LABORATORY INVESTIGATIONS

A Manual for General Biology

Michael B. Clark

Michael R. Riddle

Southwestern College
Chula Vista, CA

***Suspended Animations*, Publisher**
Jamul, CA

LABORATORY INVESTIGATIONS

A Manual for General Biology

First Edition

Cover Designs and Illustrations: Sandra Schiefer, © 1994–1998 *Schiefer Enterprises*

This book is printed at:

Commercial Printing Centre
1585 N. Cuyamaca Street
El Cajon, CA 92020

ISBN 1-885380-57-7

Printed in the United States of America by:

Suspended Animations
20275 Deerhorn Valley Road
Jamul, CA 91935

June 2002

Table of Contents

Foreword IV
Notes to the Student V
Credits VI
Covers VII
Chapter 1 Science and Human Perception *(tutorial)* 1
Chapter 2 Are Humans Doomed to Intuition? 7
Chapter 3 Measurement 23
Chapter 4 Statistics 35
Chapter 5 The Microscope 61
Chapter 6 Cells 69
Chapter 7 Chemistry Concepts *(tutorial)* 77
Chapter 8 Chemistry of Water 87
Chapter 9 Enzymes 99
Chapter 10 Photosynthesis and Respiration *(tutorial)* 107
Chapter 11 Photosynthesis 125
Chapter 12 Respiration 133
Chapter 13 Genes and Protein Synthesis *(tutorial)* 141
Chapter 14 Sameness and Variety 153
Chapter 15 Genetics 169
Chapter 16 Evolution *(tutorial)* 183
Chapter 17 Human Evolution 205
Chapter 18 Surrounded by Microbes 219
Chapter 19 Mosses and Ferns 231
Chapter 20 Dry Land Plants 243
Chapter 21 Survey of Animals 255
Chapter 22 Dimensional Realms *(tutorial)* 275
Chapter 23 Embryology 285
Chapter 24 Diet Analysis 303
Chapter 25 The Heart 317
Chapter 26 Senses and Perception 331
Chapter 27 Patterns in Nature 341
Selected Readings 351

Foreword

Laboratory Investigations covers all of the topics you expect to find in a traditional biology lab book for non-majors. In addition, there are human oriented and special-topic labs that cannot be found in other lab manuals. Finally, we have added several tutorial chapters that were specifically designed to assist students with difficult lecture concepts. All labs have been tested in the classroom, and the changes suggested by students, teachers, and lab technicians have been incorporated into *Laboratory Investigations.*

Our goal is to provide basic labs that are self-contained and totally dependable. Other lab manuals usually require that the organizational work be handled by the teacher and the lab technician. They provide only a skeletal set of instructions and guiding questions, leaving most of the technical problems and student questions unaddressed. The teacher and the lab technician are in service to these books. In our opinion, that idea is exactly backwards. A good lab book should serve the students, the lab technician, and the teacher who adopts it. *Laboratory Investigations* and its supplementary materials were created with that in mind. Because our labs are easy to understand, the students soon become self-directed. The open-ended format of each lab activity allows you to expand a favorite topic and add your own experiments to the lab without disrupting the overall organization.

The lab technician will be relieved to discover that our labs are easy to set up and maintain during the week. All experiments are written with supply budgets and safety in mind. The labs contain reminders to the students about proper lab procedures. However, those statements are not meant to replace the thorough discussions of safe lab behavior and proper equipment handling that are part of every lab class.

Finally, *Laboratory Investigations* owes its success to many people. We offer special recognition to the teachers and students and lab technicians at the following colleges whose support and suggestions have helped to make our biology lab books better every year.

Central Arizona College, AZ
Paradise Valley Community College, AZ
Pima Community College, AZ
College of the Canyons, CA
Cuyamaca College, CA
Laney College, CA
Mission College, CA
San Jose City College, CA
Sierra College, CA
Southwestern College, CA
College of Southern Idaho, ID
Lincoln Trail College, IL
IVY Tech State College, IN
Oakland Community College, MI
Washtenaw Community College, MI
Inver Hills Community College, MN
Itasca Community College, MN
Camden County College, NJ
Warren County Community College, NJ
Jefferson Community College, NY
Hofstra University, NY
Central Piedmont Community College, NC
Rowan-Cabarrus Community College, NC
Shaw University, NC
New Mexico State Univ.—Alamogordo, NM
Southwestern Oregon Community College, OR
Butler County Community College, PA
University of South Carolina—Beaufort, SC
Salt Lake Community College, UT
South Seattle Community College, WA
Madison Area Technical College, WI

Suspended Animations, Publisher
20275 Deerhorn Valley Road
Jamul CA 91935

Notes To The Student

Welcome to *Laboratory Investigations*. This book was written for you—the non-major biology student. *Laboratory Investigations* uses an approach that is easy to understand. It allows you to build your knowledge of science one step at a time, no matter what your previous background.

A lab class is quite different from a lecture class. Mostly, it is more "doing" than listening, and more "looking" than taking notes. Some students are apprehensive because of these differences, and they worry about their chances of success. The comments and suggestions below should answer most of your questions about how to succeed in your lab class.

1. We use words sparingly, constructing each sentence as a clue to something important about the topic. Read carefully, and work step-by-step through the instructions and explanations.
2. Always practice safety in the lab classroom. Make sure that you understand the instructions for properly using lab equipment, lab materials, and chemical solutions. Ask your instructor when you have a question or any confusion about safe lab procedures.
3. Be prepared. Briefly read the lab activities *before* you come to class. If you and your lab partner are prepared, you will learn more during the lab and will finish the work before the end of class.
4. Pay particular attention to the ***bold-italics*** terms and concepts. These designate the central themes and definitions of words used in the lab. They are important ideas and many will be found on your lab tests.
5. Use the blank space in the left margin of the page for lab notes and detailed answers to the questions.
6. Take time to review your work. Most students leave as soon as they finish the lab. Later, they are surprised when their answers are incorrect. Take advantage of the last half-hour of each lab. It is an excellent opportunity for checking your understanding of the topic with your instructor and other students.
7. Come early to the first day of lab class. Check out the other students. One (or more) of them is going to be your lab partner. Pick a good one! Some students are not interested in learning biology. Find someone who is serious about success. Be a good partner for them. Look for several other students who are good lab partners and form a study group for tests.
8. You will discover that many of the lab chapters will be helpful in understanding material presented in lecture class. Pay particular attention to chapters that are labeled *(tutorial)* in the Table of Contents.

Finally, have fun! Taking an active interest in the lab activities will make the time pass quickly, and will increase your chances for success in biology lab class. Good luck to you, and let us know how *Laboratory Investigations* has helped you. We welcome your comments and suggestions.

Suspended Animations, Publisher
20275 Deerhorn Valley Road
Jamul CA 91935

Credits

The Burgess Shale animals are unique, and are not available to the public except in museum collections and fossil exhibits. In order to draw the cover pictures, I had to rely on the drawings of other people—paleontologists and illustrators—as the source work for my own designs.

I would like to thank the Scandinavian University Press and the Geological Survey of Canada, for their quick response to my first permission requests. To all permission sources, I thank you for your generosity.

And, I would especially like to thank Frederick Collier, at Harvard University's Department of Invertebrate Paleontology, for his assistance in helping me to locate and contact his fellow scientists who worked on the various Burgess Shale projects and expeditions.

THE FIGURES

#1 ***Leanchoilia:*** From D. L. Bruton and H. B. Whittington, 1983. *Emeraldella and Leanchoilia*, two arthropods from the Burgess Shale, British Columbia. *Philosophical Transactions of the Royal Society, London* B 300:553-85. Adapted.

#2 ***Marrella:*** Courtesy of the *Geological Survey of Canada Bulletin* 209:1-24. (From H. B. Whittington, 1971. Redescription of *Marrella splendens*, Trilobitoidea, Middle Cambrian Burgess Shale, British Columbia.) Adapted with permission.

#3 ***Anomalocaris:*** From H. B. Whittington and D. E. G. Briggs, 1985. The largest Cambrian animal, *Anomalocaris*, Burgess Shale, British Columbia. *Philosophical Transactions of the Royal Society, London* B 309:569-609. Adapted.

#4 ***Pikaia:*** From *The Fossils of the Burgess Shale*, D. E. G. Briggs, Douglas H. Erwin, and Frederick J. Collier, 1994. Originally from Figure 3 of Briggs, D. E. G., 1991, Extraordinary Fossils, *American Scientist*, 79:130-141, illustrator Virge Kask; adapted, from Conway Morris and Whittington (1985). Adapted with permission.

#5 ***Opabinia:*** From H. B. Whittington, 1975. The enigmatic animal *Opabinia regalis*, Middle Cambrian, Burgess Shale, British Columbia. *Philosophical Transactions of the Royal Society, London* B 271:1-43. Adapted with permission.

#6 ***Canadaspis:*** From D. E. G. Briggs, 1978. The morphology, mode of life, and affinities of *Canadaspis perfecta*, Middle Cambrian Burgess Shale, British Columbia. *Philosophical Transactions of the Royal Society, London* B 281:439-87. Adapted with permission.

#7 ***Burgessia:*** Courtesy of the Lethaia Foundation & Scandinavian University Press, Olso, Norway. (From C. P. Hughes, 1975. Redescription of *Burgessia bella*, Middle Cambrian Burgess Shale, British Columbia.) Published in *Fossils and Strata* 4:415-35. Adapted with permission.

#8 ***Dinomischus:*** From *The Fossils of the Burgess Shale*, D. E. G. Briggs, Douglas H. Erwin, and Frederick J. Collier, 1994. Originally from Figure 3 of Briggs, D. E. G., 1991, Extraordinary Fossils, *American Scientist*, 79:130-141, illustrator Virge Kask; adapted, from Conway Morris and Whittington (1985). Adapted with permission.

#9 ***Branchiocaris:*** Courtesy of the *Geological Survey of Canada Bulletin* 264:1-29. (From D. E. G. Briggs, 1976. The arthropod *Branchiocaris*, Middle Cambrian Burgess Shale, British Columbia. Adapted with permission.

#10 ***Burgess Shale Fauna:*** From *The Fossils of the Burgess Shale*, D. E. G. Briggs, Douglas H. Erwin, and Frederick J. Collier, 1994. Originally from Figure 3 of Briggs, D. E. G., 1991, Extraordinary Fossils, *American Scientist*, 79:130-141, illustrator Virge Kask; adapted, from Conway Morris and Whittington (1985). Adapted with permission.

#11 ***Wiwaxia:*** (back cover) From S. Conway Morris, 1985. The Middle Cambrian metazoan *Wiwaxia corrugata* (Matthew) from the Burgess Shale and *Ogygopsis* Shale, British Columbia, Canada. *Philosophical Transactions of the Royal Society, London* B 307:507-82. Adapted with permission.

—Sandra L. Schiefer—

Covers

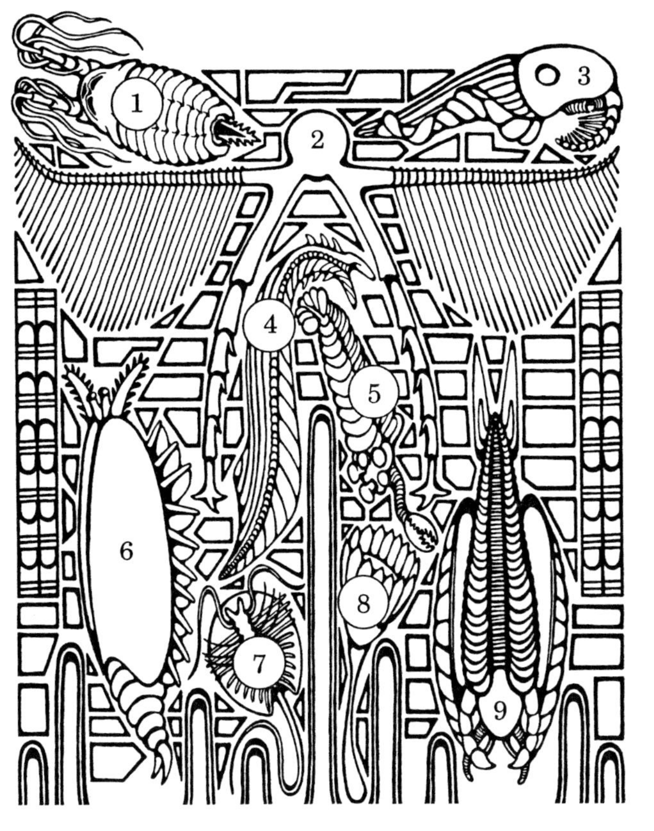

(10) All the animals.

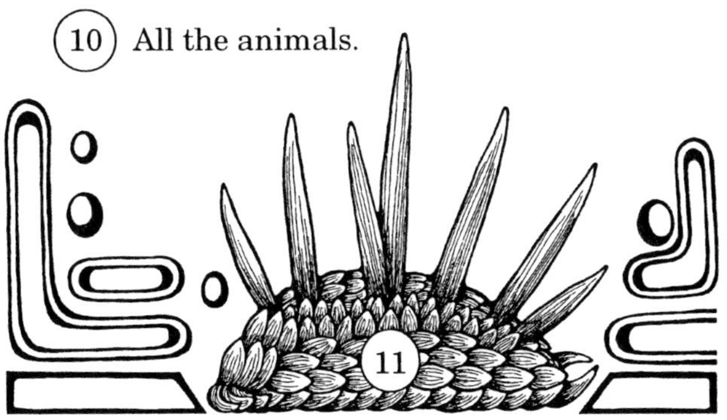

In Australian Aboriginal art and culture, "Dreamings" are Ancestral Beings. These beings are believed to exist in the present time. One of their roles is as storytellers, sharing tales of current and historical Aboriginal life.

To me, the fossils of the Burgess Shale are like the "Dreamings" paintings of the Australian Aborigines. The fossils are a current physical reality that tell a story about life's historical struggle to survive. These fossil animals are our Ancestral Beings. Fortunately for science, they lived and died under circumstances that left them perfectly preserved in layers of fine silt, to be discovered 500 million years later in the mountains of British Columbia, Canada.

The story of their initial discovery, and later re-examination, is excellently told by Stephen Jay Gould, in his book, *Wonderful Life: The Burgess Shale and the Nature of History*. It is not only a remarkable tale, it also demonstrates the scientific method at work.

In 1909, Charles Walcott found the Burgess Shale fossils. He formed a hypothesis based on his ideas about these previously unknown animals. As more samples were collected and examined, his observations eventually led to published conclusions. According to the scientific knowledge of his time, and the understanding of the process of evolution, the fossils were classified as probable ancestral forms of known modern species.

Then, in the 1960's, Charles Walcott's findings were re-examined. Modern expeditions went to the sites. More fossils were collected, and the original fossils of 50 years earlier were re-examined. Names like Whittington, Briggs, Morris, Bruton, and Hughes were associated with new discoveries. They turned the original classifications and conclusions about the Burgess Shale animals upside down. Most of the animals were not direct ancestral forms of species alive today. They were, instead, unique beings whose stories revealed: "The theme of evolution is primarily one of extinction, not survival."

The accepted idea of a branching "tree of life" that always leads to more and greater diversity over time, was felled to the ground. The Burgess Shale trunk did not fork upwards. Just the reverse. In the Burgess story, few evolutionary experiments in the Cambrian sea of life survived. However, one slender twig, which did make its way forward in time, possessed a notochord (a precursor to the internal skeleton). That twig, *Pikaia gracilens*, the first known chordate, represents the oldest recorded animal ancestor for animals classified under the phylum Chordata, which includes all vertebrates.

Behold the animals of the Burgess Shale, our Ancestral Beings, who as fossils survive to share their fascinating tales from the history of life on Earth.

The covers, *Burgess Shale Dreamings* and *Wiwaxia Dreamings*, are artistic representations of some of the unique animals discovered in the shale deposits of British Columbia. These covers were designed to enclose *Laboratory Investigations*, a unique manual for the study of general biology.

— Sandra L. Schiefer —

Are Humans Doomed to Intuition?

People often make the comment, "Problems in the modern world are out of control." Furthermore, they say, "If it weren't for science telling us what to do, everything would be better."

Those who understand the methods of science find these comments maddening. The everyday decisions controlling government, social institutions, and individual freedom are not made using the empirical method of science. Instead, society usually searches for solutions and reaches decisions by using human intuition.

This week's lab will focus on ***intuitive*** and ***counter-intuitive*** processes, and the role that science could play in our society's future decisions if we began to use it.

Exercise #1 "Questionnaire" 7
Exercise #2 "Intuition and Counter-intuition" 9
Exercise #3 "Decisions" 10
Exercise #4 "Problems of Scale" 16
Exercise #5 "Problems of Cost" 19
Exercise #6 "Knowing That You Are Right" 22

EXERCISE #1

"Questionnaire"

It is important for you to answer all of the following questions by yourself *(with no discussion at all)* before starting the other Exercises in this lab. The answers to these questions will give you valuable insight into your own thinking processes, and will be used as part of the class data for group comparisons during the lab.

? QUESTION

1. Define what you mean by the word "intuition."

2. What is the source of intuition?

3. Do you use intuition to make decisions?

4. If so, in what kinds of decisions?

5. Can intuition be wrong?

6. If so, when is it wrong?

7. Don't analyze. Make a quick decision after reading the following information. There are two buttons: **A** and **B**. You must push one of them. (circle your choice)

   Button A—If you push this button, you will be given $3000.

   Button B—If you push this button, you have an 80% chance of getting $4000, and a 20% chance of getting nothing.

8. Don't analyze. Make a quick decision after reading the following information. There are two buttons: **C** and **D**. You must push one of them. (circle your choice)

   Button C—If you push this button, you lose $3000.

   Button D—If you push this button, you have an 80% chance of losing $4000, and a 20% chance of losing nothing.

9. How many people do you usually date before forming a serious relationship?

An answer that comes quickly to mind is most likely a reflection of your intuition.

10. How many different college majors are you considering as definite possibilities?

11. In your opinion, do most people stay in bad relationships longer than they should?

12. In your opinion, do most people stay in unsatisfying jobs longer than they should?

13. Your brain makes some decisions without your conscious awareness. What could you do to discover this?

14. If you observed that all humans make the same mistake, what would you conclude about human behavior?

15. How many people live in the USA?

16. Using the population of California as a base, how many *additional people* are added to the world each year? (That is, how many "Californias" are added to the planet each year?)

17. Using the population of Los Angeles as a base, how many *additional people* are added to the Western Hemisphere each year? (That is, how many "LAs" are added to our half of the planet each year?)

18. You have decided to buy an $18,000 new car, and have been offered three methods of payment by the dealer. Quickly estimate the total cost of each payment method.

    Payment by cash = ____________

    5-year 10%-interest car loan = ____________

    Add the car payment to your existing 8% home-equity loan = ____________

19. How many months of work will it take you to pay for a new $18,000 car if the method of payment is cash?

20. Compared to bicycle travel, how much time in a year is saved when traveling by car?

21. When you make decisions, how important is the intuitive feeling of "knowing that you are right"?

EXERCISE #2

"Intuition and Counter-Intuition"

When you use your intuition to explain a counter-intuitive process, expect to be wrong.

Several concepts must be understood before we can continue our investigation. First, we need some clear definitions. The dictionary describes ***intuition*** as (1) the act or faculty of knowing without the use of rational process; (2) a capacity for guessing accurately; and (3) a sense of something not deducible.

These definitions suggest that intuition does not depend on analysis of a situation, and that intuition is usually correct. In addition, the definitions imply that certain truths may be discovered only through intuition and *cannot* be known by using logical processes.

The primary process in science is called empirical thinking. ***Empirical*** is defined as (1) relying upon or derived from observation or experiment; and (2) not guided by theory or intuition, but by verifiable measurements or observations.

When science tests intuitive ideas, there are two possible results:

1. The intuitive idea agrees with the empirical evidence.
2. The intuitive idea does not agree with the empirical evidence. When science describes a process differently than intuition does, that process is called ***counter-intuitive.***

One example of a counter-intuitive process is the *rotation of the Earth.* Our intuition tells us that the sun revolves around us. We see it rise in the east and set in the west. However, science has proven that our intuition is wrong. The planet spins on its axis from west to east, and one complete rotation takes 24 hours.

Dreaming is another example of a counter-intuitive process. Dreams usually involve strange places and situations, yet our friends tell us that we haven't physically moved from our beds at night. Our intuition leads us to conclude that there is a spirit world that we go to when we are sleeping. However, science has proven that dreams result from various spontaneous activities in the brain that can be stimulated or suppressed by machines or drugs. Dreams apparently have nothing to do with a spirit world.

There are *two* useful ways of describing the consequences you may experience when some aspect of the world operates counter to your intuition.

What are the consequences when intuition is wrong?

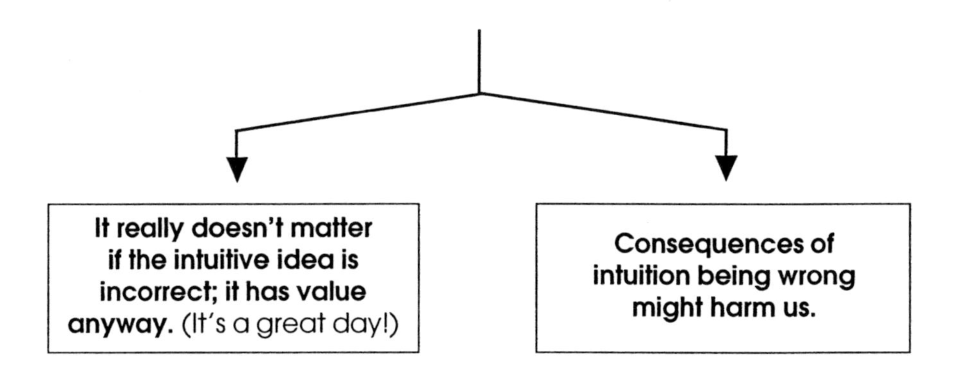

While you are doing the following Exercises, consider what serious consequences might occur when intuition is wrong. Also consider if there might be a scientific explanation for why humans don't recognize and solve certain world problems. Finally, consider what would happen if we used scientific thinking to help us when our intuition fails.

? QUESTION

1. Define intuition.

2. Define empirical.

3. Define counter-intuitive.

Remember: The concepts in this lab require your complete understanding of the definitions for intuitive, empirical, and counter-intuitive thought. Check with your instructor now if you are confused.

EXERCISE #3

"Decisions"

Can we agree that the ability to make a decision is an important human act? Can we agree that your relationships, your choice of a college major, and how you spend your money are important decisions? If you answered "yes," then let's look at the discoveries made by two scientists, Daniel Kahneman and Amos Tversky, who have studied human decision-making.

Class Data

Refer to the "Button Problems" in the Questionnaire, Exercise #1. Record your answers to questions #7 and #8 on the chalkboard. Write the class totals on the next page.

Total # of Students in Your Class = ____________

Choosing Button A = ____________

Choosing Button B = ____________

Choosing Button C = ____________

Choosing Button D = ____________

Kahneman and Tversky found that most people choose Button A and Button D. Do these findings agree with your class results? ____________
Do the findings agree with your personal choices? ____________

Low Risk and High Risk

A ***low-risk situation*** is one in which you are not going to lose something that you already have.

A ***high-risk situation*** is one in which you are going to lose something whatever choice you make.

? QUESTION

1. Which situation is low risk? (circle your choice)

A and B   or   C and D

2. Which Button choice in the low-risk situation is "playing it safe"?

3. Which Button choice in the low-risk situation is "taking a chance"?

4. So, how do most humans act when they are in a low-risk situation? (circle your choice)

They play it safe.   or   They take a chance.

In the "world of intuition," there are low-risk losers and high-risk losers.

5. Which Button choice in the high-risk situation is "playing it safe"?

6. Which Button choice in the high-risk situation is "taking a chance"?

7. So, how do humans usually act when they are in a high-risk situation? (circle your choice)

They play it safe.   or   They take a chance.

Let's Examine the Facts

The button decision problems were designed to offer you a choice, and determine whether you have an intuitive bias. The results reveal that intuition directs us to choose A and D, both of which are ***losing*** strategies in life. The ***winning*** choices are actually B and C.

In a low-risk situation there is no real personal loss to you if you took a chance. You were being given money. The odds favored more money if you chose Button B. Your choice of Button A means that you stayed with the safe bet. It's just like staying with one major in college or dating only one person. ***You are limiting your options for better returns.*** After all, who would insist that there is only one deserving person in the world with whom you could have a successful relationship? Likewise, our economy proves that there are many successful choices for college majors. But, in money, love, and vocation we usually find ourselves playing it safe in a world of opportunity.

By the way, if you happened to choose Button B and you're feeling pretty smug right now, then forget it. Research has shown that when the problem is restated in terms other than money, you will make the same mistake as everyone else (Button A). For example: You are hungry. Button B offers you, free of charge, a seven-course dinner at a posh restaurant with a 20% chance that the restaurant is closed. Button A offers you, free of charge, a meal at your favorite fast-food emporium that is open 24 hours a day, seven days of the week.

A more serious problem is revealed by the selection of Button D. Both Buttons C and D involve loss. However, the intuitive bias towards Button D means that you are willing to expose yourself to *even more risk* in an effort to avoid any pain. Choosing Button D compounds the risk. Choice C, while still a loss, represents the best strategy—*take the smallest loss and move on.* Moving on to a new relationship has a greater chance of success than waiting for a bad relationship to improve. Getting a new job has more possibility of encouraging the development of your talents than staying at an unsatisfying job.

Procedure

1. Divide the class into discussion groups. Assign one person the responsibility of coordinating the discussion. (This person is not supposed to come up with all of the ideas—just keep the rest of the group on task.) Assign another person to write down the group's answers

2. For each of the three topics, determine how your using strategy A or D to solve the problem could actually make the situation worse.

   A = ***Playing it safe when there is opportunity to take a chance without loss.***

   D = ***Taking even more risk in an attempt to avoid loss.***

3. For each of the three topics, determine how your using strategy B or C to solve the problem might improve the situation.

   B = ***Taking a chance for a new opportunity when there is no real loss at stake.***

   C = ***Taking the loss and moving on to new opportunities.***

4. Work on the three topics for 10 minutes each. There will be a class discussion following your group discussions.

Group Report

Topic #1
Human Overpopulation

1. Using **Strategy A**, what political policy might be adopted towards human population growth?

   What are possible *negative* consequences of using this strategy?

2. What is an example of a **Strategy D** political policy towards overpopulation?

   What are possible *negative* consequences of using this strategy?

3. What is an example of a **Strategy B** political policy towards population growth today?

   What are possible *benefits* of using this strategy?

4. What is an example of a **Strategy C** political policy towards human overpopulation?

   What are possible *benefits* of using this strategy?

Topic #2
Unemployment

Try another example using a separate sheet of paper. What are examples of the four strategies an unemployed person could have used prior to a factory shutdown, or while they are on unemployment benefits?

Topic #3
Personal Relationship

Finally, try personal examples using the four strategies to find a good relationship, or while in a bad relationship.

Random Decisions During Problem-Solving

Many aspects of decision-making and problem-solving have been investigated by science. These problem-solving situations have been tested and retested by researchers, and the conclusions are similar whether a single person, a family, a company, or a governmental agency is studied. Scientists have discovered the influence of ***random thinking processes*** during problem solving.

The first step the research team took was to methodically interview a test group. This was to determine what different *decision-making criteria* the group used when solving a particular problem. The City Council flow chart below is a typical example of the decision-making process.

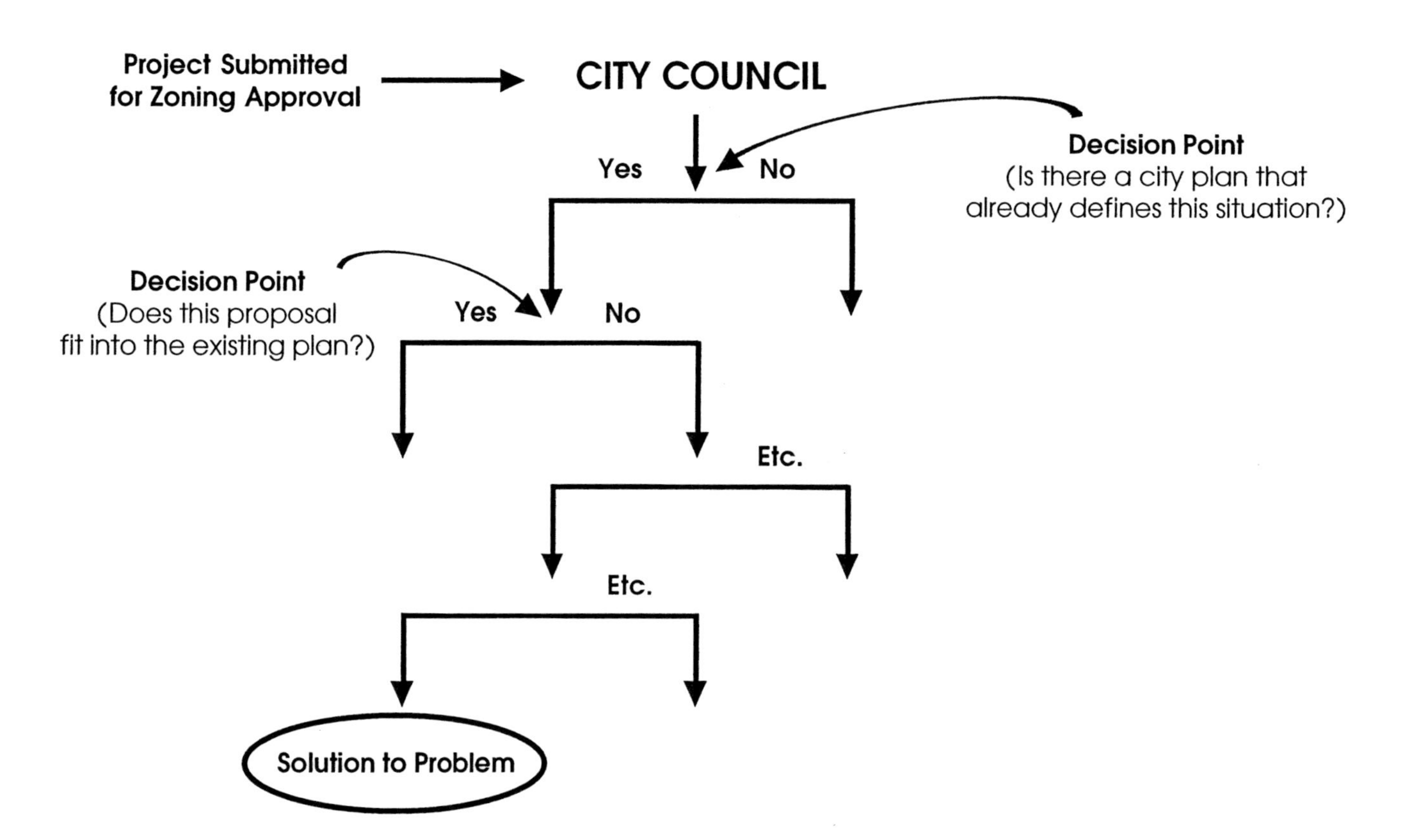

Once all of the test group's decision-making criteria had been determined, the researchers began submitting problems to the group to see if they actually follow their own rules.

Results

1. All groups developed solutions that included *violations* of their original decision-making criteria.
2. When the decision-points were carefully monitored, the researchers discovered that the test groups were completely *unaware* that they had "violated" their own rules.
3. The violated decision-points were retested. Researchers determined that there was no pattern. The test groups *randomly chose* between "yes" and "no," and they were unaware that their choices were random.
4. Further investigation revealed that the test groups made random choices whenever there was *ambiguous* or *conflicting information* surrounding a particular decision-point.

Conclusion

There is something inherent in the human thinking process that automatically creates random decisions in the brain. And, whenever there is ambiguous or conflicting information, the number of random choices increases, and ***the brain doesn't realize it!***

Procedure

[1] Divide the class into discussion groups. Answer the questions below keeping in mind what researchers learned about decision-making. Assign one person the responsibility of coordinating the discussion. Assign another person to write down the group's answers to each of the eight questions. Be very specific about answering question #7.

[2] Your group has 20 minutes to discuss all the questions. Then the class will have a general discussion for 10 minutes.

? QUESTION

1. When people have strong disagreements in personal relationships or involving group decisions, is it likely that there will be ambiguous or conflicting information at the decision-points?

2. What is going to happen in both people's minds as a result of the answer to question #1?

3. Will they be aware of this process during the discussion?

4. How are they likely to treat each other during the problem-solving?

5. Based on what you have learned in this exercise, would you expect humans to be able to work out their personal relationship problems by using the usual human thinking processes?

6. Would you say that your society has the same problems as you do?

7. What is your group's solution to the dilemma illustrated in questions #1–6?

8. Use the same approach as you used to answer question #7 to project how our society might solve the problem of overpopulation.

EXERCISE #4

"Problems of Scale"

We have a great deal of difficulty understanding processes that are many times larger or smaller in scale than those in our personal lives. You were asked three questions about population (questions #15, #16, and #17) in the questionnaire at the beginning of this lab.

The bigger the numbers, the harder you fall.

The answers to the three questions might surprise you.

#15: There are 265 million people in the United States.

#16: Three times the population of California are added to the world each year.

#17: Three times the population of Los Angeles are added to the Western Hemisphere each year.

Most people miss these estimates unless they have recently seen them in a magazine article. Apparently, we do not comprehend population numbers because they are not in the scale of our normal, everyday experience. Therefore, the process of analyzing population growth is *counter-intuitive*. One way of revealing this fact is to examine our intuitive inadequacy in understanding the ***rate of growth*** and ***doubling times***.

There is a rule and a diagram that can help us to go beyond our intuition when we try to solve a problem involving a rate of growth (business planning, investments, debt, population, etc.).

Rule of 70

If you divide the number 70 by the growth-rate percent of a process, then you will have estimated the number of years required for that process to double. For example: If the national debt is increasing at a rate of 7% per year, then the debt will double in 70 ÷ 7 = 10 years.

The Doubling Rectangle

There is a visual aid to help you understand the doubling process. It is called the ***Doubling Rectangle***, and it looks like this:

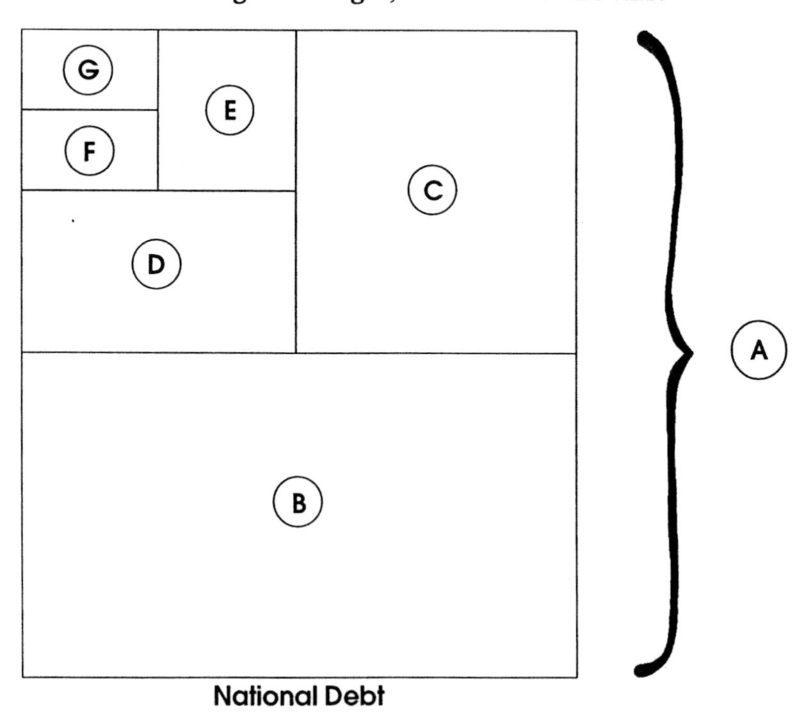

National Debt

Use the Doubling Rectangle on the previous page as a visual aid to understand the doubling process of the National Debt. The easiest application of the *Doubling Rectangle* is when the growth rate has been constant over a long period of time. Let's assume that the growth in National Debt has been constant and that each rectangle represents 10 years of time. Label rectangle Ⓓ as the 1980s decade. That means that Ⓒ, Ⓑ, and Ⓐ are the doubled debts for the 1990s, from 2000–2010, and from 2010–2020. Going back in time, Ⓔ becomes the decade of the 1970s, Ⓕ the 1960s, and Ⓖ represents all debt in the United States before 1960. Label all of the rectangles in the diagram with these decades. *You will be using this Doubling Rectangle to compare with the one that you do next.*

Doubling of GDP

Procedure

1. Try a problem of scale using the tools you have just learned. The GDP (Gross Domestic Product) has been growing at an average of 3% per year since 1900. Your job is to calculate the doubling time, and then label the *Doubling Rectangle* below.
2. Start with rectangle Ⓒ, and label it 1970 ⟶ ?. The **?** will be known after you calculate the GDP doubling-time using the *Rule of 70*. GDP doubles in ____________ years.
3. Rectangle Ⓑ will start with the year that rectangle Ⓒ ended with. Label the doubling time for Ⓑ.
4. Finish the diagram after doing the calculations. Label Ⓐ, Ⓓ, Ⓕ and Ⓔ. Rectangle Ⓖ represents the GDP before 1900.

Compare the doubling of the GDP with the doubling of the National Debt.

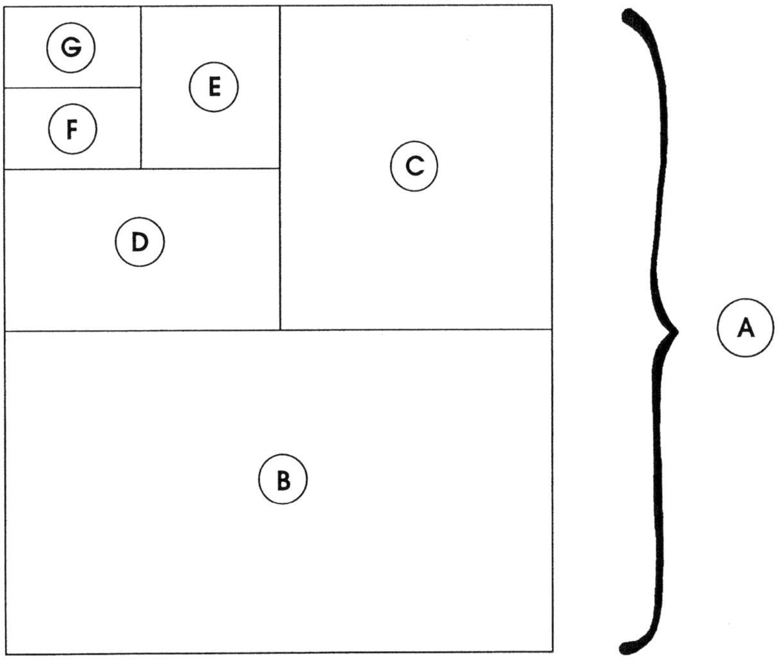

G
F
E
D
C
B
A

Gross Domestic Product

Check with your instructor to make sure that your answers for doubling times are correct before going on to the next questions.

? QUESTION

1. Which process has the quickest doubling time? (circle your choice)

GDP or National Debt

2. You should have calculated that the GDP rectangle (B) starts in 1993. During the time when the GDP is doubling in (B), how many doublings occur in the National Debt? (Refer to your first doubling rectangle.)

3. The period between 1940 and 1960 was the time of fastest growth in personal income and lifestyle for Americans. Economists tell us that the GDP grew at a rate of 4.4%, which is faster than the average for the century. What is the GDP doubling-time based on a 4.4% growth rate?

Was the growth in the GDP closer to the growth of debt during the 1940–1960 period than it has been since then?

4. We have been increasing our use of oil at a rate of 7% per year since the turn of the century. What's the doubling time?

When your intuition fails, the Doubling Rectangle and the Rule of 70 are powerful tools for solving problems of scale.

Pick any decade (such as 1980–1990). The amount of oil consumed in that decade is how much compared to all of the oil used before that decade?

5. The world population is growing at about 2% per year. What's the doubling time?

6. The doubling time in food production for both the USA and Mexico has been about 35 years. Mexico's population doubling-time has been about 25 years and the USA's has been about 50 years. What has Mexico had to face during the past three decades that the USA has not had to contend with?

7. Since 1990 the average wage has been increasing at 1.5% per year. (This does not consider people who lost higher-paying jobs and were rehired at lower-paying ones.) The rate of inflation has been increasing at about 3% per year. At these rates, how long will it be before your real spendable income becomes one-half what it is today?

In Conclusion

Doubling times cannot be accurately comprehended by using human intuition. As you saw in the examples above, the counter-intuitive problems are complex, and we haven't considered that (1) doubling times are not constant, and (2) there are many other variables besides the GDP and debt that need to be considered at the same time.

Perhaps you have heard of the computer game called *SimCity*, based on city planning. The fun of this game is that the dynamics of the game are based on rates of change (doubling times). While playing the game, we discover that our intuition is continually being "fooled."

What is fun in a game can be a catastrophe in real life!

EXERCISE #5

"Problems of Cost"

An unfortunate life is created by opportunities forgone in making choices.

Modern economic theory states that the free-market capitalistic system operates efficiently because of rational self-interest. ***Rational self-interest*** means that we make decisions that are in our best interest. Economists tell us that rational self-interest creates a natural buying and selling market that balances supply and demand so that stable prices result. One test of these assumptions is to ask the question: "Can our intuition balance the costs and benefits of our purchasing decisions?"

Cost

Economics textbooks define ***cost*** as ***the value of opportunities forgone in making choices***. One simple test of our ability to determine monetary costs is found in question #18 of the questionnaire. Let's check your accuracy.

The cost of a car when paying in cash is $18,000.

The cost of using the 5-year 10%-loan is $24,000.

The cost spread over a 30-year 8% home-equity loan is $54,000.

? QUESTION

1. How close were your questionnaire estimates to the actual costs?

2. Was your intuition adequate to the task?

Procedure

Convert the cost of a new car into the number of months you must spend working to pay for it.

? QUESTION

1. What wage do you expect to be making? ($ per hour) = __________.
2. Assuming 160 hours of work per month, your Gross Monthly Wage equals __________.
3. Subtract your taxes (about 35% of income). Corrected Monthly Wage equals __________.
4. Subtract your estimated rent and food. Corrected Monthly Wage equals __________.
5. Subtract routine miscellaneous expenses. Corrected Monthly Wage equals __________. This is your *Spendable Income.*
6. Divide the cost for each of the three methods of car purchase by your Spendable Income, and put your answers in the table below.

Method of Payment	Months of Work to Pay for Car
$18,000 ÷ Spendable Income	__________
$24,000 ÷ Spendable Income	__________
$54,000 ÷ Spendable Income	__________

7. Do the calculated values for the amount of work required to pay for your car surprise you?

8. If so, what does this mean about your intuitive ability to understand costs?

Remember: The time spent working for the car is a measure of cost (opportunities forgone in making the decision to buy a new car).

Traveling by Car or by Bicycle

Could driving a car cost you as much time in your life as if you traveled by bicycle? Set your intuition aside and give the environmentalists a chance to convince you.

Information

1. The average travel by car in the USA is 12,000 miles each year. The environmentalists are correct when they say that many European cities are designed for bicycle travel. Their cities are efficiently planned so that an average person needs to travel only 6,000 miles per year to do everything we do in 12,000 miles.

   If you drive a combination of city and open-road, then divide 12,000 by 25 mph (your average speed), and put this value into the Travel Comparison Table (travel time by car). If you drive only in the city, then add another 320 hours of time because your average speed drops to 15 mph in a big city.

   Traveling by bicycle averages 10 mph. Divide this into the 6,000 miles that you would need to commute in a well-designed city. Record that number in the Travel Comparison Table (travel time by bicycle).

2. The average expense of operating a car is $3,900 per year (car cost, repair, insurance, gas, etc.) In addition, government estimates reveal that about $1,000 of tax money is spent subsidizing the automobile each year for each driver (highway construction, etc.). Divide the total yearly expense for operating a car ($4,900) by the hourly wage you chose for the car cost problem on the previous page. Record your answer in the Travel Comparison Table (expenses).

3. The average minimum expense of operating a bicycle is estimated to be $500 per year (bike cost, repair, road upkeep, insurance, etc.) Divide the total yearly expense for operating a bicycle ($500) by your hourly wage. Record your answer in the Travel Comparison Table (expenses).

4. Calculate the totals for each method of travel.

TRAVEL COMPARISON TABLE
(Hours of Your Life Spent on Travel)

	Car	Bicycle
Travel Time		
Expenses		
Total		

Are you surprised that travel by bicycle compares so well to travel by car?

Bicycle travel is *counter-intuitive* for Americans. This comparison does not prove that car travel is inferior, nor does it prove that we should all travel by bicycle. However, our study does demonstrate that by using our intuition, the estimate of actual cost of travel will be incorrect (another opportunity forgone).

EXERCISE #6

"Knowing That You Are Right"

The final Exercise of this lab examines the intuitive feeling of *knowing that you are right* in a particular situation. Your group will solve a very common problem in city redevelopment. ***An area of the city does not have enough jobs, and there are not enough cheap houses for the lower-income people who live there.***

Procedure

Your group is to quickly write down your solutions for city redevelopment, and predict the positive results of your plan. After you have done this, read the information below.

Don't read the Empirical Findings until after your group has devised a solution to redevelopment. If you cheat, you won't get as much value from testing your intuition (an opportunity foregone).

Empirical Findings

The most common solution to the city redevelopment question is to build more cheap housing. This solution always fails. More lower-income people are attracted to the area, and *the local unemployment actually increases.* Another common solution is to build factories in the area. This approach works only when building the factories also reduces the existing housing and number of residents at the same time. An unexpected finding is that destroying low-income housing will increase the percentage of employed people remaining in the area. These results have been known for several decades, yet the same failed approaches are continually recommended by politicians and planners.

? QUESTION

Review your group's solution again. Do you still believe that your plan would work if only it was given a little time and a fair chance?

ANSWER

Most people answer "yes" to this question. It is almost universal that when people are presented with the empirical evidence that disproves their intuitive idea, they still "feel" that they are correct. The intuitive feeling of *knowing that you are right* apparently does not change even when the facts oppose you.

In Conclusion

This is the final challenge to humanity as it considers those aspects of the modern world that are counter-intuitive. To be successful as a species, we may have to ignore our intuitive feelings when solving certain problems—and that could be very difficult.

How will you play the game of life: Take your losses and move on, or risk even more? There is a lot to be said for understanding the real world choices behind Buttons A, B, C, and D. Ultimately, your future will be determined by the buttons you choose.

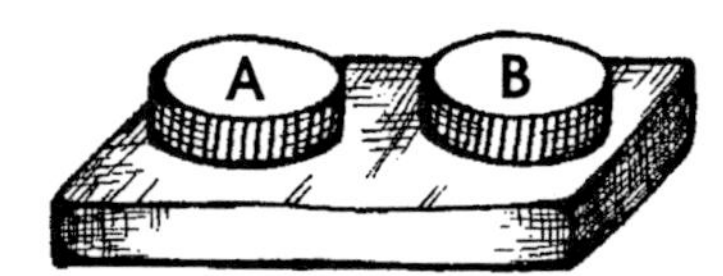

Measurement

Some people insist that measuring things is the only way to get a fair deal from another human being. But it is also said that the measuring of things is at the heart of all mistrust between people. Whatever the correct judgment may be, historians tell us that measurement systems are based on political and economic needs, and can be found wherever humans have a network of dealing with each other.

Originally, measurement systems depended on the type of material being exchanged. For example, a farmer selling apples would price them by the cart-load, not by the bucket. But when selling milk, a cart-load of liquid would have been ridiculous, and a bucket more appropriate. These kinds of systems developed and changed as they were used. To prevent squabbles among merchants and buyers, ***standard*** "cart-loads" and "buckets" were determined, and these became the basis for the systems of measurement we use today.

As science developed in civilized countries, the need for scientific measurement began to overlap the measurement systems used for trade. But there are problems with adapting the common system for use in science. In grade school children are taught the English System which uses inches, feet, yards; cups, pints, quarts, gallons; ounces, pounds, tons. However, a quart in Canada is equal to 1.136 liters. A quart in the United States is equal to 0.946 liters. And in Mexico there are no quarts, only liters. For trade purposes, these inequities can be worked out, but in science or industry where precision is important, the English System simply will not do.

Then, 100 years ago, the International Metric System was devised to standardize measurement around the world. This system provides exact precision in ***powers of 10***. There is a standard reference unit used in each measurement category. On either side of the reference unit, the units increase by 10, 100, and 1000 or decrease by $\frac{1}{10}$, $\frac{1}{100}$, and $\frac{1}{1000}$.

The universal standard unit for weight is the ***gram*** (g). The standard for volume is the ***liter*** (l). The standard unit for length is the ***meter*** (m).

Science embraced the Metric System and so did many European countries and Mexico. It was anticipated that the United States would be converted to the Metric System by now. It is unfortunate that so much confusion is created by converting quarts to liters, pounds to grams, and inches to meters. The solution is familiarity and practice until the United States completely adopts the Metric System.

In today's lab you will be introduced to the Metric System. You will learn to use simple measuring equipment that will help you remember the different units of the Metric System.

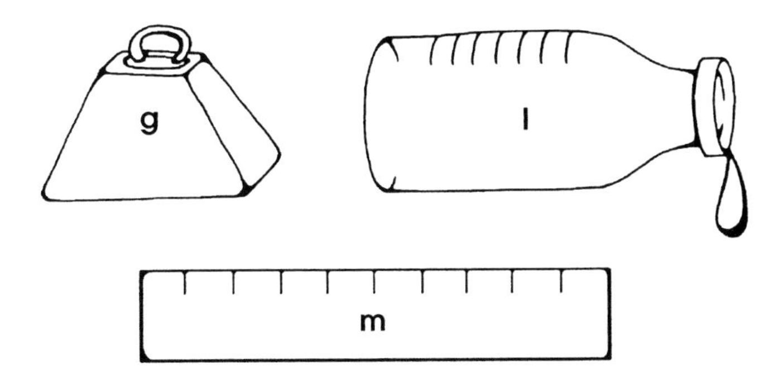

Exercise #1 "Some Basics of the Metric System" 24
Exercise #2 "Conversions: Metric ←→ English" 28
Exercise #3 "Personal List of Metric References" 32
Exercise #4 "A Test of Your Measuring Skills" 33

EXERCISE #1

"Some Basics of the Metric System"

Fractions

Fractions represent parts of a whole, and there are two methods of expressing them:

First—Simple Fractions. Examples are $\frac{1}{4}$, $\frac{1}{2}$, $\frac{9}{25}$, etc.

Second—Decimal Fractions. Examples are 0.25, 0.50, 0.36, etc.

Fraction Conversion

Rule 1: ***If you want to convert a simple fraction into a decimal fraction, then divide the top number of the simple fraction by the bottom number.*** Therefore, the simple fraction $\frac{9}{25}$ is converted into decimal form by:

$$25 \overline{\smash{)}\,9.00}^{\;0.36}$$

Rule 2: ***If you must convert a decimal fraction into a simple fraction, then put the decimal fraction over 1.00 (this represents the whole), and reduce the top and bottom numbers to the simplest fraction.*** Therefore, the decimal fraction 0.36 converts into:

$$\frac{0.36}{1.00} \text{ or } \frac{36}{100}\;\frac{\div 4}{\div 4} \text{ or } \frac{9}{25} \text{ expressed as a simple fraction}$$

? QUESTION

Show your work in figuring out the following problems.

1. $\frac{3}{4}$ = ____________ (decimal fraction)
2. $\frac{9}{20}$ = ____________ (decimal fraction)
3. 0.4 = ____________ (simple fraction)
4. 0.15 = ____________ (simple fraction)

Percents %

In the previous discussion we saw that fractions are ***parts*** of the number ***1***. Percents (%) are ***parts*** of the number ***100***. Therefore, in order to convert a decimal fraction into a percent, we must first *multiply* the decimal fraction by 100.

The decimal fraction 0.23 is equal to 23%. (See the factors of 10—Multiplication Rule if you need help with this.)

? QUESTION

Convert the following decimal fractions into percents, and show your work.

1. 0.79 = ____________ %
2. 0.07 = ____________ %
3. 0.18 = ____________ %
4. 0.001 = ____________ %
5. 0.809 = ____________ %

Factors of 10

Because the Metric System is based on units that differ from each other by factors of 10, we need to review how the decimal position *moves* when converting within metric units.

Multiplication Rule

When multiplying a number by 10, 100, 1000, etc., move the decimal position ***to the right*** by the number of 0's (zeros) in the multiplier. For example:

25 x 10 = 25.0 = 250

25 x 100 = 25.0 0 = 2500

By the way, you add 0's as you move the decimal point to an empty space.

25 x 1000 = ______

2.5 x 100 = ______

0.25 x 1000 = ______

0.002507 x 100,000 = ______

Try these for practice.

This rule allows you to convert ***large*** metric units like meters into ***small*** metric units like millimeters.

Division Rule

When dividing a number by 10, 100, 1000, etc., move the decimal position ***to the left*** by the number of 0's (zeros) in the divisor. For example:

25 ÷ 10 = 2.5. = 2.5

25 ÷ 100 = .2 5. = 0.25

25 ÷ 1000 = .0 2 5. = 0.025

Again, you add 0's as you move the decimal point to an empty space.

0.25 ÷ 10 = ______

2.5 ÷ 10 = ______

2.5 ÷ 1000 = ______

Try these for practice.

This rule allows you to convert ***small*** metric units like millimeters into ***large*** metric units like meters.

In science, the rule about decimal fractions like .025 is that you always put a "0" (zero) in front of the decimal point so that there is no confusion about the number being smaller than one. Therefore, we write the above example as 0.025.

Metric Prefixes

The Metric System is a standardized method for defining ***length, volume,*** and ***weight.*** The standard metric units are ***meter*** (m), ***liter*** (l), and ***gram*** (g).

There are over a dozen prefixes used with each of these standard units that define "portions" of units and differ from each other by factors of 10. In this lab you will need to use only three of the metric prefixes. *Memorize them now!*

Prefix	Abbreviation	Compared to Standard	Examples
Milli	m	$\frac{1}{1000}$	mm, ml, mg
Centi	c	$\frac{1}{100}$	cm, *, *
Kilo	k	1000	km, *, kg

* cl, cg, and kl won't be used in this lab.

? QUESTION

1. If you convert from a centi metric unit to a milli metric unit, are you going to a smaller unit or a larger unit in size?

2. How much difference is there between a *centi* unit and a *milli* unit?

3. How many decimal positions would you move for a conversion from *centi* to *milli*?

4. Converting from *centi* to *milli*, would you be moving the decimal position to the left or to the right? __________ You would be using the ____________________ rule.

5. Let's review your answers with an example.

   24.5 centimeters = __________ millimeters

6. Now, let's try it in reverse.

   53 millimeters = __________ centimeters

There will be more practice on these conversions as you go through the next sections on *standard* metric units.

Length

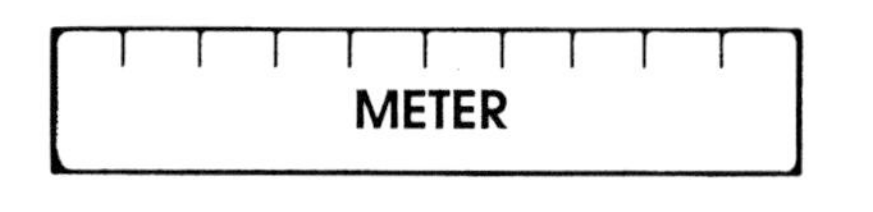

The standard reference for ***length*** in the Metric System is the ***meter*** (abbreviated as m). You will be using three prefixes with this standard unit: milli ($\frac{1}{1000}$), centi ($\frac{1}{100}$), and kilo (1,000).

Remember to use the rules pertaining to the movement of the decimal point.

? QUESTION

1. The abbreviation for millimeter is **mm**. How many millimeters are in a meter? __________
2. The abbreviation for centimeter is **cm**. How many centimeters are in a meter? __________

3. The abbreviation for kilometer is **km**. How many kilometers are in a meter? ______ (Did you get tricked by this question?)
4. How many meters are in a kilometer? ______
5. How many mm are in 1 cm? ______
6. How many cm are in a mm? ______
7. 40 cm = ______ m.
8. 40 cm is what fraction of a meter? ______
9. 40 cm is what % of a meter? ______
10. $\frac{32 \text{ m}}{10 \text{ mm}}$ = ______ **Hint:** You must have the same units on the top and bottom before doing the division.
11. 1.2 m x 30 = ______ m = ______ cm.

Volume

The standard value for ***volume*** in the Metric System is the ***liter*** (abbreviated as **l**).

The prefixes used for metric length units also apply to volume units. However, milli is the only prefix that we will use during this lab. The milliliter (**ml**) is a very common unit for the scientific measurement of small volumes of liquid.

? QUESTION

1. How many milliliters are in a liter? ______
2. How many liters are in a milliliter? ______
3. 355 ml = ______ l (This is a familiar volume for a canned soda.)
4. 750 ml = ______ l (This is a familiar volume for wine bottles.)
5. 15 ml = ______ l (This volume is one tablespoon.)
6. What % of a liter is 15 ml? ______
7. $\frac{1}{2}$ liter = ______ ml

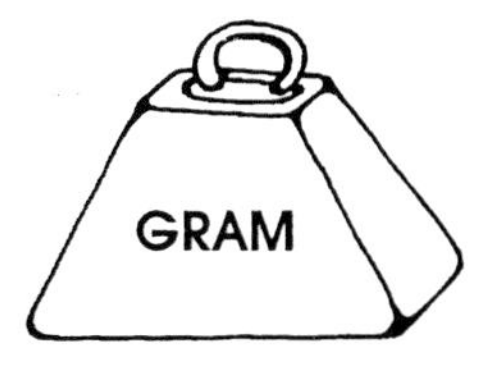

Weight

The standard reference for ***weight*** in the Metric System is the ***gram*** (abbreviated as **g**).

The metric prefixes that we will use during this lab are milli and kilo.

? QUESTION

1. How many grams are in a kilogram? ______
2. How many kilograms are in a gram? ______
3. How many milligrams are in a gram? ______
4. How many milligrams are in a kilogram? ______
5. 454 g = ______ kg (This is the weight of a small loaf of bread.)
6. What % of a kilogram is 454 g? ______
7. 2.265 kg is the weight for a small bag of sugar. You are baking cupcakes for a school fund drive. It takes 100 g of sugar to make one batch of cupcakes. How many batches of cupcakes can you make with one small bag of sugar? (Show your work.)

EXERCISE #2

"Conversions: Metric ⟷ English"

You are familiar with the English System of measurement. In this Exercise we will review some conversions between the English and Metric Systems.

Length

Materials

- A combination meterstick/yardstick.

Procedure

[1] Make a line on a piece of paper exactly 10 inches long.

[2] Measure that same line in centimeters. 10 inches = __________ cm.

? QUESTION

1. How many centimeters are in one inch? __________ cm = 1 inch
2. How many centimeters are in one foot? __________ cm = 1 foot
3. How many centimeters are in one yard? __________ cm = 1 yard
   (Check your answer with the measuring stick.)

Volume

Materials

- A 1-liter graduated cylinder.
- A 10-milliliter graduated cylinder.
- A 1-quart graduated cylinder.
- An eyedropper.
- A teaspoon.

Procedure

The curve of the water line is called the ***meniscus***.

[1] Fill the container marked "1 quart" with water. (There may be a painted line indicating the exact 1 quart amount.)

[2] Pour the 1 quart of water into a graduated cylinder for measuring liters.

? QUESTION

1. How many ml are there in a quart? __________ ml = 1 quart.
2. Which is the greater volume? (circle your choice)

   4 Liters   or   1 Gallon

Procedure

[1] Count the number of drops of water it takes to fill the small graduated cylinder to the 1-milliliter mark.

1 ml = __________ drops

[2] Fill the 10-milliliter cylinder to the 5-milliliter mark. Pour that amount into the teaspoon.

5 ml = __________ teaspoon

Weight

? QUESTION

Try this weight problem. Home Depot bought sacks of cement from a company in Mexico. The sign above the cement display reads: "100 lb $4.95." The 100 lb cement sacks were also marked "45 kg."

1. How many sacks of cement can you buy for $4.95? ____________

2. How many pounds are in one kilogram? ____________

3. How much do you weigh in kilograms? ____________

Temperature

The English measurement of temperature is in degrees Fahrenheit (°F). Using this scale, water freezes at 32°F and boils at 212°F.

There is a scientific temperature scale that is more like a metric scale. It uses Celsius (°C). On this scale water freezes at 0°C and boils at 100°C.

Procedure

Use the dual scale thermometer illustration to answer the questions below.

? QUESTION

1. How many °F are there between the 0°C mark and the 100°C mark?

____

Therefore, how many °F are there in one °C? ____________

2. Water freezes at what temperature in °F? ____________

At what temperature in °C? ____________

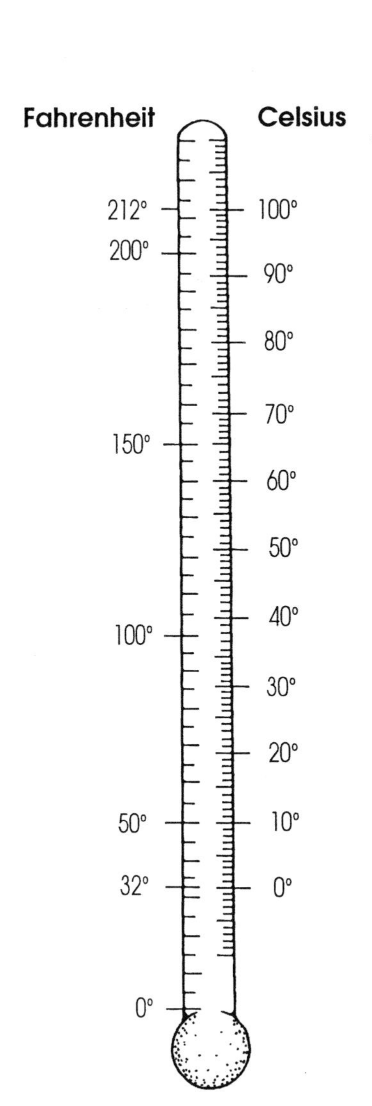

3. Water boils at what temperature in °F? ____________

At what temperature in °C? ____________

4. What is your favorite air temperature in °F? ____________

What would that be in °C? ____________

5. What is your idea of a "hot day" in °F? ____________

What temperature is that in °C? ____________

6. Your normal body temperature is 98.6°F. Your child's forehead seems to be hot. You grab a Celsius thermometer by mistake and take her temperature. It reads 37°. Should you rush her to the hospital?

____

(The formula for converting °C into °F is: **°F = °C x 1.8 + 32**.) Why must the number 32 must be added?

7. Your cookbook says that roast beef is rare at 140°, medium at 160°, and well done at 170°. You like your beef cooked between medium and rare and only have a meat thermometer in °C. What temperature will the thermometer reach for the roast to be done the way you like it?

8. The water temperature gauge on your new Volkswagen reads 85°C. Are you overheating your engine? ____________

Conversion Factors

(Optional Exercise)

Check with your instructor to see if you are responsible for doing this section before going on.

Conversion Factors (also called dimensional analysis in science) are a bit more complex than the methods used in the previous exercises. However, Conversion Factors are capable of solving both simple and complex problems in science, and they have an automatic self-checking feature when the rules are followed.

Rule 1

The top unit of a Conversion Factor must be equal to the amount of the bottom unit of that factor.

There are many Conversion Factors that you can use. Examples:

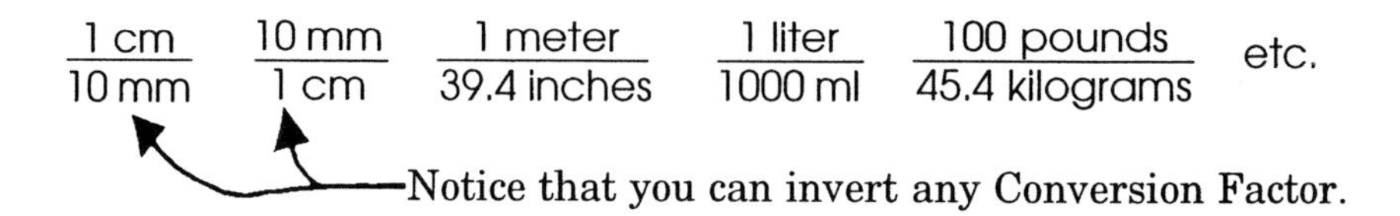

All of the above factors are correct (OK factors) because the amount on the ***top*** of the factor is the *same amount* as that on the ***bottom*** of the factor.

$$1 \text{ cm} = 10 \text{ mm, so } \frac{1 \text{ cm}}{10 \text{ mm}} = \frac{\text{same}}{\text{same}} = \text{OK factor}$$

Rule 2

The Conversion Factor that you choose must automatically cancel out the starting unit and leave the desired unit as your answer.

For example, if you were starting with a kilometer unit and wanted to convert it into a yard unit, then the correct Conversion Factor would be:

$$\frac{1094 \text{ yards}}{1 \text{ kilometer}}$$

Remember: Starting Unit x Conversion Factor = Desired Unit

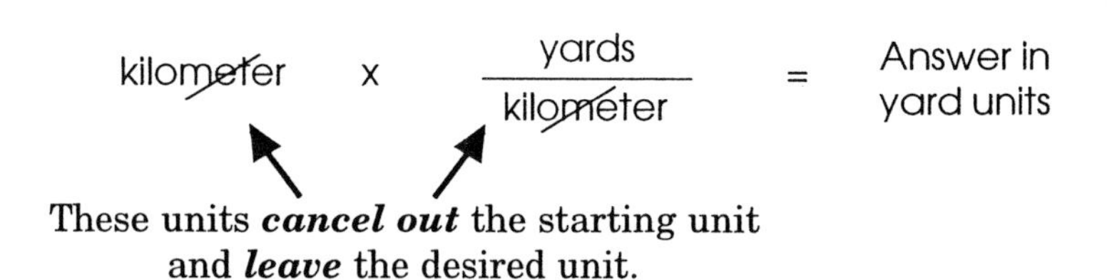

RULE 3

You may have to use more than one Conversion Factor in order to solve a particular conversion problem.

Let's use the previous example, and follow it through to see how Conversion Factors work. As you will see, knowing a little information and using a Conversion Factor can get you where you want to go.

In the example under Rule 2, we wanted to convert kilometers into yards. We decided that the necessary Conversion Factor must be:

$$\frac{\text{yards}}{\text{kilometers}}$$

But, what if you couldn't remember how many yards there are in a kilometer? The solution is: Use as many Conversion Factors as are necessary, starting with the one that you *do remember*.

Perhaps you remember that 1 meter = 39.4 inches. And you know that there are 1,000 meters in a kilometer. How can you use these Conversion Factors to answer the above question?

Start with what you know:

$$1 \text{ kilo}\cancel{\text{meter}} \times \frac{1000 \text{ me}\cancel{\text{ters}}}{1 \text{ kilo}\cancel{\text{meter}}} \times \frac{39.4 \text{ in}\cancel{\text{ches}}}{1 \text{ me}\cancel{\text{ter}}} \times \frac{1 \text{ yard}}{36 \text{ in}\cancel{\text{ches}}} =$$

. . . and ***cancel*** the "units" as you work through the problem.

$$\frac{1000 \times 39.4 \times 1}{1 \times 1 \times 36} = \text{\_\_\_\_\_\_\_\_}$$

Your Answer in yards in a kilometer!

? QUESTION

1. How many inches are there in 50 cm?
Hint: There are 39.4 inches in a meter.
What is the starting unit? ______________
What is the desired unit? ______________
What is your answer? ______________
(Show your Conversion Factors.)

2. How many liters are there in 10 gallons?
Hint: There are 946 ml in 1 quart.
What is the starting unit? ______________
What is the desired unit? ______________
What is your answer? ______________
(Show your Conversion Factors.)

EXERCISE #3

"Personal List of Metric References"

Think of something easy to remember that you can associate with each of the metric units below. Perhaps a centimeter might be the width of one of your fingernails. Be specific.

Whatever you choose as a reference, *make sure it's something you won't forget!*

Name: ______________________________

Metric Unit **My Personal Reference**

Length:

mm ______________________________

cm ______________________________

km ______________________________

Volume:

ml ______________ (How many drops?) ________

l ______________________________

Weight:

mg ______________________________

g ______________________________

kg ______________ (How many pounds?) ________

EXERCISE #4

"A Test of Your Measuring Skills"

How To Weigh an Object

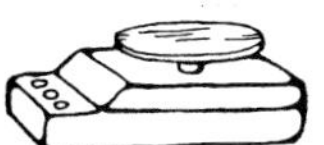

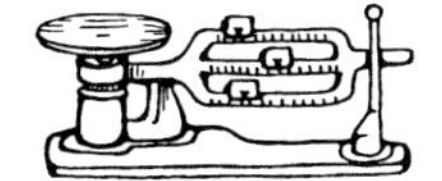

Weighing balances are used to measure the weight of an object. It is important that the object to be weighed is put inside a weighing container so that the material is not spilled on the balance pan.

Step 1 "Zero" the scale. (Your instructor will show you how to do this.)

Step 2 Then, weigh the weighing container. Why is it important to weigh the container first?

Step 3 Put the substance or object into the weighing container, and weigh them together.

Step 4 Determine the difference between the weights for Step 1 and Step 2.

Materials

- Some table salt.

Procedure

Your instructor will give you specific directions for using each type of weighing scale. Make sure that you "zero" the scales before weighing.

[1] Weigh 2.7 g of table salt, and put that amount onto a piece of paper. This is the recommended daily intake of salt in your diet.

[2] Weigh 11.6 g of table salt, and put that amount next to the other pile. This is the typical daily salt intake by people in our society.

A Test of Your Weighing and Volume Skills

Experimental Questions

It has been said that pennies before 1982 are heavier than pennies after 1982. We have 20 pennies in each category on the lab table.

1. Are the post-1982 pennies lighter or heavier?
2. If there is a difference in weight, then is that difference because . . .
   - **a.** they are not made of the same metal, or
   - **b.** they are not the same size coin.

Procedure

[1] Design and perform an experiment to answer the questions above.

[2] Compare groups of at least 10 coins in each category to measure the difference in weight or volume.

[3] Ask to see one of the post-1982 pennies that has been cut in half to show its metal composition.

LAB REPORT

Question:

Hypothesis:

Experimental Design:

Results:

Conclusions:

Statistics

There's statistics and there's damn statistics.
—Attributed to Benjamin Disraeli
(British Prime Minister 1874–80)

People have a strange relationship with statistics. When someone beats us in an argument by using their facts and figures, we hate statistics. But when we use statistics to win our side of an argument, we love those facts and figures. Statistics is much more than numbers. It is the mathematics for the scientific collection, organization, and interpretation of numerical data.

But statistics is not just calculation activities. During scientific research, significant choices are made about the analysis of data, and these choices influence the quality of the research. In this regard, statistics is a bit like art; it is best learned by the experience of doing it.

In the following Exercises you will learn a few of the basic ideas about probability and statistics and their relationship to scientific investigation.

Exercise # 1 "Probability and Statistics" 35
Exercise # 2 "Data Patterns" 40
Exercise # 3 "Describing Data Patterns" 44
Exercise # 4 "Two Statistical Tests of Experimental Data" 51
Exercise # 5 "The Hypothesis" 55

EXERCISE #1

"Probability and Statistics"

Probability begins with knowing everything. Statistics begins with knowing nothing.

Probability is the mathematical tool for analyzing situations in which there is a description of all individuals in a population. The term ***population*** refers to all the organisms or objects within a group. For example, when playing a card game, you know all of the individual cards in the deck of cards (the population). Therefore, you would use *probability* to calculate the odds of getting a particular hand in a game of poker.

Statistics is the mathematical tool for figuring out a situation in which you don't have a description of the entire population, but only have data from a small subgroup of the population. The small subgroup is called a ***sample***.

Probability Used Here

Bottle full of 100 black marbles and 100 white marbles.

- **Problem:** What is the chance of picking a black marble?
- **Solution:** Do calculations using the laws of probability.

Statistics Used Here

Bottle full of black and white marbles, but you don't know how many of each.

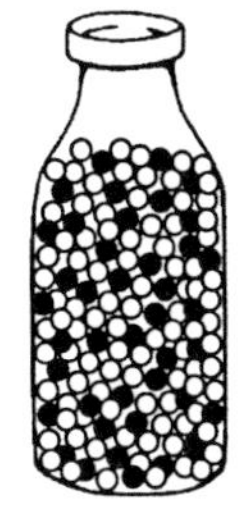

- **Problem:** What portion of the marbles is black?
- **Solution:** Take a sample, and draw a conclusion based on the results.

Probability

Probability is a mathematical description of how likely it is that an event will happen. Since any event that you can imagine is either impossible, certain, or somewhere in-between, the mathematical probability of an event is either 0 or 1, or a value between 0 and 1.

Probability = 0

Probability = 1

Multiplication Rule for Compound Events

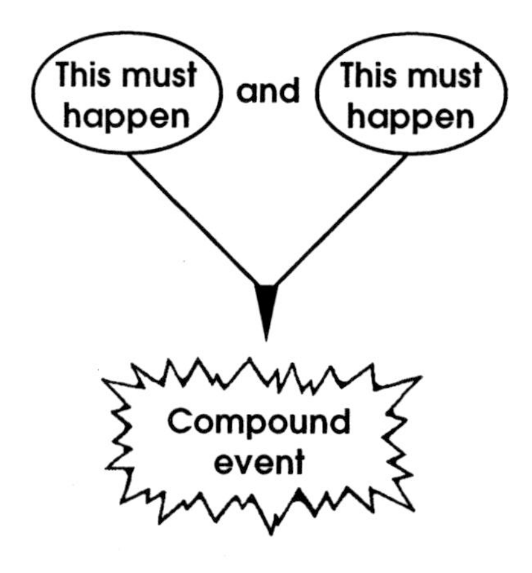

A ***compound event*** is a situation in which more than one independent event must happen simultaneously. For example, if you are interested in the probability of reaching into the bottle (of 100 black and 100 white marbles) without looking, and randomly picking out a black marble twice in a row, then this situation is a compound event. ***If a situation requires more than one independent event to happen, then you* multiply *the separate probabilities of the required events.*** The answer is the probability for the compound event.

The probability of picking one black marble is $\frac{1}{2}$. But, picking the same color twice is a compound event. Two events (black and black) must happen for the compound event (two black marbles in a row) to occur. The probability of picking two black marbles = $\frac{1}{2} \times \frac{1}{2}$.

? QUESTION

1. When do you use probability? (circle your choice)

When you know all of the population
or
When you have a sample

2. When do you use statistics? (circle your choice)

When you know all of the population
or
When you have a sample

3. The probability of a nearly certain event is close to the number ______.

4. The probability of a very rare event is close to the number ______.

5. When do you use the multiplication rule in calculating probability?

6. What is the probability of picking a black marble on the first try from a bottle of 100 black and 100 white marbles?

7. What is the probability of picking a black marble three times in a row (replacing the marble after each "pick")?

8. What is the probability of picking a black marble and then a white marble? (Assume that the first marble is replaced before picking the second marble.)

9. Show how you would calculate the probability of picking the Ace of Spades out of a deck of cards five times in a row. (The Ace is replaced into the deck after each draw.)

Addition Rule

Situations that require the Addition Rule are not the same as those that require the Multiplication Rule.

When calculating probabilities, it is necessary to determine if there is more than one way that an event can happen. For example, what if you wanted to know the probability of there being a boy and a girl in a two-child family? You would first ask yourself, is there more than one way for this situation to happen? The answer is yes! A daughter could be born first and then the son, or a son could be born first and then the daughter.

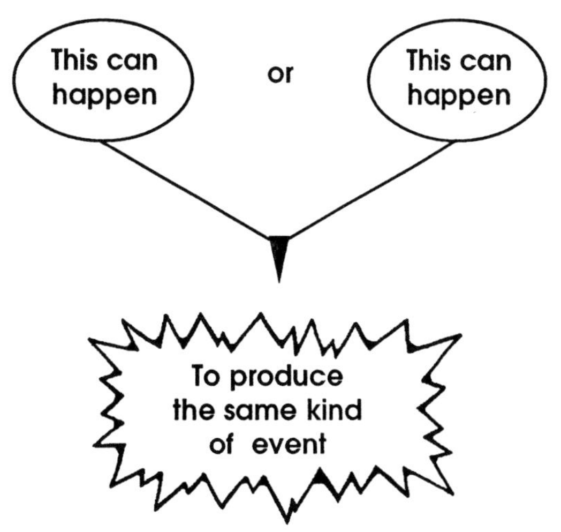

If there are two or more independent ways for an event to happen, then the probability of that event occurring is calculated by adding the probabilities of the two or more independent ways. (In a statistics course, you would learn another variation of the addition rule that is used when the different ways for an event to happen aren't completely independent. This variation is not covered here.)

Returning to our example, there are two ways that a boy and a girl can be born into a family. Use the addition rule! But first, you must use the *multiplication* rule to calculate the probability of each way—girl-then-boy and boy-then-girl. Then *add* the probabilities of the two different ways.

? QUESTION

1. What is the probability of a daughter being born first and then a son?

2. Which rule did you use to answer question #1?

3. What is the probability of a son being born first and then a daughter?

4. Now, what is the probability of a family having both a boy and a girl (not specifying which is born first)?

5. Which rule did you use to answer question #4?

Probability and Goals in Your Life

The laws of probability were revealed as people discovered how events actually happen in the world. Let's see if the laws of probability can suggest how you might try to make something happen in your life.

? QUESTION

1. Imagine that you have planned a business venture that requires a series of ten independent events to occur for the business to succeed. Use your knowledge of the multiplication rule in probablility to change your business plan in a way that will definitely increase your chance of success.

2. What does the addition rule suggest you could do to increase your chance of success?

3. When you increase the number of events that must happen, the total probability of success moves closer towards . . . (circle your choice)

   0 or 1

4. Which approach has a better chance of achieving success? (circle your choice)

   Increase the number of events required for success
   or
   Add alternate ways of reaching success

5. How could you apply this same knowledge to planning your college career?

Statistics

There is no way to keep track of all the numbers.

Data are numerical values related to a sample, collected from a population during an experiment. These numbers represent observations or measurements. It is difficult to compare or see a pattern in many numbers, so science has invented ways of summarizing the data. These summary numbers are called ***statistics***, and they include terms such as *mean* and *range*. Specific statistics are discussed in Exercise #3. For now, remember that there are two types—***descriptive*** and ***inferential***.

An example of a descriptive statistic about circadian rhythms might be the average of the lowest daily body temperatures of a group of students. Statistical inference occurs when you use a pattern of data to predict how different factors are related to each other. For example, a graph of the daily change in body temperature might predict the ideal time for students to study. One type of statistics *describes,* the other *predicts.*

? QUESTION

1. Describe the basic difference between descriptive statistics and inferential statistics.

2. Which type of statistic do you think would be most controversial? (circle your choice)

   Descriptive or Inferential

3. Discuss why you would expect to hear more arguments about the validity of using statistics than the validity of using probability.

EXERCISE #2

"Data Patterns"

In a statistics class, you would investigate a variety of data patterns, and learn the differentiating characteristics that apply to each one. This Exercise will introduce you to four common patterns.

Patterns of Correlation

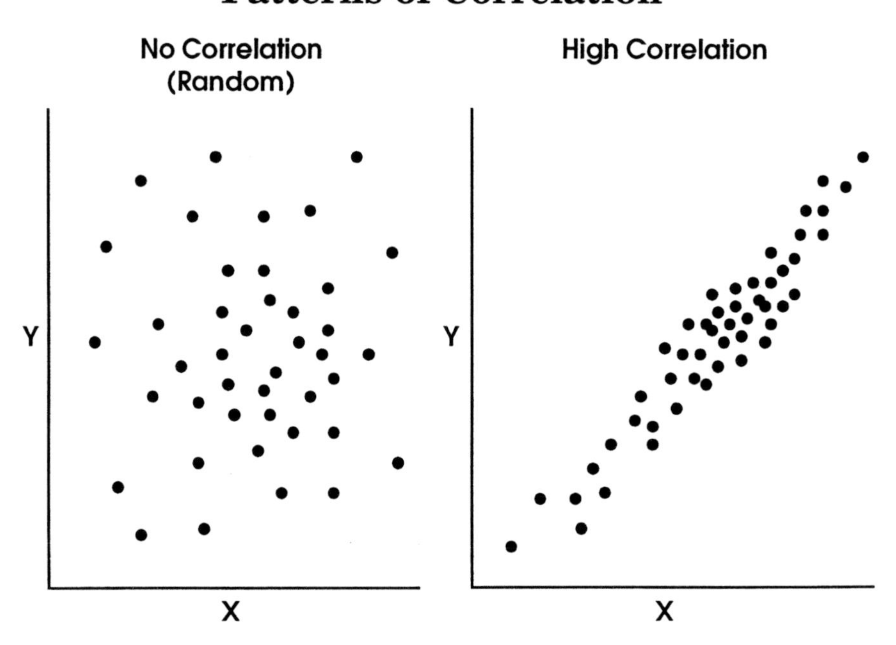

A pattern of correlation can reveal cause and effect.

The above scatter diagrams represent two different patterns of association between variables in an experiment. The Y axis is called the ***dependent variable*** because it "depends on" (or is controlled by) the X axis factor—the ***independent variable***. For example, in an investigation of the effects of watering on lawn growth, the "height of the lawn" is plotted along the Y axis, and the "amount of water" is plotted along the X axis.

A quick examination of the pattern on the right reveals that there is a strong relationship between the two factors (X and Y) in the experiment. However, our eyes can trick us, especially when we have a prior opinion and the pattern is not quite so obvious. There are statistical tools that can help us avoid this problem, but those tools are more appropriate in a statistics course. In this lab class we will limit our discussion to the obvious data patterns.

? QUESTION

1. Define *dependent variable.* On which axis is it plotted? ________

2. Define *independent variable.* On which axis is it plotted? ________

Patterns of Description

The data collected during an investigation are often presented in graphic form. Measurements are put into categories, and the number of individuals in each category is marked on a graph. Such a graph could be used to summarize the entrance examination scores for college freshmen.

A pattern of description can indicate that something is affecting the data spectrum.

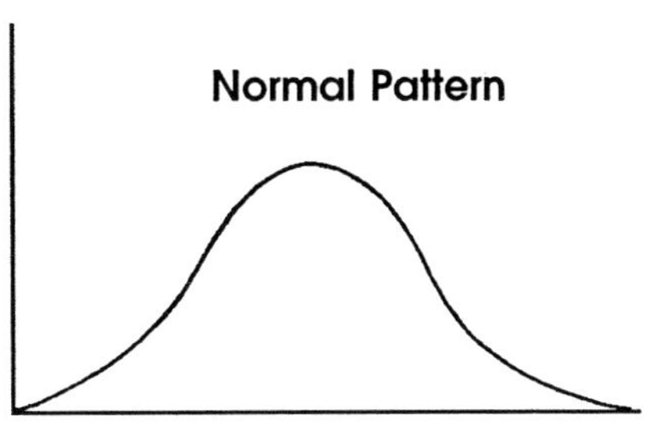

- Clustered around a central point.
- Curve drops off in both directions.
- Symmetrical.

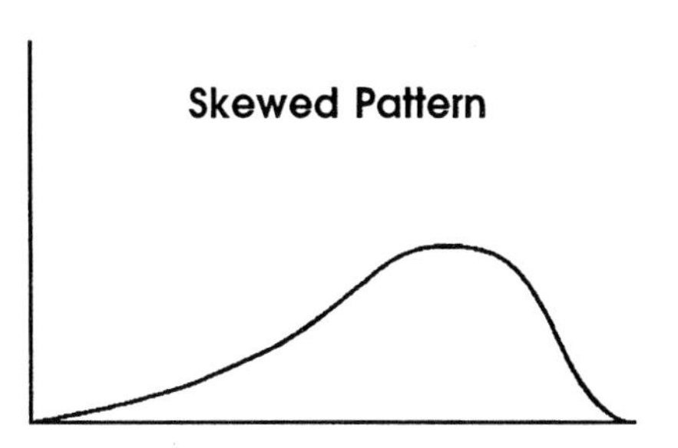

- One end of distribution is favored over other end.
- Can be skewed to the left or to the right.

These two patterns of description are also called *probability distributions* by statisticians because the patterns show the percent of individuals occurring in each measurement category of the population studied.

It is important to determine the pattern of data distribution. Some statistical tests apply only to situations in which the data show a normal distribution pattern like the one on the left. On the other hand, data that show a skewed pattern indicate that something is affecting one end of the data spectrum. This "something" might be the next step in your investigation.

? QUESTION

List the four data patterns presented in this Exercise. Describe what each pattern tells you.

1.

2.

3.

4.

Making a Descriptive Graph

Make a descriptive graph of the data in this table. Follow the directions below.

1994 Per Capita Personal Income by State

State	Income	State	Income
Alabama	18,010	Montana	17,865
Alaska	23,788	Nebraska	20,488
Arizona	19,001	Nevada	24,023
Arkansas	16,898	New Hampshire	23,434
California	22,493	New Jersey	28,038
Colorado	22,333	New Mexico	17,106
Connecticut	29,402	New York	25,999
Delaware	22,828	North Carolina	19,669
District of Columbia	31,136	North Dakota	18,546
Florida	21,677	Ohio	20,928
Georgia	20,251	Oklahoma	17,744
Hawaii	24,057	Oregon	20,419
Idaho	18,231	Pennsylvania	22,324
Illinois	23,784	Rhode Island	22,251
Indiana	20,378	South Carolina	17,695
Iowa	20,265	South Dakota	19,577
Kansas	20,896	Tennessee	19,482
Kentucky	17,807	Texas	19,857
Louisiana	17,651	Utah	17,043
Maine	19,663	Vermont	20,244
Maryland	24,933	Virginia	22,594
Massachusetts	25,616	Washington	22,610
Michigan	22,333	West Virginia	17,208
Minnesota	22,453	Wisconsin	21,019
Mississippi	15,838	Wyoming	20,436
Missouri	20,717		

Source: Bureau of Economic Analysis—U.S. Department of Commerce

Step 1 Label the Y axis of each graph "Number of States in Each Income Category." Label the X axis "Income Categories (in thousands)."

Step 2 Decide on the ***interval*** for income categories (this has been done for you). This step takes a bit of practice to do well because each interval should not be too large or too small. You will plot the data using two different interval categories. Both are pretty good choices.

Step 3 Count the data in each income category, and make a bar graph. Plot the data twice using each of the two interval categories.

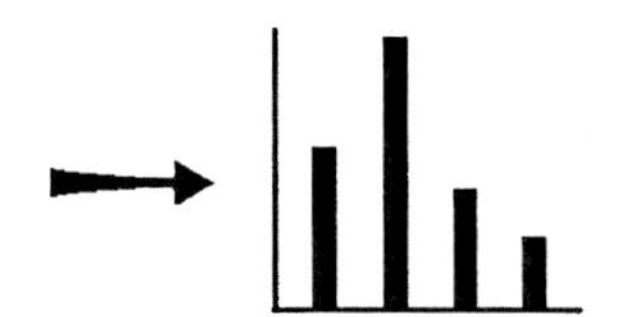

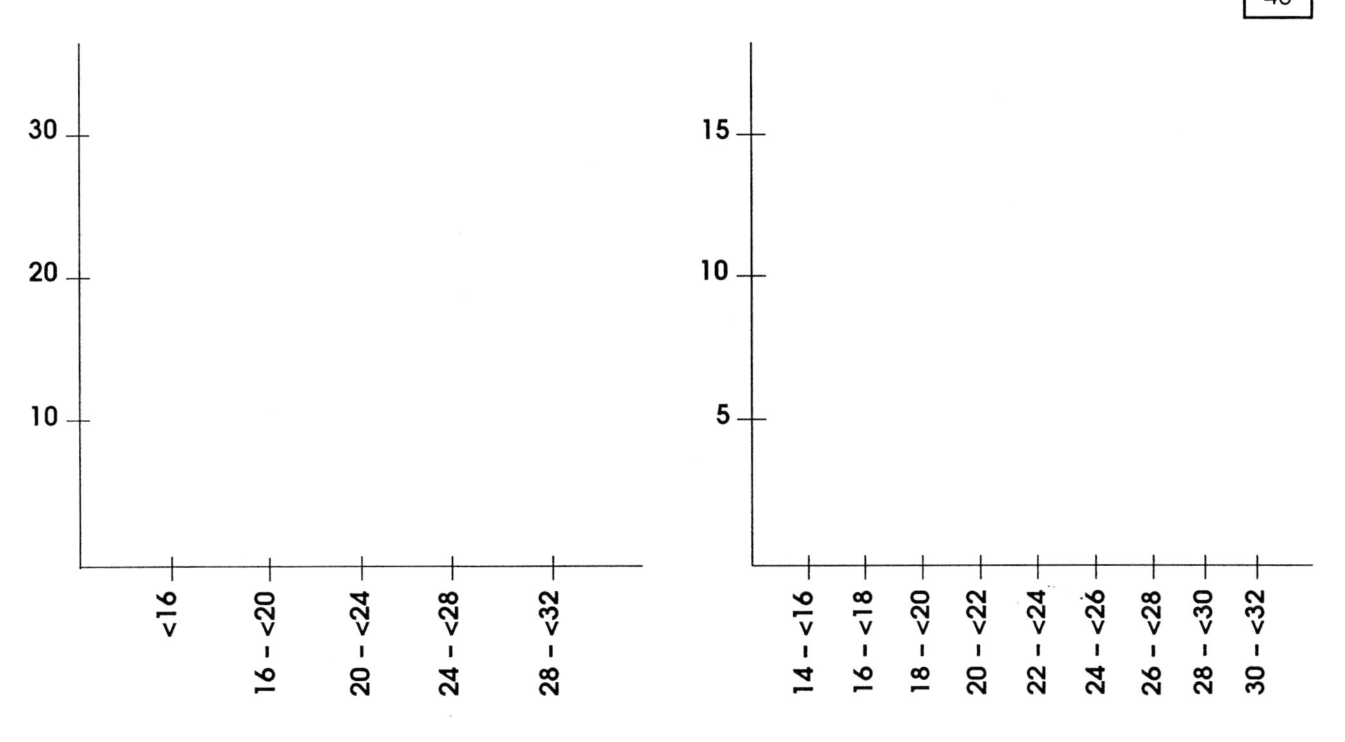

? QUESTION

1. For each descriptive graph, draw a curve that represents the distribution of incomes. Which pattern is it?

2. You should see a slight skew in the pattern. Is the "tail" of the curve on the right side or the left side?

3. Discuss a possible explanation for the slight skew in the distribution of state incomes.

EXERCISE #3

"Describing Data Patterns"

There are three good ways to describe the pattern of data collected during a scientific investigation:

1. Graphic form of the data (normal, skewed, or other),
2. Most representative data value (mean or median), and
3. Spread of the data (variation).

Graphic forms of data were discussed in Exercise #2. The ***most representative data value*** and ***spread of the data*** are presented next.

Most Representative Data Value

The mean is subject to "the oddball effect."

The graphic form of a data pattern usually shows more data clustered around one part of the pattern and the numbers trailing off in both directions. The Normal Pattern and the Skewed Pattern are examples of this kind of clustering. There are two kinds of statistics for describing the most representative data value in the middle of a cluster of data—***mean*** and ***median***.

Mean

The ***mean*** is another word for average.

$$\text{mean} = \frac{\text{sum of all of the data measurements}}{\text{number of data in the sample}}$$

However, there are situations in which the mean can be distorted by what we call "the oddball effect." The following distribution of data represents the amount of money (in $100s) in the saving accounts of fourteen students:

			5	6			
		4	5	6	7		
2	3	4	5	6	7	8	9

? QUESTION

1. What is the mean of the above data?

2. Now, assume that the "oddball effect" happens, and you collect a 34 and 35 instead of the 8 and the 9. What is the new mean of the data?

3. What kind of distribution best describes the second set of data that includes the "oddball effect"? (circle your choice)

   Normal or Skewed

4. In which distribution is the mean the most representative data value? (circle your choice)

   Normal or Skewed

Median

The ***median*** is the data value that is exactly in the middle of the ranked data. First, you put the data into order from low to high measurements. The data 25, 9, 21, 16, 12 would be reordered as 9, 12, 16, 21, 25. Pick the middle number. That number is the *median*. If there is an even number of data, average the two "middle" values in the ranking. The median of the example data is 16. If the data are collected randomly and the distribution of numbers is not skewed, then the median and mean statistics will be the same number. The mean of the above five numbers is 16.6—amazingly close to the median 16, isn't it?

? QUESTION

1. Go back and calculate the medians for the two sets of numbers presented under the discussion of the mean. Does the median change when there is an "oddball effect"?

Median of first set of data = ____________

Median of data with the "oddball effect" = ____________

2. In which type of distribution is the median better than the mean as the most representative data value? (circle your choice)

Normal or Skewed

3. If the data show a normal distribution pattern, then the mean is ________________ the median.

4. Calculate the mean for each set of data below.

STUDENT HEART RATES	
Normal Day	**Test Day**
68	80
80	80
72	80
78	80
84	75
82	80
82	86
63	84
80	85
80	88
68	80
66	68
68	72
64	64
78	91
80	96
80	98
75	92
80	86
52	64
mean = _______	mean = _______

5. Put each set of data from the previous question into order from low to high measurements, and then find the median. (The median will be the average between #10 and #11 in the ranking.) You will use these ranked data values again in Exercise #4.

RANKED HEART RATES

High ⟷ Low

Normal Day	Test Day
median = ______	median = ______

Variation

Variation is a measure of the spread of data values. There are two common statistical terms for variation—*range* and *standard deviation.*

Range

The ***range*** of a set of data includes the highest data value and the lowest data value. For example, the range of heart rates on a normal day is from 52 to 84. However, because the range can be exaggerated by the "oddball effect," it is usually considered to be a misleading estimate of variation. The range is not used in scientific statistics.

Standard Deviation

The ***standard deviation*** is another way to describe the variation of data in a sample. The standard deviation measures the spread of data compared to the mean. It is something like calculating the average deviation (from the mean) for all of the data. It is worth learning because there are powerful uses for the standard deviation that can't be derived from calculating the average deviation or the range.

Step 1

List all data in your sample and calculate the mean. (You already did this for the normal-day and test-day heart rates. Refer to the tables on the next two pages.

Step 2

Determine the difference between each heart rate and the mean for the sample. These calculations are the ***deviations*** of the data values. Complete the calculations for the second column in each of the two tables.

At this step, if you add all of the deviations and divide by the sample size, then you have calculated the *average deviation* of the sample. However, mathematicians tell us that a more powerful statistic (the standard deviation) can be calculated if you put in just a little more work after Step 2.

Step 3

Square the deviation for each of the data values. Then determine the sum of all of the squared deviations. (Ask your instructor whether you should continue doing the calculations in the third column of each table, or if the calculations will be done using a computer.)

Step 4

Calculate the standard deviation. In this equation, n = sample size, and the sum of squared deviations is the answer to Step 3.

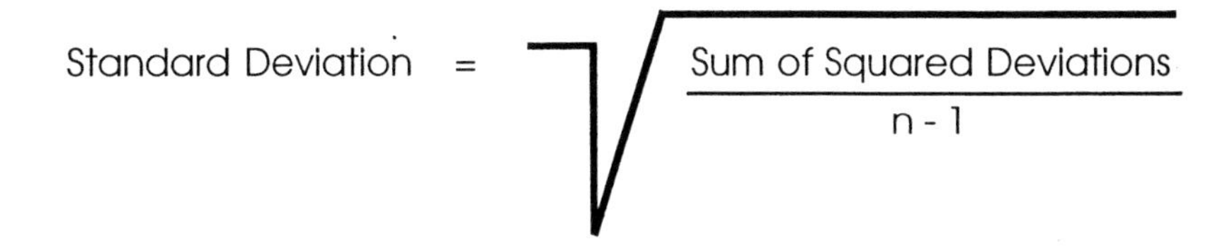

Technical note to advanced students: In Step 3, the deviations are squared which converts any negative deviations to positive numbers. But the numbers are then in "squared units," which is why the square root calculation in Step 4 returns the units back to "normal units."

There are two ways of calculating the standard deviation for the test-day and normal-day heart rate samples. You can do the math step by step, or you can use a scientific calculator that automatically does all of the math after the data have been entered. Your instructor will suggest which method to use in today's lab. Do the math now!

STANDARD DEVIATION FOR NORMAL-DAY HEART RATES

Heart Rates	Deviations (Difference Between H.R. and Mean)	Square of Deviations
68	68 - 74 = 6	36
80	80 - 74 = 6	36
72	72 - 74 = 2	4
78	78 - 74 = 4	16
84		
82		
82		
63		
80		
80		
68		
66		
68		
64		
78		
80		
80		
75		
80		
52		

mean = 74

Sum of Squared Deviations = __________

Standard Deviation = __________

STANDARD DEVIATION FOR TEST-DAY HEART RATES

Heart Rates	Deviations (Difference Between H.R. and Mean)	Square of Deviations
80		
80		
80		
80		
75		
80		
86		
84		
85		
88		
80		
68		
72		
64		
91		
96		
98		
92		
86		
94		
mean = 81		

Sum of Squared Deviations = ________

Standard Deviation = ________

Importance of Variation

You have learned that the standard deviation is the best statistic for describing the variation of data. But, why is it important for you to know the variation? Two uses of variation are shown next, and a third use is presented in Exercise #4.

The mean may be meaningless without the variation.

Case 1

Variation in the data should be examined to reveal possible differences between samples that have the same mean.

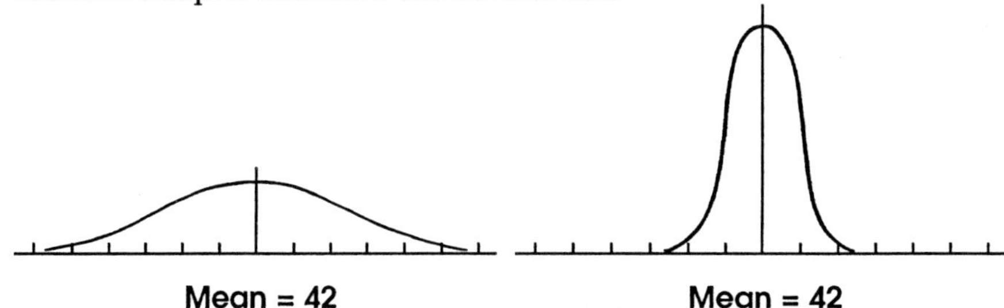

Case 2

The standard deviation can also be used to show what percent of the population is at different distances from the mean.

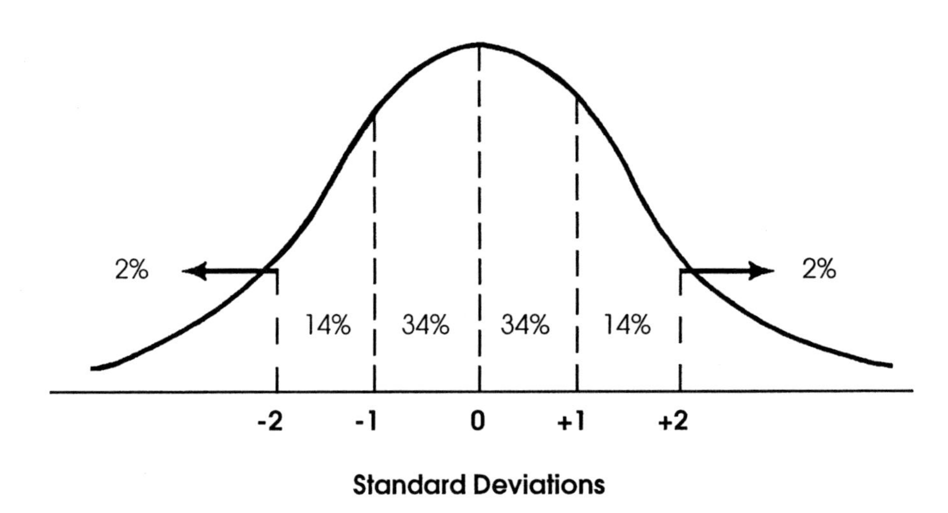

Technical Note: This distribution curve is a simplified version of what you would learn in a statistics class.

EXERCISE #4

"Two Statistical Tests of Experimental Data"

If you prove that two situations are not the same, haven't you proven that they are different?

Statistical tests are developed by mathematicians to help investigators determine whether two sets of data are the same or are different from each other. For example, if you discover that the heart rates on a normal day are not the same as those on a test day, then you might conclude that taking exams affects the heart rates of students. In this Exercise, you will learn two simple statistical tests to check your conclusions. The first example, the "Standard Error Test," works best when you have a lot of data, and those data show a normal pattern of distribution. The second example, the "Comparison of Ranks Test," works in all situations, and it is especially useful when the data has a skewed pattern or has a lot of variability.

Standard Error Test

$$\text{Standard Error} = \frac{\text{Standard Deviation}}{\sqrt{n}}$$

n = sample size

The Standard Error Test is a simplified version of the parametric statistical tests that are presented in a statistics course. Two of those other versions are the *Student t Test* and the *Z Test*. Your instructor may want you to use those tests. If so, realize that they are based on the same logic as the Standard Error Test.

When Sample A has a mean different from Sample B, then you might suspect that there is a real difference between the two samples. But, how can you be sure? A special mathematical tool, called the ***standard error***, will help you to make that decision.

Mathematicians have determined that if you select an infinite number of samples from a population, then you would get a standard normal distribution of sample means. For example, 68% of all samples should fall within ± 1 standard error unit of the true population mean. And 96% of all samples should fall within ± 2 standard error units of the true population mean.

Remember that standard error is a special case of standard deviation. It describes the variation among different sample means.

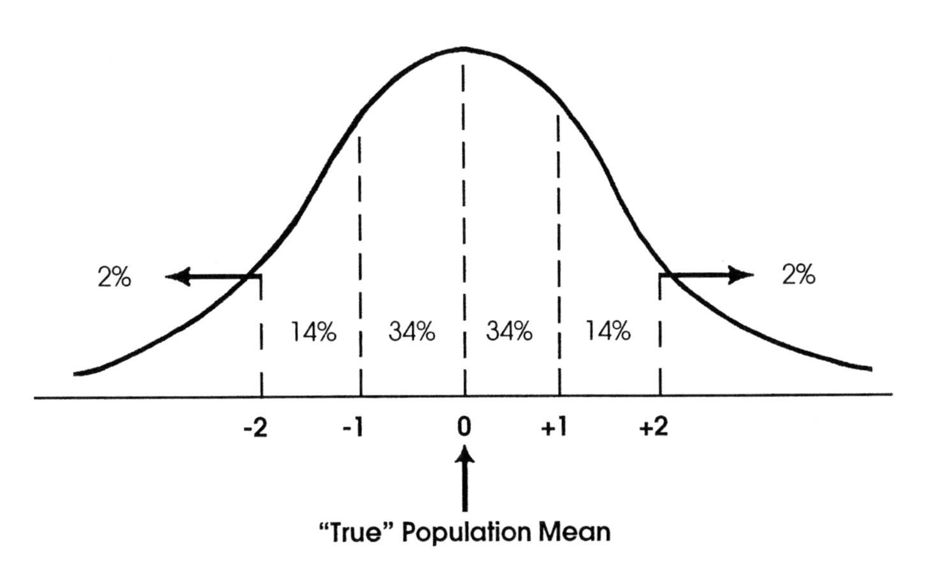

The standard error statistic tells you how much "error" you can expect when sampling a population. Furthermore, mathematicians have determined that we can use the standard error to test whether there is a real difference between two samples (test-day heart rates and normal-day heart rates). The logic goes something like this: If Sample A is a truly different situation from Sample B, then the mean of Sample A should differ from the mean of Sample B by at least two standard error units. Remember, this Standard Error Test is a simplified version of what a statistics course would present. A more accurate approach would be to use statistical tools such as *confidence intervals, probability values of T scores, maximum error,* etc.

Step 1

Calculate the standard error for each of the two samples you want to compare. (Refer to your data tables in Exercise #3 and the formula for standard error on the previous page.)

Test-Day Heart Rates	Normal-Day Heart Rates
mean = ____	mean = ____
standard error = ____	standard error = ____
mean ± 1 S.E. = ____ to ____	mean ± 1 S.E. = ____ to ____

Step 2

For each sample, calculate the range of value from -1 S.E. of the mean to +1 S.E. of the mean. If the mean heart rate is 60 and the standard error is 2, then the range of values is 58 to 62.

Step 3

If you discover that the mean ±1 S.E. of Sample A does not overlap the mean ±1 S.E. of Sample B, then there probably is a real difference between these two samples. An example of two non-overlapping samples would be:

Sample A mean ±1 S.E. = 58–62

Sample B mean ±1 S.E. = 63–66.

? QUESTION

Based on the Standard Error Test, is there a real difference in heart rates on test days compared to normal days? ____________

Comparison of Ranks Test

The Comparison of Ranks is an example of a non-parametric test presented in a statistics class. These tests are used when the data shows a skewed pattern or there is a lot of variability.

Step 1

Refer to the ordered ranking of heart rates you performed in question #5 under "Median" in Exercise #3. Arrange the data in each sample from low to high values. (You already did this for the heart rates.)

Step 2

Rank the data considering *both* samples. Start with the lowest heart rate, and assign number ranks to each data value. There are forty data values in the two samples of heart rates. This means that the data will be ranked from 1 to 40. In our example, the heart rates of 52 and 63 from the Normal Day get the first two rankings because they are the lowest data values of the two samples. The next three heart rates are all 64 and are tied for ranks 3, 4, and 5. To be fair to each sample, you assign the rank of 4 to each of the three heart rates. (Four is the average of the tied rankings.) The heart rate of 66 from the Normal Day gets the ranking of 6. Continue assigning ranks until you reach rank 40.

Write down the numbers 1 to 40, and check them off as you assign ranks. Be careful when there is a tie between samples! *Check with your instructor if you get confused.*

Step 3

Calculate the sum of the ranks for each of the samples.

Test-day sum of ranks = ________

Normal-day sum of ranks = ______

Step 4

Determine which sample has the lower sum of ranks. You will be using that number in Step 5.

Lower sum of ranks = ______

Step 5

The lower sum of ranks is compared to a particular value in the Comparison of Ranks Table on the next page. You enter the table using the sample size for Sample A along the bottom and Sample B up the side. In this case, both of our heart-rate samples have n = 20. Therefore, the ***critical value*** is 348.

Step 6

If the lower sum of ranks (Step 4, above) is less than the critical value (348), then the samples are assumed to be *statistically different.*

? QUESTION

Based on the Comparison of Ranks Test, is there a real difference in student heart rates on exam days as compared to normal days? ______

CRITICAL VALUES FOR COMPARISON OF RANKS TEST*

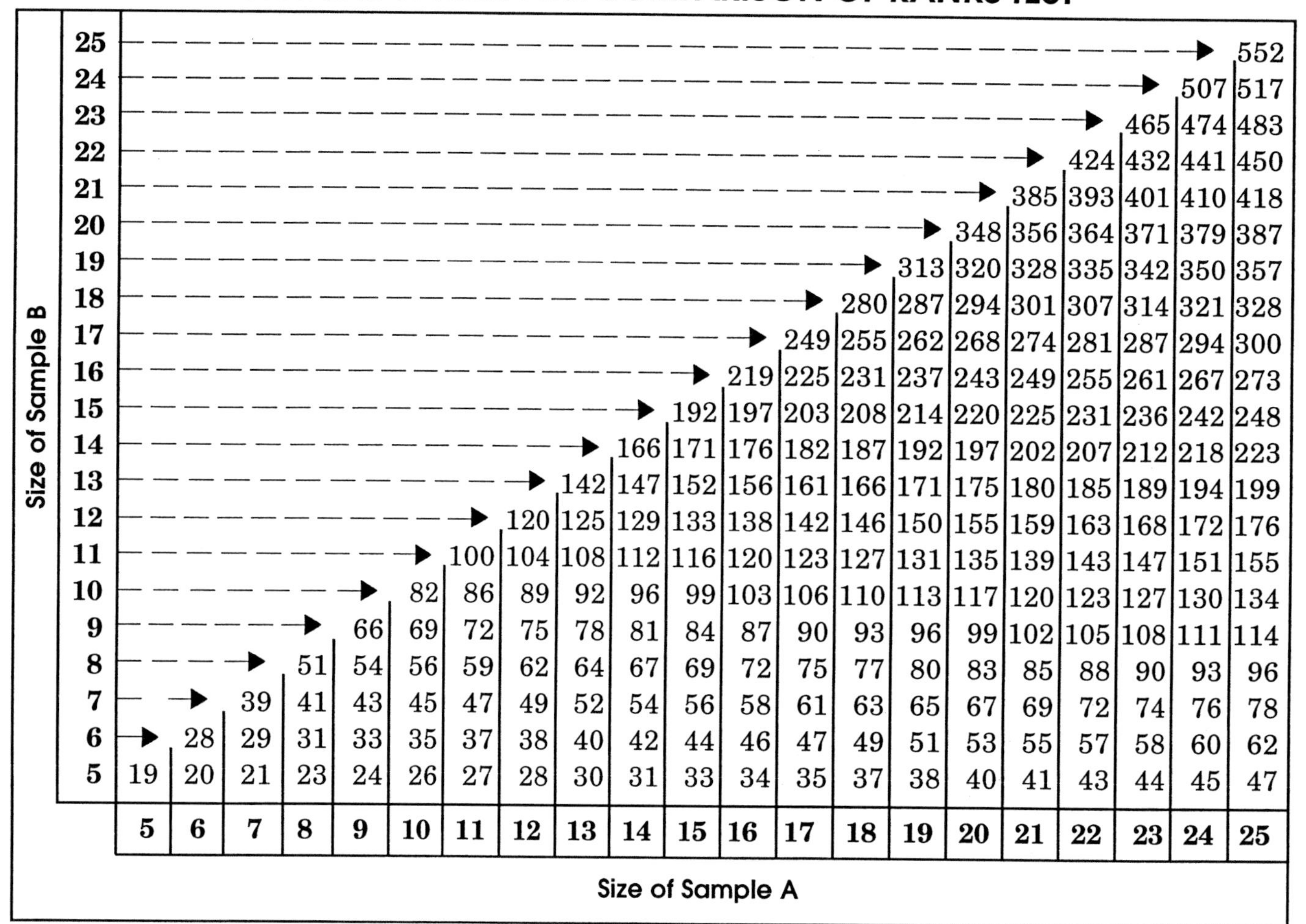

Size of Sample B	5	6	7	8	9	10	11	12	13	14	15	16	17	18	19	20	21	22	23	24	25
25																					552
24																				507	517
23																			465	474	483
22																		424	432	441	450
21																	385	393	401	410	418
20																348	356	364	371	379	387
19															313	320	328	335	342	350	357
18														280	287	294	301	307	314	321	328
17													249	255	262	268	274	281	287	294	300
16												219	225	231	237	243	249	255	261	267	273
15											192	197	203	208	214	220	225	231	236	242	248
14										166	171	176	182	187	192	197	202	207	212	218	223
13									142	147	152	156	161	166	171	175	180	185	189	194	199
12								120	125	129	133	138	142	146	150	155	159	163	168	172	176
11							100	104	108	112	116	120	123	127	131	135	139	143	147	151	155
10						82	86	89	92	96	99	103	106	110	113	117	120	123	127	130	134
9					66	69	72	75	78	81	84	87	90	93	96	99	102	105	108	111	114
8				51	54	56	59	62	64	67	69	72	75	77	80	83	85	88	90	93	96
7			39	41	43	45	47	49	52	54	56	58	61	63	65	67	69	72	74	76	78
6		28	29	31	33	35	37	38	40	42	44	46	47	49	51	53	55	57	58	60	62
5	19	20	21	23	24	26	27	28	30	31	33	34	35	37	38	40	41	43	44	45	47

Size of Sample A

* If the lower sum of ranks of the two samples is equal to or less than the table value, then there is a 95% chance that the two samples really are different.

EXERCISE #5

"The Hypothesis"

Experiments are designed to test the hypothesis.

The scientific method begins with a possible explanation for something we are observing. That explanation is called the ***hypothesis***. The hypothesis generates questions that we attempt to answer with an experiment.

As it turns out, the very specific way that you state a hypothesis leads to a particular experimental design. It also determines the kind of possible error that you might make in your study. Today, science generally agrees that all experiments should be designed to test what is called the ***Null Hypothesis***. The simplest statement of a null hypothesis is:

Sample A = Sample B.

Error

Mistakes do happen!

Scientists accept that research errors can happen. Therefore, error is one of the first topics discussed when scientists are trained to use the scientific method. There are two kinds of error that can be made when a researcher tests a null hypothesis.

Two Possible Errors When Testing the Null Hypothesis	
Type I Error	The null hypothesis is actually true (Sample A = Sample B), but the experimenter mistakenly concludes it is false.
Type II Error	The null hypothesis is actually false (Sample A ≠ Sample B), but the experimenter mistakenly concludes it is true.

? QUESTION

1. State the null hypothesis in your own words.

2. The way that you state the null hypothesis determines the kind of ____________ that you might make in your conclusions.

3. Define Type I error in your own words.

4. Define Type II error in your own words.

Avoiding The Catastrophes of Error

Protect yourself from a Type I error, and you automatically expose yourself to a Type II error.

You have learned that science can make two kinds of mistakes—Type I and Type II. The question is, "How do we avoid each of these errors?"

In science, Type I error can be reduced to as low a risk as you choose. For example, if you want to reduce possible Type I error when using the Standard Error Test, then you could require that the two sample means must differ by 3 S.E. units in order to conclude that they are different. (You used a threshold of only 2 S.E. units in Exercise #4.) If you want to reduce Type I error even more, you could require 4 S.E. units as the threshold.

The dilemma is that when you reduce the chance of a Type I error, the chance of a Type II error automatically increases. That is because the factors creating these two kinds of error are linked.

When either one of them is reduced, the other one tends to increase. Therefore, when scientists reduce Type I error, they must make adjustments in their experimental design (such as greatly increasing the sample size) to prevent Type II error from increasing out of their control.

Another way of avoiding the catastrophe of Type I and Type II errors lies in the way that the null hypothesis is stated. There are two very important rules for stating the null hypothesis.

Rule #1 The null hypothesis is written in the *equality form.* (Heart rates on a test day are *the same as* heart rates on a normal day.)

Rule #2 The null hypothesis should be stated in the "safest and most conservative way" to minimize danger. It must be stated in such a way that a Type II error won't result in someone dying or something just as serious.

Humans love to prove an idea wrong.

The main reason that scientists use the equality form of hypothesis is because this sets up a situation in which they can attempt to prove the hypothesis wrong. (Humans excel at thinking up experiments designed to try to prove an idea *wrong.*) It is much more difficult to devise ways of proving that an idea is correct, especially when not much is known about the phenomenon. Scientists can deliberately set the Type I error to a very low risk, and they are very accurate at proving a hypothesis wrong. However, when scientists set the Type I error at a low risk, this increases the chance that they will make a Type II error.

Now, what about Type II error? Remember, Type II error means that the null hypothesis actually is false (Sample A $\neq$ Sample B), but the experimenter mistakenly concludes it is true. Science is vulnerable to Type II error. Because of this vulnerability to making a Type II error, scientists use Rule #2 (safe and conservative) to avoid the "catastrophe of error."

A safe and conservative null hypothesis minimizes the danger coming from a Type II error. The following story illustrates how the "safe and conservative" null hypothesis became established in science.

During the early years of World War II, London was being battered by daily bombing attacks. Many people had been killed and the city was in chaos. It was a bleak time for the British people. Then a truly amazing man, the Prime Minister, took control of the situation. His name was Winston Churchill, and he inspired his people to persevere.

One of Churchill's brilliant inspirations occurred when he ordered a group of the best British scientists and mathematicians to meet with him. This stern and pugnacious man walked into the room of experts and stared at each of them without saying a word.

If they were wrong, people would die.

After what must have seemed like an eternity, he bellowed, "We will not cower to the scoundrels! London must go on!"

Then he proceeded to give these men and women an incredible challenge. They were to determine which areas of London would be safe from the bombing—safe for work, safe for transportation, safe for schools and the nation's children.

The experts were shocked. They had been ordered to assume an enormous responsibility. What if they were wrong? People would die! This was their state of mind as they worked feverishly to develop a method of testing various hypotheses so they could identify which areas would be safe from bombing. Soon they came to realize that there were two very different kinds of null hypotheses.

The Safest Null Hypothesis

Scientists test a hypothesis by comparing data collected from two samples, or in some experiments, a sample may be compared to an expected value. Remember, the simplest statement of the null hypothesis is Sample A = Sample B. But, how we *define* Sample A and B becomes crucial! In the bombing of London, the safest null hypothesis became:

"The zone in question (Sample A) ***is the same as known bombing zones*** (Sample B).***"***
(Zone A is a bomb area.)

Why is this the safest way to state the null hypothesis? Remember, science tries to prove this hypothesis wrong. Science is at its best and most certain when proving a hypothesis wrong. Statistical comparisons were designed to test this hypothesis. The statistical threshold for making a Type I error was set at an extremely low risk. If England's scientists were successful at disproving this safe hypothesis ("Zone A is a bomb area"), then they could be reasonably certain that they had actually identified a place where bombs would not fall—a safe area for England's people.

Most important of all, the scientists were able to protect themselves against the danger of making Type II error. (Remember, science is most vulnerable to Type II error.) Type II error in a safely stated null hypothesis would have led them to mistakenly conclude that their hypothesis was true (that Zone A was a bomb area, when actually it was a safe area). But, no one would die from this mistake!

The Risky Null Hypothesis

"The zone in question (Sample A) ***is the same as known unbombed zones*** (Sample B)."
(Zone A is a safe area.)

This is the risky way to state the null hypothesis. It leads to catastrophe if a Type II error is made. (A Type II error for this hypothesis would mistakenly conclude that Zone A is a safe area when it is actually a bomb area.) People would die because of this mistake!

This dramatic example from World War II demonstrates why science uses only the "safe and conservative" way of investigating problems.

? QUESTION

1. How is Type I error reduced in science?

2. What is the value of stating the hypothesis in an equality form (Sample A = Sample B)?

3. Why do scientists try to prove ideas wrong rather than trying to prove them correct?

4. What kind of error is science most vulnerable to? (circle your choice)

Type I or Type II

5. What is it about science that naturally leads to Type II error?

6. Explain the importance of stating the null hypothesis in the "safest and most conservative way" to minimize danger.

7. Using the same approach as science, decide which is the "safest and most conservative" null hypothesis. (circle your choice)

A. Defendant is assumed to be innocent, and the trial tries to prove him guilty.

or

B. Defendant is assumed to be guilty, and the trial tries to prove him innocent.

8. If you made a Type II error using hypothesis **A**, what would be the result? (circle your choice)

Guilty man goes free   or   Innocent man goes to jail

9. If you made a Type II error using hypothesis **B**, what would be the result? (circle your choice)

Guilty man goes free   or   Innocent man goes to jail

Summary

Statistics, probability, data patterns, data tests, hypotheses, error—there are many fascinating and important aspects in the topic of statistics that you can't cover in one lab. Consider taking a good statistics class. Most people don't know what you've already learned in this lab. What you have learned will give you a more thoughtful reaction the next time you hear someone say . . .

There's statistics, and there's damn statistics.

Chemistry Concepts

(Tutorial)

The first principles of the universe are atoms and empty space; everything else is merely thought to exist.
—Democritus (~ 400 B.C.)

Chemistry is the science that deals with the nature of and the changes in the composition of matter. The relationship of particles within the atom, and the interactions between the atoms account for everything that we call "matter" in our world. At first it seems impossible to believe that air, liquid, and solid could be made of the same basic particles. But when we watch a cube of ice melt in a pool of water, then drip onto a hot plate and become vapor, we are forced to conclude that there is something going on that we can't explain in any other way. Our eyes can't see the particles but we know they must be there—in another world—in a world of atoms.

This set of Exercises is meant to be an aid to your understanding of some of the chemical processes and interactions that will be discussed throughout your study of biology. Here we will start the discussion, and we hope it will encourage you to continue your exploration into the science of chemistry.

Exercise #1 "The Periodic Table of Elements" 77
Exercise #2 "Atoms and Isotopes" 79
Exercise #3 "Molecular Formation" 82
Exercise #4 "The Rule of Eight in Molecular Formation" 85

EXERCISE #1

"The Periodic Table of Elements"

The word ***element*** originally came from a Latin word meaning "first principle." In Roman times, people thought that the universe was made up of four basic elements: earth, air, fire, and water.

Today, the modern chemist defines an *element* as the most basic kind of substance which cannot be broken into simpler substances by ordinary chemical processes. Over the years, more than 100 elements have been discovered and described.

Water is the most abundant material on the surface of the earth. However, using our definition, we don't call water an element because it can be broken down into two simpler substances: *hydrogen* and *oxygen*. Hydrogen and oxygen cannot be broken down into simpler substances, so we conclude that they are elements.

Study the "Abbreviated" Periodic Table of Elements on the next page for a few minutes. Notice how the different elements are listed, grouped, and numbered.

All living things are chemical combinations based on the lighter elements. These are shown in this "Abbreviated" Periodic Table. (A complete Table of Elements can be found in your textbook or the lab room.) During the formation of the earth, heavy elements sank into the inner molten core and were not available for use in chemical processes taking place on the surface of the planet. Look again at the Table of Elements. These are the common elements that were available in the "Soup of Life" from which you came.

? QUESTION

1. Do you know what the numbers on the periodic chart mean?

2. Do you know why the elements are listed in certain groups?

3. Do you recognize the names of any elements on the chart by their alphabetical symbols?

4. If you have access to a complete list of element names (in your textbook or on a lab chart), then you will notice that the element symbol is not necessarily an abbreviation of the word we use for that element. Can you find the common name for the element symbols Na and K?

"ABBREVIATED" PERIDODIC TABLE OF ELEMENTS

Group I	Group II	Group III	Group IV	Group V	Group VI	Group VII	Inert Gases
1 **H** 1							2 **He** 4
3 **Li** 7	4 **Be** 9	5 **B** 11	6 **C** 12	7 **N** 14	8 **O** 16	9 **F** 19	10 **Ne** 20
11 **Na** 23	12 **Mg** 24	13 **Al** 27	14 **Si** 28	15 **P** 31	16 **S** 32	17 **Cl** 35	18 **Ar** 40
19 **K** 39	20 **Ca** 40						

EXERCISE #2

"Atoms and Isotopes"

Atomic Structure

One of the most important questions in chemistry is: What makes one element different from another element?

The smallest particle of an element that can exist and still retain the chemical properties of that element is called an ***atom***. Today we know that there are particles even smaller than atoms called ***protons***, ***neutrons***, and ***electrons***. It is the various combinations of these subatomic particles that control the chemical properties of the different elements.

All atoms consist of an inner core that contains most of the mass of the atom, and an outer zone in which there are very light particles, called electrons, that are either like energy waves or like particles, depending on how they are studied. If you monitor electrons with a machine that looks for particles, they will appear to be particles. And if you try to find them as energy waves, that machine will record them as energy waves. Their nature is both particle and wave.

Electron
(a particle and an energy wave at the same time)

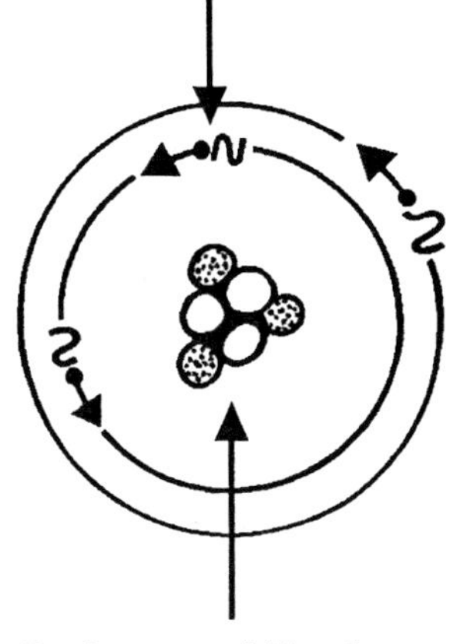

Protons and Neutrons of the Nucleus

The inner core is called a ***nucleus***, and it is made up of two kinds of particles: *protons* and *neutrons*.

The outer zone of the atom is filled with those tricky, high energy particles called *electrons*.

Atom Facts

1. A proton has a mass of 1 atomic mass unit.
2. A neutron also has a mass of 1.
3. An electron has a mass of almost 0 (zero).
4. A proton has an electrical charge of +1.
5. A neutron has an electrical charge of 0 (neutral).
6. An electron has an electrical charge of −1.

In this discussion we must simplify definitions and explanations. A chemistry class would look at the concepts of weight and mass with important differences. Here, we consider them as equals. (This is, in fact, not true.)

Atomic Number and Atomic Mass

The Periodic Table of Elements contains the atomic mass and atomic number of each element. These numbers indicate the number of protons, neutrons, and electrons in each element's basic structure.

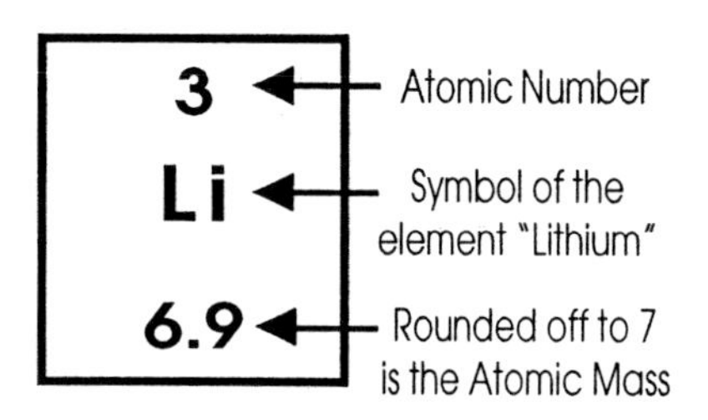

- Each square in the Periodic Table represents an *element.*
- There are two numbers in each square.
  - The smaller number at the top of the box is the ***atomic number***.
  - The number at the bottom of the box is the average of all isotopes of the element. If you "round off" this value to the nearest whole number, then that number is the ***atomic mass***. On your "Abbreviated" Periodic Table we have provided the rounded off atomic masses.

Atomic Number = # of Protons

Atomic Mass = # of Protons + # of Neutrons

A single atom has an overall neutral charge because the number of + charges (protons) is equal to the number of – charges (electrons) in the atom. **Hint:** If you know the atomic number, then you automatically know how many protons or electrons are in that atom. This is because the number of protons is equal to the number of electrons.

? QUESTION

1. Using the "Abbreviated" Periodic Table of Elements, put the appropriate symbol by the element's name, and determine the number of protons, electrons, and neutrons.

Element	Symbol	# of Protons	# of Electrons	# of Neutrons
Lithium				
Beryllium				
Boron				
Carbon				
Nitrogen				
Oxygen				
Fluorine				

2. Use the chart you just completed to answer the following questions.

   a. What is the atomic number for carbon? ________

   b. What is the atomic mass for lithium? ________

   c. What is the atomic number for nitrogen? ________

   d. What is the atomic mass for oxygen? ________

Isotopes

Elements exist in different forms called ***isotopes***. You may have heard of radioactive iodine which is used in medicine, heavy water which is used in nuclear reactors, or carbon 14 which is used in paleontology to date fossils. All of these are isotopes of the elements iodine, hydrogen, and carbon, respectively.

Isotopes: Same properties, but different weights.

Isotopes of an element usually have the same chemical properties. Isotopes have the *same* number of *protons,* but have *different* numbers of *neutrons*. That means they will have slightly different *atomic weights*.

The atomic weight as recorded in the Periodic Table is the average weight of all the various isotopes of that element. That value is recorded at the bottom of each element box.

The mass (protons + neutrons) of individual isotopes is not included in the Periodic Table, but instead is indicated by putting a number to the upper left of the element's symbol. For example, ^{12}C is an isotope of carbon and is called carbon twelve because its atomic mass is twelve. ^{14}C has an atomic mass of 14; it is called carbon fourteen. Because of its radioactive properties, ^{14}C can be put into a sugar molecule, and that molecule can be "followed" through your metabolism to determine how you process your food Calories.

$$^{14}C_6H_{12}O_6$$

SUGAR

Isotopes are often used as a chemical "tag" to follow the molecules of a particular substance through a biochemical process. For example, in medicine, radioactive iodine is injected into the the bloodstream and the isotope is tracked with special machines as it filters into the patient's kidney and urine. It is used to discover kidney disease or kidney malfunction.

? QUESTION

1. How many neutrons are in ^{14}C ? ________ Its atomic mass is ________.

2. How many neutrons are in ^{12}C ? ________ Its atomic mass is ________.

3. ^{2}H is called deuterium. It is the isotope of hydrogen that is used to make heavy water for nuclear power plants.

How many neutrons are in ^{2}H ? ________ Its atomic mass is ________.

4. Why do the isotopes of an element have the same basic chemical properties?

5. What is the primary use of isotopes in medicine and biological research?

EXERCISE #3

"Molecular Formation"

The Eight Groups of Elements

Look at your "Abbreviated" Periodic Table of Elements. About 150 years ago, chemists discovered that all of the known elements could be arranged from low atomic weight to high atomic weight. And, when they were put in this order, it was noticed that there were repeating physical and chemical properties. It was amazing! There was some kind of "order" in the universe that could be seen through the study of chemistry.

So, the Periodic Table of Elements was constructed on those repeating patterns. The elements in Groups I, II, and III are all ***metals*** (except for hydrogen, which chemically reacts like a metal but is a gas). The other groups are called the ***non-metals***.

Metals easily *release* electrons to other groups. The ***group number*** of the metals indicates *how many electrons* can be released: 1, 2, or 3.

The non-metal elements in Groups IV through VII *attract* or *share* electrons from other elements. Now, if you take the group # of a non-metal element and subtract it from the number *eight*, then you will know the *number of electrons* that this element will attract or share.

For example, the element oxygen will attract two electrons when it combines with other elements to form molecules.

8 - Group VI = 2

To complete our discussion of the groups in the Table, the ***inert gases***—the final group—get their name because they *do not* react with any other element.

The key to understanding the process of ***chemical bonding*** comes from understanding the interactions between the electrons of different elements when they bump into each other. Do they create a chemical reaction, or is there no chemical reaction? The number "8" turns out to be a very important clue in predicting molecular formations in chemistry.

? QUESTION

Keeping in mind what you have just learned about the Periodic Table of Elements, answer the following questions.

1. Does sodium (Na) attract or release electrons? ____________________

How many electrons are involved in the transfer? ________________

2. Does chlorine (Cl) attract or release electrons? ________________

How many electrons are involved in the transfer? ______________

3. What do you think would happen if a sodium atom and a chlorine atom bumped into each other?

4. Welders use helium (He) and argon (Ar) gases to blow over the metal during the welding process. Why would they want to use these gases?

5. If 10 magnesium (Mg) atoms and 10 chlorine (Cl) atoms were allowed to bump into each other, could they combine to form a substance?

How many molecules would be made? ________

Would there be any atoms left over? __________

Which ones? ______________________ How many? ________

6. My friend, the inventor, says that she has just made a translucent, light-weight metal by combining aluminum with helium and silicon. She wants me to invest $10,000 in the process, and promises that we will be millionaires in only six months. What do you think I should do?

Why?

Ions

An ion is charged because it has either gained electrons or lost electrons.

An atom has the same number of positive charges (protons) as negative charges (electrons). Some atoms can lose electrons (Groups I, II, and III), and other atoms can gain electrons (Group VII). An ***ion*** is an atom that has lost or gained electrons, and thereby becomes charged (either + or–) An ion that has *gained* an electron will acquire an excess *negative* electric charge, and the ion formed by *losing* an electron will have a *positive* charge. These ions are very important in our life processes, and we would all die without them. (More about that later.)

Ions act differently than uncharged atoms. Think of them as being like magnets + and – , and other atoms as being like non-magnets. Does that start to explain how molecular formation can happen?

The "Abbreviated" Periodic Table of Elements provides some information about the formation of ions. As we discussed before, Group I elements can lose *one* electron, and Group II elements can lose *two* electrons. Guess what Group III elements can lose? *Three* electrons!

So, when atoms in Groups I, II, and III become ions, they will have a ________________ charge.

Groups IV, V, and VI elements usually *share* electrons with other elements and don't form ions. These elements will have a neutral charge.

Group VII elements attract and *gain one* electron. They can form ions, and they will have a ________________ charge.

? QUESTION

1. What is the electrical charge of an electron? ________________

2. What would be the electrical charge of a fluorine (Fl) atom that gained an electron? ________________

3. What would be the electrical charge of a sodium (Na) atom that lost an electron? ________________

4. Refer to your "Abbreviated" Table of Elements, and determine whether the following atoms would form ions by losing or gaining electrons.

	Lose Electrons	Gain Electrons	Ion Electrical Charge
Cl			
Be			
Al			
Mg			
H			

Chemical Bonds

Chemical bonds hold molecules together in living organisms. Furthermore, energy for the biological processes in living things is provided by the breaking and forming of chemical bonds in the molecules that drive our metabolism.

Two important types of chemical bonds are produced by the interactions between the electrons of different atoms—ionic bonds and covalent bonds.

Ionic Bonds

What do you think would happen if a + ion of lithium got very close to a – ion of fluorine? (Remember to think about ions as behaving like the + and – ends of magnets.)

The atoms of some molecules are held together by the attraction between oppositely charged ions. That force of attraction is called an ***ionic bond***.

Ions are essential in many of the physiological processes of organisms such as nerve conduction and muscle contraction. Because ionic bonds are easily broken by water, the ions can be dissolved and moved throughout tho fluids of an organism and delivered to every part of the body.

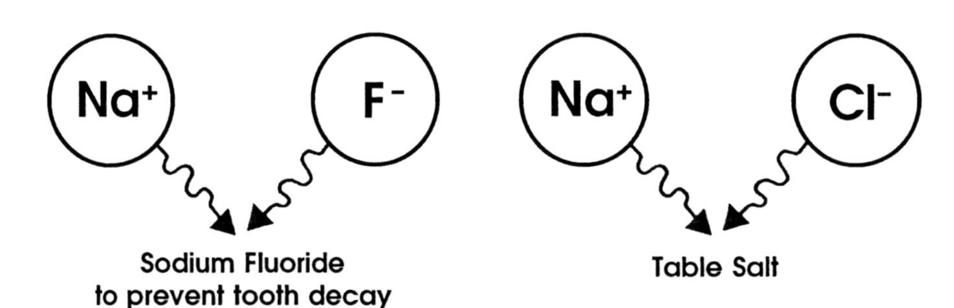

Covalent Bonds

Elements in Groups IV, V, and VI *share* electrons with other atoms when bonding. The atoms of some molecules are held together by this mutual sharing of electrons. This cohesive force is called a ***covalent bond***.

Covalent bonds are important to living organisms because they are strong enough to hold together the molecular structure of the organism. However, these bonds aren't so strong that they can't be broken and "remodeled" into other molecules that the organism needs.

Glucose Molecule

- Covalent bonds in glucose are strong enough to make wood (cellulose).
- Covalent bonds in glucose have enough energy to "drive" our metabolic processes (cell respiration).
- Covalent bonds in glucose can be "remodeled" into other organic molecules that the organism needs.

Based on what you know about elements in the Periodic Table, determine whether the following elements would interact to form ionic or covalent bonds.

Elements Reacting	Type of Bond
Na and **Cl**	
H and **O**	
H and **Cl**	
C and **H**	

EXERCISE #4

"The Rule of Eight in Molecular Formation"

Remember: The elements in the Periodic Table of Elements are listed in vertical *groups* that have Roman numbers I through VII plus a group called the "inert gases." We have been discussing each group # with a special description of the behavior of electrons in the atoms of that group.

Procedure

1. Whenever chemists observed atoms reacting with each other, they noticed that the *sum of their group numbers* usually equalled eight, or multiples of 8 (16, 24, 32, 40, etc.). This is called ***"The Rule of Eight."*** Let's see how this "Rule of Eight" works in understanding *chemical formulas.*

One Molecule of Sugar

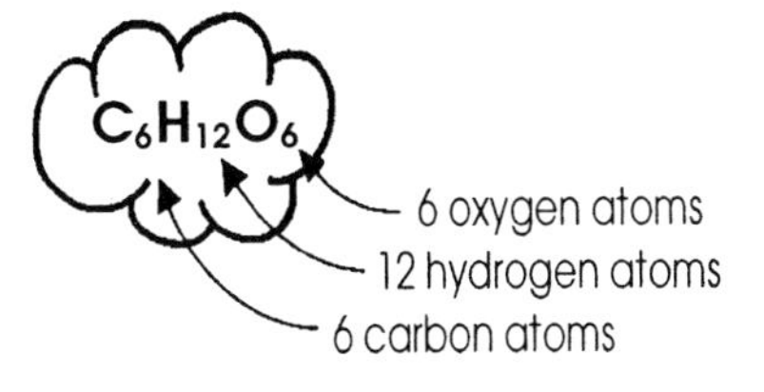

2. Chemical formulas are written to show how many atoms of each element are bonded together to make a molecule of a substance. For example $C_6H_{12}O_6$ is the chemical formula for one molecule of sugar.

3. Let's look at water: H_2O. By looking at this formula, you know that it takes ________ hydrogen atom(s) to bond with ________ oxygen atom(s).

4. Now, add the group numbers. Hydrogen has a group # of I (1). Oxygen has a group number of VI (6). Two hydrogens (1 + 1) and one oxygen (6) equals a total of 8. *"The Rule of Eight!"*

5. Try another simple one—table salt (sodium chloride): NaCl. The formula tells you that it takes ________ sodium atom(s) to bond with ________ chlorine atom(s). Add the group numbers: The group # of sodium is I (1) and the group # of chlorine is VII (7). 1 + 7 = 8. There's that *"Rule of Eight"* again!

[6] Try this one on your own. Go back to the sugar molecule: $C_6H_{12}O_6$

6 atoms of carbon (group # ____) 6 x ________ = ______

12 atoms of hydrogen (group # ____) 12 x ________ = ______

6 atoms of oxygen (group # ____) 6 x ________ = ______

Total = ______

Is the total of ____ a multiple of 8? ____

"The Rule of Eight" again!

? QUESTION

1. Determine the sum of the group numbers for the atoms in each of the molecules below, and see if your answer still agrees with the "Rule of Eight."

	Sum of Group #'s	Rule of 8?
NH_3 (ammonia)		
$CaCl_2$ (calcium chloride)		
Al_2O_3 (aluminum oxide)		

2. Use the "Rule of Eight" to write a chemical formula by determining how many *atoms* of one element would be necessary to combine with how many *atoms* of another element to form a particular molecule.

1 Mg + ____ Cl ⟶ MgCl ____ (magnesium chloride)

2 H + 1 S + ____ O ⟶ H_2SO ____ (sulfuric acid)

1 C + ____ H ⟶ CH ____ (methane)

2 H + ____ O + 1 C ⟶ H_2CO ____ (carbonic acid)

In Conclusion

Chemists explain this "Rule of Eight" phenomenon by saying that all atoms (except for the inert gases) have *incomplete* outer electron shells. When atoms react with each other, each atom ends up with a compete outer shell of 8 electrons. They call this principle "The Rule of Eight" or "The Octet Rule."

The Rule of Eight does not explain all reactions of atoms. Some atoms are stable with other arrangements of electrons. For instance, if you consider the different molecules of gas that exist separately as the air you breathe: H_2, O_2, or N_2, do their sum of group numbers equal 8? What's going on? There is much more to this idea, but we'll have to leave that to your further explorations into chemistry.

Chemistry of Water

Life first evolved in water, and most life on this planet lives in water. Furthermore, plant and animal forms are 70% to 90% water. These facts clearly establish the importance of this substance to all living beings. Water is rare in the universe, and because of this, life is also rare.

In this lab we will investigate some of the unique chemical properties of water: the most essential substance for living organisms.

Exercise #1 "Properties of Water" 87
Exercise #2 "What the Heck is pH?" 92
Exercise #3 "Molecular Motion" 94
Exercise #4 "Diffusion of Water Into and Out of Cells" 96

EXERCISE #1

"Properties of Water"

In this Exercise you will investigate some of the physical properties of water. A more thorough discussion of water's characteristics can be found in your textbook.

Hydrogen Bonds

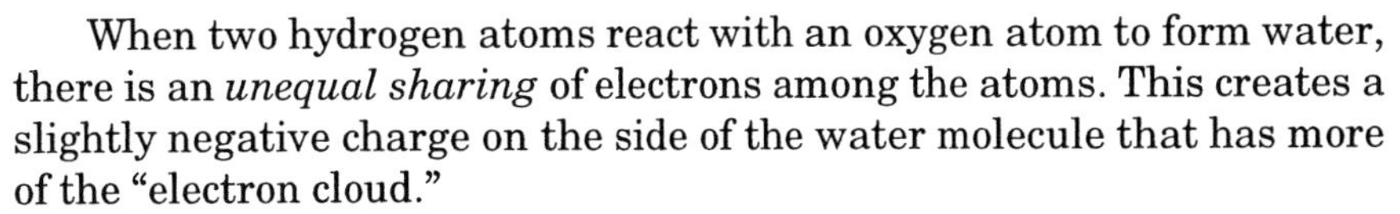

When two hydrogen atoms react with an oxygen atom to form water, there is an *unequal sharing* of electrons among the atoms. This creates a slightly negative charge on the side of the water molecule that has more of the "electron cloud."

Water molecules act like a bunch of magnets holding on to each other by the attractions between the + and – ends. The force created by those attractions is called a ***hydrogen bond***, and it is the secret to all of water's special properties.

? QUESTION

1. As water falls from the clouds, what force keeps the water in drops?

2. In order for the liquid water to evaporate and become steam, heat must be added. In a pan of boiling water, what bond is being broken by the heat of the stove?

3. So, if heat is required to evaporate water, then what is released when water condenses?

4. On a calm, but rainy day, the temperature rises slightly when it starts to rain. Explain.

Water Adhesion

Adhesion occurs when two or more different substances are stuck together as if by glue. Water has some interesting adhesion properties.

Materials

- A small container of water.
- An eyedropper.
- Four microscope slides.

Procedure

[1] Put several drops of water on one slide, then place the other slide directly on top of the wet slide.

[2] Try to pull apart the glass slides *without* sliding them past each other.

[3] Repeat this experiment with the two dry slides.

? QUESTION

1. How strong is the force of attraction between the two dry slides?

2. How strong is the force of attraction between the two wet slides?

3. What is the name of the force that holds the two slides together?

4. A freeze-dried anchovy is fairly easy to break between your fingers. Yet, when the fish is allowed to sit in water for a while, it only bends with the same effort. Explain.

5. Based on your answer to #4, what is one important role of water in living organisms?

Capillarity

Materials

- A glass capillary tube.
- A small container of water.

Procedure

Hold the capillary tube vertically between your fingers, and put the bottom end just below the surface of the water. The process you have just observed is called ***capillarity***.

? QUESTION

1. What happened when you did this experiment?

2. Draw a simple sketch of the results.

3. Explain your results. What is it about water that makes it do this? (Be specific about the forces of attraction.)

4. How is this property of water important to plants?

Capillarity

Heating Properties of Water

The difficulty of heating water reveals the strength of the hydrogen bonds and the *temperature stabilizing role* of water within living organisms. When a substance is heated, the rate of temperature increase depends on how easy it is for heat energy to increase the speed of molecular motion in that substance. If it doesn't require much heat to increase the motion, then we say that the substance is "easy to heat up."

You can apply this idea to the heating of water and answer the question: ***"Are the hydrogen bonds between water molecules strong enough to make water a substance that is hard to heat up?"***

Materials

- A hot plate.
- A chunk of steel.
- A beaker (the 100-ml size is best).
- A container of water.

Procedure

The cooling rate of an object is directly related to the amount of heat energy absorbed by that object. If an object cools quickly, then it didn't absorb much heat energy to start with.

1. Weigh the piece of steel.
2. Put an amount of water equal to the weight of the steel object into the beaker.
3. Put the piece of steel into the water, and heat the container until it just begins to steam.
4. Immediately remove the piece of steel from the beaker, and put both the beaker of water and the piece of steel onto the table. (Your results are more accurate if you pour the hot water into a room-temperature beaker at the same time you put the steel on the table.)
5. Repeatedly touch both the water and the piece of steel until both are approximately the same temperature. Keep track of how long it takes for each to cool.

Time for the water to cool = ____________________

Time for the steel to cool = ____________________

? QUESTION

1. Which substance cooled the slowest? (circle your choice)

Steel or Water

2. Which substance would require more heat energy to heat it up? Remember that the amount of heat given off by a substance equals the amount of heat absorbed by that substance when it was heated. (circle your choice)

Steel or Water

3. We know that sitting in 70°F water is more chilling to the body than sitting in room air at 70°F. Explain why.

   If you were made of steel, would it be more chilling or not? Explain.

4. Based on your experimental results, what do you conclude about the importance of the hydrogen bond on the heating up of water?

5. Is water a temperature-stabilizing substance for living organisms?

Evaporation of Water

The ***heat of vaporization*** is the amount of heat energy required to vaporize a substance (like water).

We can estimate the *heat of vaporization* for water by comparing that process with what we saw in the previous experiment.

Materials

- A hot plate.
- Three equal-size beakers (the 250-ml size is best).
- A thermometer. *Be careful, please!* This equipment is fragile.

Procedure

[1] Fill beakers (A) and (B) with the proper amounts of water, and put them in the freezer or into special "ice tubs" for beakers (A) and (B) to stay cold.

[2] Prepare a large container of boiling water. *Use the special measuring pipettes for safely removing 10 ml of boiling water.* You must prepare beaker (C) at the last minute just before you start the experiment. Read on.

[3] Preheat your hot plate at a setting that you know will boil water moderately. (*Not the highest setting!*)

[4] It is important that you start all three beakers at *exactly the same time*, without time for the beakers to change temperature before heating on the hot plate. Record the starting time as soon as all beakers are on the hot plate.

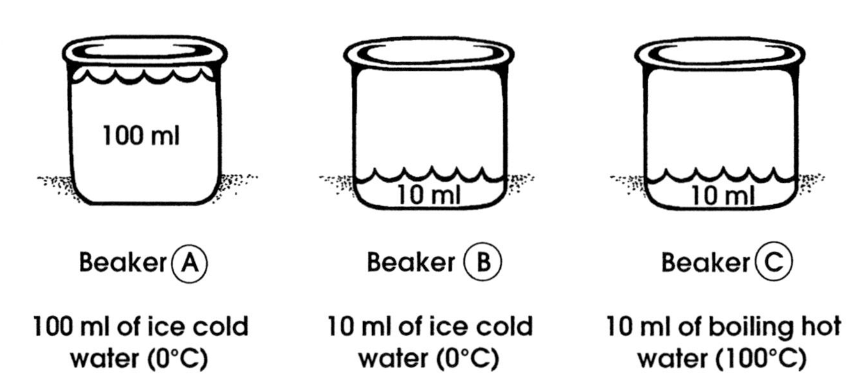

[5] There are *three* events that you must record during this experiment.

a. How many minutes does it take for Beaker (B) to just start to form little bubbles at the bottom?

____

b. How many minutes does it take for all of the water in Beaker (C) to boil away?

____

c. You must watch very carefully to determine the exact temperature of the water in Beaker (A) at the exact time when all of the water finally boiled out of Beaker (C). (Subtract the starting temperature if it was above 0°C.)

__________ °C in Beaker (A)

? QUESTION

1. What takes more energy? (circle your choice)

To heat 10 ml of water
from 0°C to 100°C.

or

To boil away (*evaporate*) 10 ml of water
that is already at boiling temperature (100°C).

2. Based on your results, how effective is the evaporation of water (sweating) at removing excess heat from your body?

3. We can calculate the actual number of calories of heat required to evaporate water by examining the results of Beaker (A). One calorie of energy is the amount of heat required to increase the temperature of 1 ml of water 1°C. Beaker (A) is a measure of the heat energy coming from the hot plate and going into Beaker (C). How much did the temperature change in Beaker (A)? __________ °C.

4. Now, how many calories of heat from the hot plate were required to evaporate the 10 ml of water in Beaker (C)? **Remember:** You must consider both how much water is in Beaker (A) and how much its temperature increased.

__________ calories

5. So, how many calories would be required to evaporate only 1 ml of water?

__________ calories

6. What does this experiment tell you about calories, exercise, and sweating?

EXERCISE #2

"What the Heck is pH?"

Hydrogen Ions

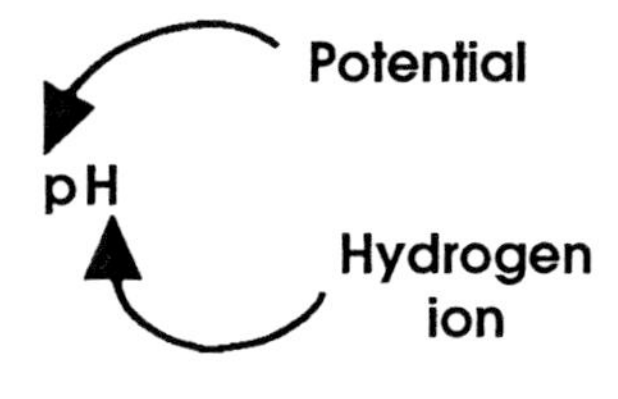

An ion is an atom that has lost or gained electrons, and thereby has become electrically charged (either + or –). They act differently than uncharged atoms. (It's like comparing magnets with non-magnets.) These ions are very important in our life processes, and we would die without them. Table salt is an example of two essential ions—sodium and chloride.

Of all the ions in your body, none is more important than the ***hydrogen ion***, H^+. The term *"pH"* refers to the concentration of H^+ ions in water. Biologists are interested in H^+ concentration because it affects chemical reactions so greatly.

A small change in the hydrogen ion can dramatically affect life. Acid rain and acid stomach are two expressions of the concentration of H^+ ions. Also, the blood of a human is so sensitive to H^+ concentration that a small pH change from your normal of **7.4** can result in your death.

We monitor the pH of our fish aquariums and our swimming pools in order to avoid potential problems. In the case of the aquarium, we are trying to maintain a good environment for micro-organisms, whereas, in the swimming pool, we are trying to prevent micro-organisms from growing.

? QUESTION

1. Knowing how important pH can be to living organisms, what would be the effect of acid rain on the ecosystem?

2. What does the "p" stand for in the term pH?

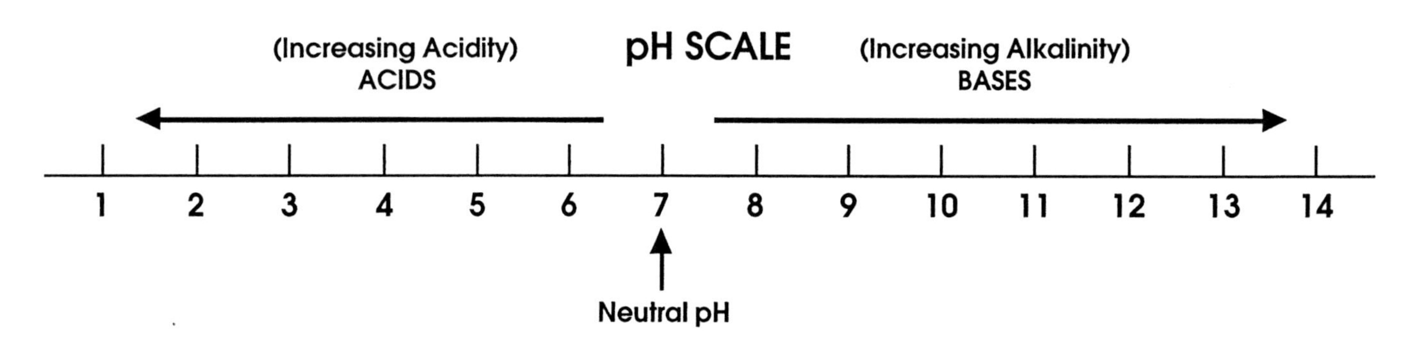

Information

- The pH scale ranges from 1 to 14.
- Each step *up* the pH scale means that there is 10x more OH^- (base) and 10x less H^+ (acid) than the step below.
- Each step *down* the pH scale means that there is 10x more H^+ (acid) and 10x less OH^- (base) than the step above.
- A water solution with a pH of 7 is ***neutral***. The concentration of the H^+ ions (acid) is equal to the concentration of the OH^- ions (base).
- If the pH is less than 7, then there are more H^+ (acid) ions than OH^- (base) ions, and the solution is called an ***acid***.
- If the pH is more than 7, then there are more OH^- (base) ions than H^+ (acid) ions, and the solution is called a ***base***.

? QUESTION

1. pH is a measure of ______________ ion concentration.

2. A pH of 3 would be . . . (circle your choice)

   Acid or Base

3. A pH of 11 would be . . . (circle your choice)

   Acid or Base

4. What is the relationship between H^+ and OH^- at a pH of 7?

5. How much more H^+ is in water at a pH of 3 when compared to a pH of 6?

6. How much more OH^- is in water at a pH of 11 when compared to a pH of 7?

How to Measure pH

The pH can be measured with a machine or with special color indicators. A pH machine directly reads the H^+ concentration and displays the pH on a screen. A color indicator is a special molecule that changes color at a particular pH level.

Follow the directions given by your instructor if you are using a pH machine. Otherwise, follow these instructions.

Procedure

[1] Use the pH paper test kit to determine the pH of the *three unknown solutions* on the demonstration table.

[2] Tear off a 1" strip of pH test paper, and squirt a drop on it from the test solution. Compare the color of the pH test paper to the pH color chart. That's the pH!

? QUESTION

1. Is solution A acid or base or neutral? ______________

2. Is solution B acid or base or neutral? ______________

3. Is solution C acid or base or neutral? ______________

Buffers

A ***buffer*** is a chemical substance that can be added to water, and will make that solution *resist* a pH change.

Materials

- 25 ml of *Sample X* in a small beaker.
- 25 ml of *Sample Y* in a small beaker.
- A dropper-bottle of phenol red.
- A dropper-bottle of acid.

Procedure

[1] Put 5 drops of *phenol red* into each beaker. Phenol red is a pH color indicator. It turns *yellow* in *acid*, and it turns *red* in *base*.

[2] Counting the drops, add acid one drop at a time until each beaker turns yellow. *Gently shake* the beakers after each drop in order to mix the acid into the test solution. (If either solution hasn't changed to yellow after 25 drops of acid have been added, then stop adding acid and assume that it will take many more drops to change the pH.)

? QUESTION

1. Which solution contains a buffer? (circle your choice)

Sample X or Sample Y

2. How many *more* drops of acid did it take to change the buffered solution compared to the nonbuffered solution?

__________ more drops

3. Why do you think that one of the brands of aspirin is called "Bufferin"?

EXERCISE #3

"Molecular Motion"

All atoms and all molecules move! They are bouncing off of each other at an incredible speed. (It's a good thing that O_2 and N_2 molecules are so small, because they would "sandblast" your skin if they were bigger.)

Chemists discovered that the speed of molecular motion is influenced by several factors, and we will investigate two of them. Also, we will look at a couple of special effects created by molecular motion.

Can You See Molecular Motion?

Actually, molecules are too small to see. But a clever physicist calculated the energy in moving water molecules, and has determined that those molecules have enough energy to bump into and move some very small particles, like carmine dye. When viewed under a microscope, this movement can be seen.

Materials

- A slide and coverslip.
- A compound microscope.
- A drop of *carmine dye* particles.

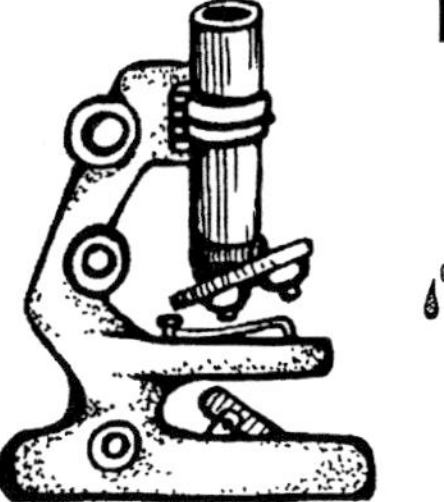

Procedure

[1] Make a wet mount of the carmine particles.

[2] Look at the *very very* smallest particles that you can see. Show your instructor.

[3] The vibrating motion of these particles suspended in water is called ***Brownian Motion***. (Named after guess who?) The tiny particles move whenever a water molecule (which you can't see) bumps into the carmine particle. Observe.

? QUESTION

1. When you watch *Brownian Motion* are you actually seeing *molecules* move? Explain.

2. What do you think would happen if you held a flame under the slide? Explain.

3. What would happen if you held an ice cube under the slide? Explain.

Light Molecules vs. Heavy Molecules

Methylene Blue dye has a molecular weight of 374. The purple dye *Potassium Permanganate* has a molecular weight of 158. If molecular weight makes any difference in the *speed* of motion, then we should be able to measure that difference.

You will need to work as a class for this experiment. Assign one member of your group to work with members from other groups.

Materials

- Two agar plates.
- Put *one* crystal of *Methylene Blue* in the middle of the agar plate. (Your instructor may have you put a drop of Methylene Blue solution into one of the small depressions in the agar.)
- Put *one* crystal of the *Potassium Permanganate* in the middle of the other agar plate. (If you are using solutions, put a drop of Potassium Permanganate into the other agar depression.)

Procedure

[1] Record the starting time for this experiment.

[2] At 30 minutes and one hour, come back to the agar plates and measure the *diameter* of the spreading colors. Record your measurements in the chart below.

RATE of SPREAD (Diameter of Color)

Test Molecule	In 30 Minutes	In One Hour
Potassium Permanganate		
Methylene Blue		

? QUESTION

1. Which molecules move faster? (circle your choice)

Potassium Permanganate or Methylene Blue

2. What does this experiment tell you about the speed of movement of different-sized molecules?

EXERCISE #4

"Diffusion of Water Into and Out of Cells"

The movement of molecules from where they are in high concentration to an area where they are in low concentration is called ***diffusion***. Because *all* molecules move, they also *diffuse*!

When a cube of sugar is put into a cup of coffee we know that the sugar will diffuse from where it is in high concentration (the cube) to where it is in low concentration (the hot coffee). However, we don't normally think about what the water molecules are doing in that same cup of coffee.

The rules of diffusion apply to water concentration just like they do to the sugar. Water will diffuse from where it is in high concentration (the hot coffee) to where it is in low concentration (the sugar cube).

We can observe the effects of water diffusion in living cells. The diffusion of water through a cell membrane is called ***osmosis***. This is a very important process involving the movement of water throughout the cells of all living organisms.

Materials

- A slide and coverslip.
- A leaf from the *Elodea* plant.
- An eyedropper of salt water.

Procedure

[1] Make a wet mount of the *Elodea* leaf and look at the leaf cells under high power. Draw a picture of the distribution of chloroplasts within a typical cell.

[2] Work with another lab group. One group is to leave their normal *Elodea* slide under the microscope. The other group is to add salt water to their *Elodea* slide.

[3] Put one drop of salt water at the edge of the coverslip. Use a piece of tissue paper on the opposite side of the coverslip to absorb and pull the salt water across the slide. This salt water will soon surround all of the leaf cells.

[4] Wait 10 minutes. Look back and forth between the two microscopes. Draw another picture of the distribution of chloroplasts within a typical cell in salt water.

Elodea Cell

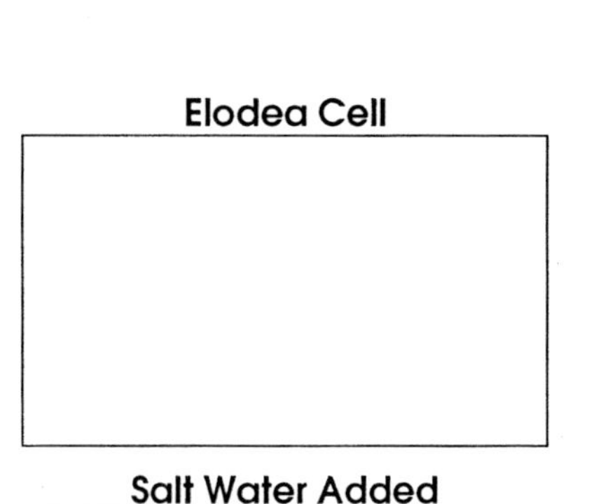

Salt Water Added

Be sure to clean the microscope stage if you got any salt water on it.

? QUESTION

Let this picture represent one *Elodea* cell surrounded by salt water. The small circles are the water molecules and the larger circles are the salt molecules.

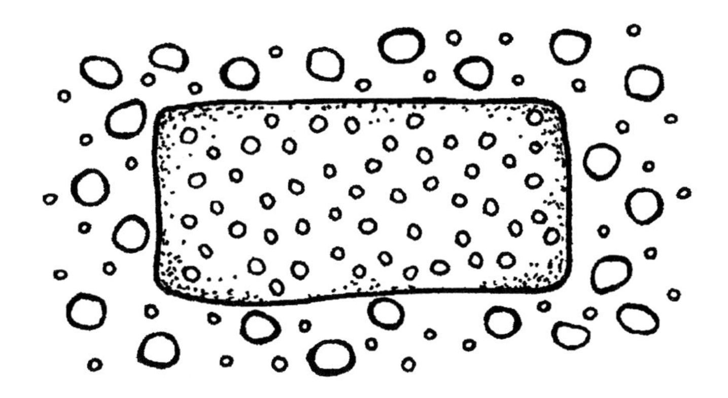

Salt Molecule
Water Molecule

1. Where is the *water* in high concentration? (circle your choice)

   Inside of the cell or Outside of the cell

2. Where is the *water* in low concentration? (circle you choice)

   Inside of the cell or Outside of the cell

3. In what direction will *osmosis* (water diffusion) occur? (circle your choice)

   Into the cell or Out of the cell

4. Explain what you saw in your second picture of the *Elodea* leaf cells. (What is the membrane surrounding the grouped chloroplasts? What happened to the central vacuole?)

5. When you eat a lot of highly salted food (a bag of potato chips, for example), what happens?

   Why?

6. People with high blood pressure or heart problems are told to be careful about their intake of salt. Why?

In Conclusion

Under normal conditions root hair cells are relatively low in water molecules and high in other kinds of molecules (cell salts and nutrients). *Draw arrows* to show water movement between the soil water, the root hair cell, and the water transporting tube.

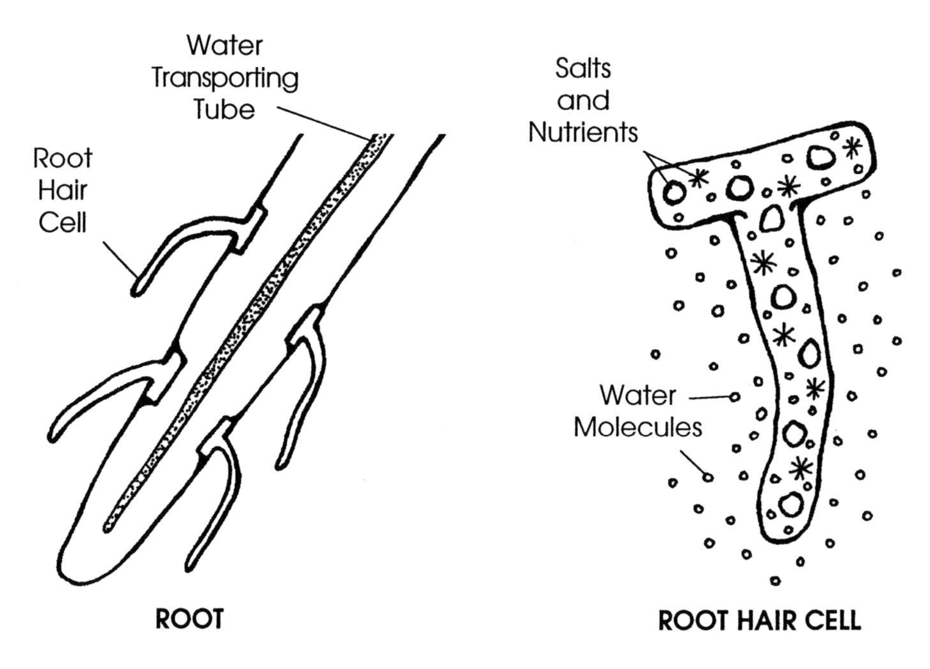

? QUESTION

1. What would happen to the water movement if you put salt or a lot of fertilizer in the soil?

2. Explain how the properties of water investigated in Exercises #1 and #4 assist a plant in the movement of water from the soil, into the root hair, and throughout the structure of the plant.

OPTIONAL

Water is a most unusual substance. It has some strange changes in volume at different temperatures. Try this surprising experiment.

PROCEDURE

1. Very accurately mark the water level in a beaker of cool water.
2. Heat the beaker of water to just before steaming (when bubbles start to appear) and record the water level.
3. Refill the beaker with ice cubes and enough water to match your first water level mark. (Push all the ice below the water surface.)
4. Allow the ice to melt, and record the water level.
5. Compare your results. Surprised?

Photosynthesis and Respiration

(Tutorial)

Physicists have developed two ways of describing the universe: One of these is energy, and the other is matter. Matter is anything that has mass and occupies space. Energy is non-material, travels in waves, and has the capacity to change matter.

Life can be described from both of these points of view. Life is an assemblage of atoms and molecules. It is also an energy process. The following Exercises present highlights of energy conversions and substance changes in living systems.

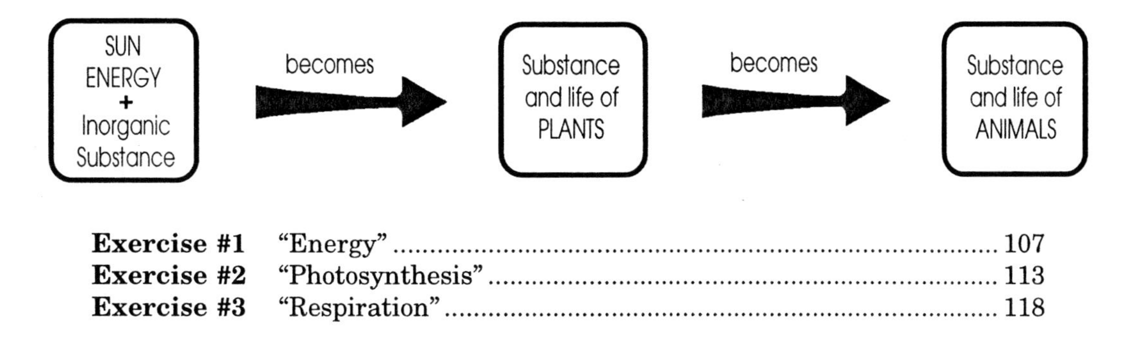

Exercise #1 "Energy" 107
Exercise #2 "Photosynthesis" 113
Exercise #3 "Respiration" 118

EXERCISE #1

"Energy"

To understand the process of life, follow the energy.

Energy can move and change matter, and can exist in different forms. Some of these forms include heat, light, electrical, chemical, and mechanical energy. In addition to describing the form of energy, we can measure the *amount* of it. The amount of energy can be measured using various experimental methods. For example, we can estimate the amount of heat energy in a flame by observing how fast that flame can "move" water molecules (heat them up).

Laws of Thermodynamics

Several basic principles of energy change have been discovered. Two of these, the First and Second Laws of Thermodynamics, are of particular value in understanding life processes.

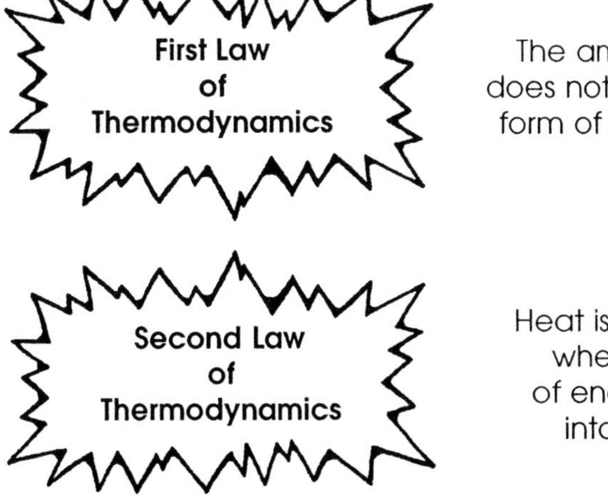

The amount of energy does not change; only the form of energy changes.

Heat is always produced whenever one form of energy is converted into another form.

Imagine that the Engineering Department decided to build a campus wind generator for capturing and storing energy in batteries that could be used later to run electric motors. The size of the energy boxes in the diagram below represents the amount of energy available at each step along the way. The numbers in the boxes represent energy units.

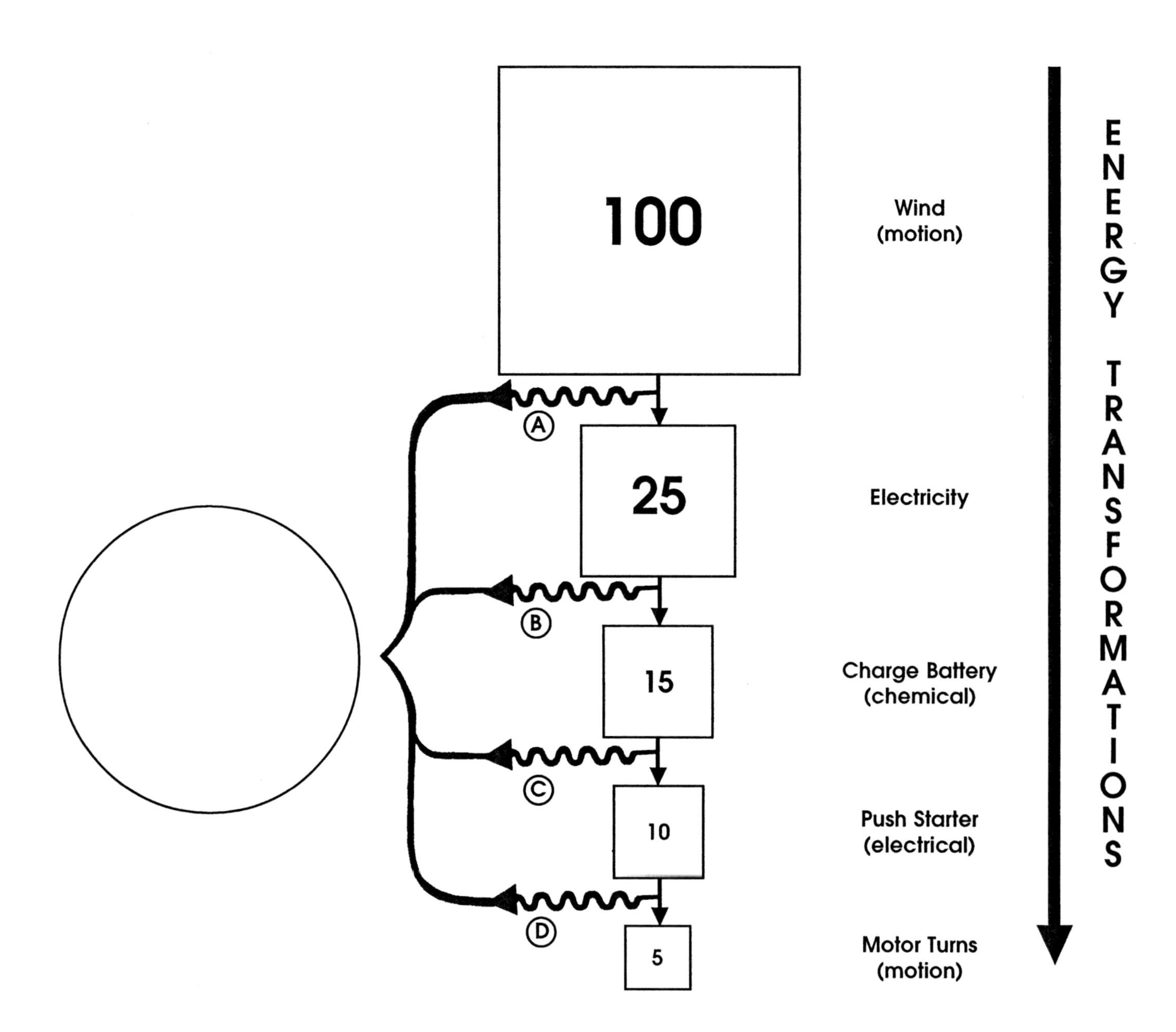

? QUESTION

1. What form of energy is represented by Ⓐ, Ⓑ, Ⓒ, and Ⓓ?

   Write the name of this form of energy inside the circle on the left side of the diagram.

2. Which Law of Thermodynamics is used to answer question #1?

3. Using the First Law of Thermodynamics, calculate the amounts of energy released at:

   Ⓐ ________ Ⓑ ________ Ⓒ ________ Ⓓ ________

4. What is the total amount of heat in the big circle? ________________

5. A biology student who learned about the Laws of Thermodynamics, suggested that the school's Engineering Department should have designed a wind machine that directly turned the campus motors. How would her suggested change improve efficiency?

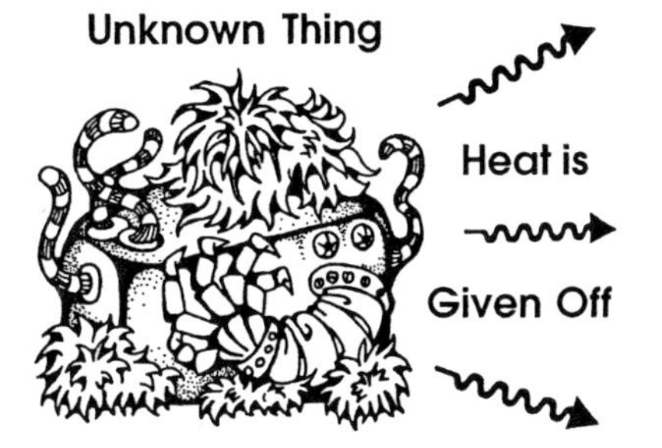

6. Imagine there is an "unknown thing" giving off heat. What can we conclude is going on inside the "unknown thing"?

Energy in Animals

The process of FOOD ENERGY ⟶ YOU ⟶ HEAT is called ***Respiration***. We will outline the substances that are changed during respiration in Exercise #3.

We can measure the amount of energy in food. Also, there are machines designed to record the heat an organism gives off during the day. Assume that the *size* of the boxes in the diagram below represents the amounts of energy for one day.

Respiration is the transformation of food energy into the chemical energy necessary to run cell processes.

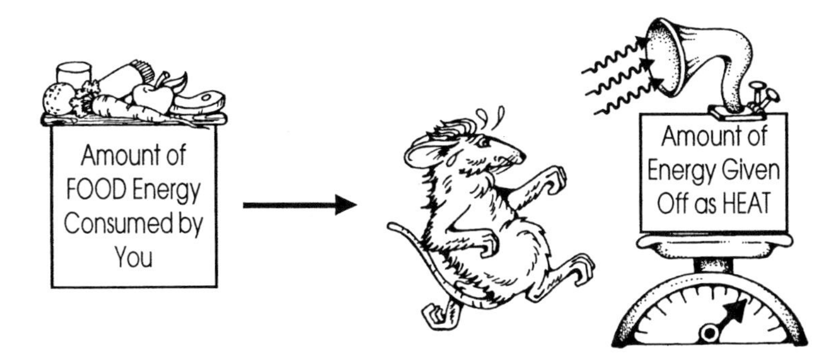

? QUESTION

1. Use the First Law of Thermodynamics to analyze the energy diagram on the previous page. What is obvious about the amounts of energy represented in the diagram?

2. Where is the rest of the energy?

Energy in Plants

Photosynthesis is the transformation of light energy into chemical energy.

The process of transforming SUN ENERGY ⟶ PLANT is called ***Photosynthesis***. We will describe the substances that are changed during photosynthesis in Exercise #2.

We can measure both the amount of light energy absorbed by a plant during a day and the amount of heat released by the plant. Assume that the size of the boxes represents amounts of energy.

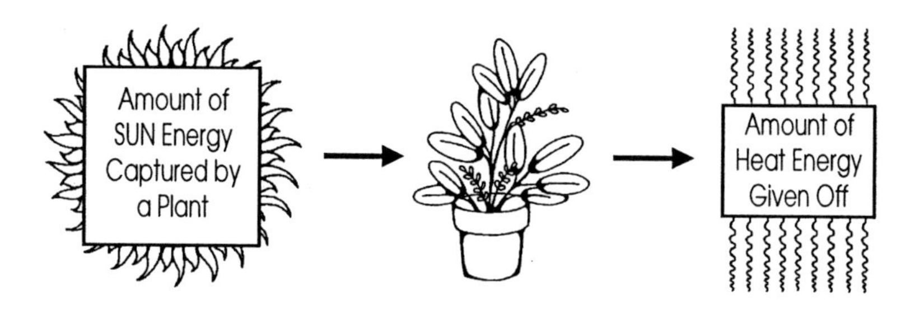

? QUESTION

Considering the First Law of Thermodynamics, where is the rest of the energy?

Energy in Ecosystems

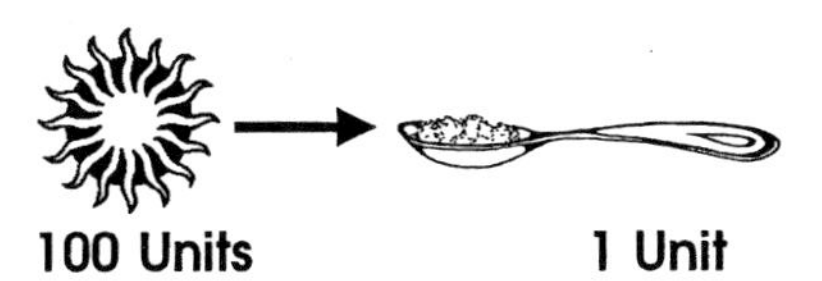

The Laws of Thermodynamics help us to understand conversions of energy within ecosystems. Energy of sunlight is first converted into the chemical energy of plants. This conversion occurs with about 1% efficiency. This means that photosynthesis requires about 100 units of sun energy to produce one unit of sugar energy.

The animal that eats plants is called a ***herbivore***. The animal that eats herbivores is called the ***first carnivore***. The ***second carnivore*** eats the first carnivore, and so on. At each step from the plant outward, the energy conversion efficiency is about 10%. (For example, 100 kg of plant are required to produce 10 kg of rabbit.) The diagram below shows the amounts of energy at each food level in the ecosystem.

Energy flows through the ecosystem.

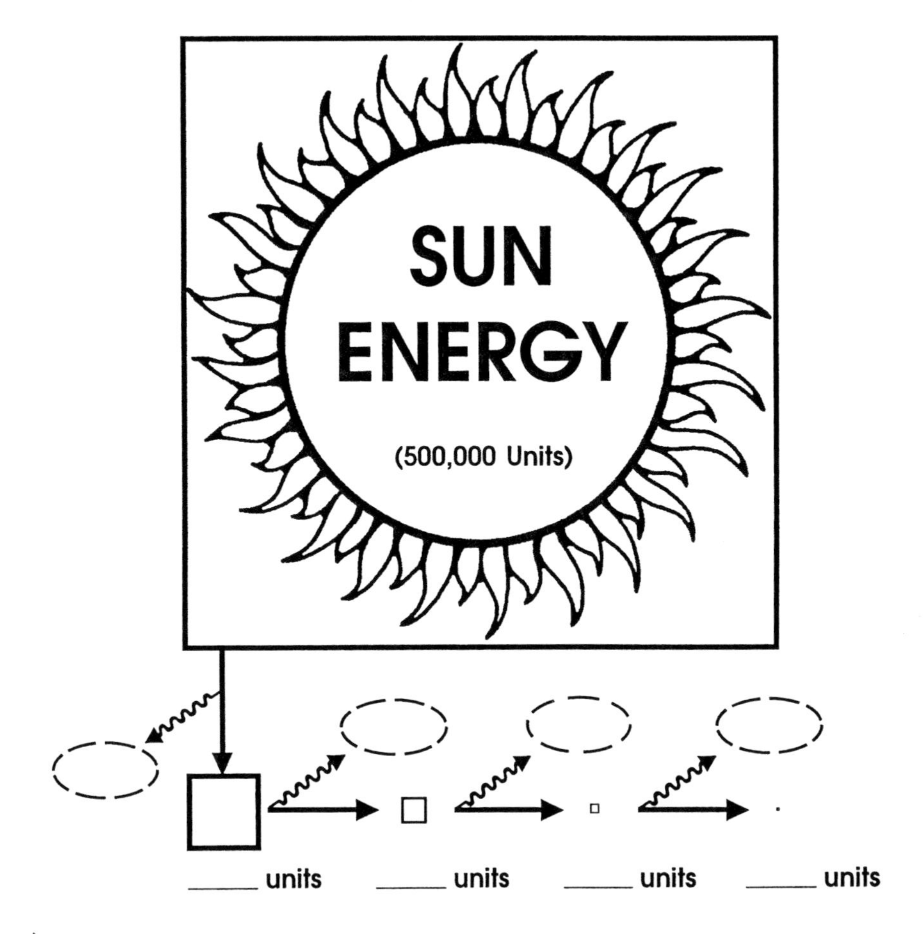

? QUESTION

1. Each of the four boxes represents a food level. Write the name of each food level (from plants ⟶ second carnivores) below the boxes.

2. Starting with 500,000 units of sunlight energy, calculate and record the number of energy units (on the lines below each box) for each of the energy levels in the diagram.

3. What form of energy is given off during the conversion from one ecosystem level to the next? **Hint:** Remember the Second Law of Thermodynamics.

4. Record the amount of energy (question #3) given off at each step in the ecosystem diagram. (Put your answers in the dotted ovals at the ends of the wavy arrows.)

5. Assume that humans are the second-level carnivores, and that each energy unit in their level represents 25 people. How many people can be supported in the ecosystem?

6. Assume that we eat herbivores only. As first-level carnivores, how many humans can now be supported in the ecosystem?

7. If we ate plants only, how many humans could be supported by the ecosystem?

8. If we ate plants only, what would happen to the amounts of energy available for the other herbivores and carnivores in the ecosystem?

9. Let's look at energy conversion another way. It takes about 0.5 kg of meat to support a human for one day. Assuming that beef is the only available nutrient, how much meat energy (kilograms of meat) will it take to get you to 20 years old?

10. How much plant energy (kilograms of plant) will it take to feed the beef that gets you to 20 years old? (Refer to question #9.)

11. Your digestive system is very efficient at digesting meat, but is not as efficient at getting all of the available nutrition from plants. Now, assume that you were raised on plants instead of meat. Plant material has about 40% of the per-kg food value as meat. How much plant energy would it take to get you to 20 years old?

Do you get the general idea of how the Laws of Thermodynamics give us a better understanding of ecosystems? The next Exercises present how matter is changed by energy in both photosynthesis and respiration.

EXERCISE #2

"Photosynthesis"

Exercise #1 presented photosynthesis from the energy perspective. In this Exercise we consider the changes in *substances* (matter) during photosynthesis.

Historical Discovery Process

The chemical changes that occur during photosynthesis have been investigated for the past 300 years. The general highlights of these discoveries help us to understand the basic process of photosynthesis.

Three centuries ago people wondered where plants came from. They knew that plants grew out of the ground, but *how* that happened was a complete mystery to them. The first step in answering the question was to plant a small tree in a large pot supported off the ground. They did this so that the soil of the container was separated from the soil of the earth. Only the dirt in the pot was available to the plant.

The people cared for the plant during one year. At the end of the year the small tree had gained 100 kg. *(The actual data from this experiment have been changed to simplify the discussion.)*

? QUESTION

1. What do you think the experimenters considered as two possible sources for the substances (matter) that became incorporated into new tree growth? (This experiment was performed nearly 300 years ago. People didn't know very much about chemistry, and air was thought to be a non-substance.)

____________________________ ____________________________

2. When the experimenters measured the weight of the two substances (question #1) given to the plant, they found 35 kg of one and 2 kg of the other. Where do you think they thought the rest of the plant's weight came from?

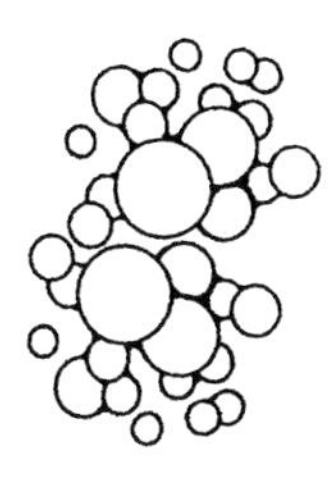

People guessed that there must be some invisible substance in light or somewhere else that was added to the plant. It took another century for humans to discover the chemical makeup of air. Once humans invented accurate scales, they could measure small changes in the weight of air.

Experimenters put a plant into a sealed jar with air that had been weighed. After the plant was exposed to light for a few hours, the air lost weight. They had discovered another source of matter for the new plant growth—the air! The basic equation for photosynthesis was complete.

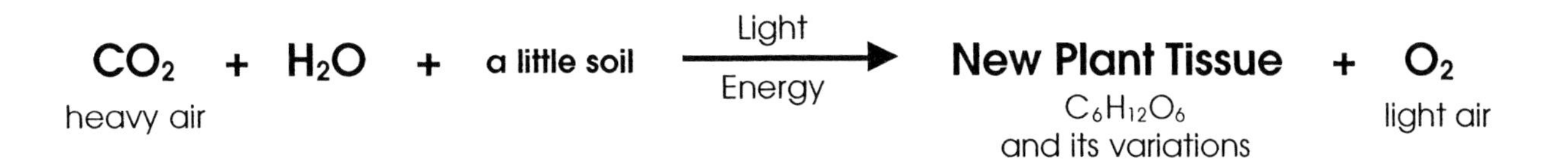

? QUESTION

1. Look at the molecular formula for plant tissue. Where does the carbon come from?

2. Where does the hydrogen in $C_6H_{12}O_6$ come from?

3. When radioactive isotopes of oxygen atoms are put into CO_2 molecules and the plant is allowed to photosynthesize, only new plant tissue ($C_6H_{12}O_6$) is radioactive and the oxygen gas given off is not radioactive. Draw a dotted line in the photosynthesis equation to show where the oxygen atoms in CO_2 go.

4. Draw a dotted line in the equation to show where the oxygen atoms in H_2O go.

Chlorophyll

All chemical reactions involve changes in the *electrons* of atoms. The key to photosynthesis is a special molecule called ***chlorophyll***. This large and complex molecule has electrons that become "activated" when light shines on the molecule.

This reaction is unusual because most substances only heat up when exposed to sunlight; that is, their electrons aren't "activated" by light energy. But the electrons in chlorophyll are activated by light, and this electron energy is used to make new plant tissue ($C_6H_{12}O_6$).

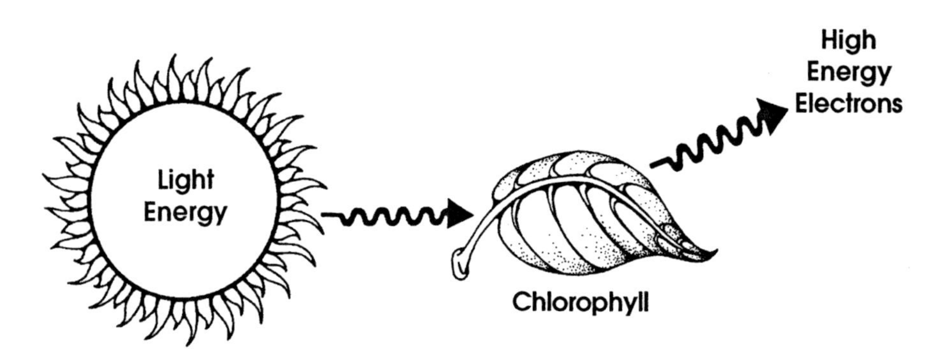

? QUESTION

1. Sunlight contains all of the colors of light in the visible spectrum. Which color do you see reflecting from the surface of plants?

2. Which two primary colors do you *not* see when looking at the plant leaves?

3. Which colors of light probably activate the electrons of the chlorophyll molecule? Explain.

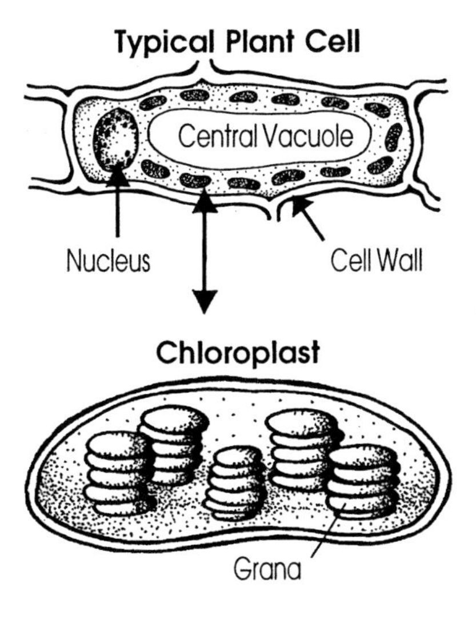

Chloroplasts

All of the unicellular algae and multicellular plants are eukaryotic and have specialized organelles called ***chloroplasts***. The chloroplasts contain chlorophyll and enzymes for the photosynthesis process. The exceptions to this rule are the photosynthetic bacteria (cyanobacteria). They are prokaryotic; that is, they don't have cell organelles like chloroplasts. However, cyanobacteria *do* photosynthesis, and they *have* chlorophyll.

The chloroplast has many specialized structures. Stacks of discs, called ***grana***, contain the chlorophyll and act like photoelectric cells. When light shines on these grana, the electrons of the chlorophyll are "activated." This is the first stage of photosynthesis.

The second stage of photosynthesis involves the building of sugar molecules from carbon dioxide and water. These two ingredients will not combine unless chemical energy is provided. That energy comes from the "activated" electrons in the chlorophyll reaction. Sugar molecules are made in the fluids of the chloroplasts that surround the grana. Simple sugars made during photosynthesis are modified into all of the other organic molecules needed by the plant.

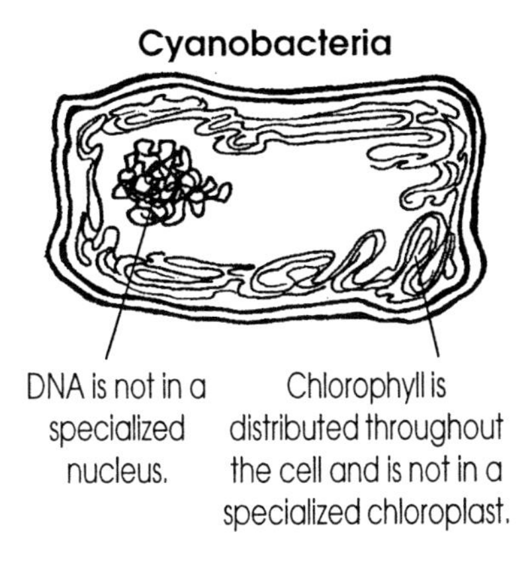

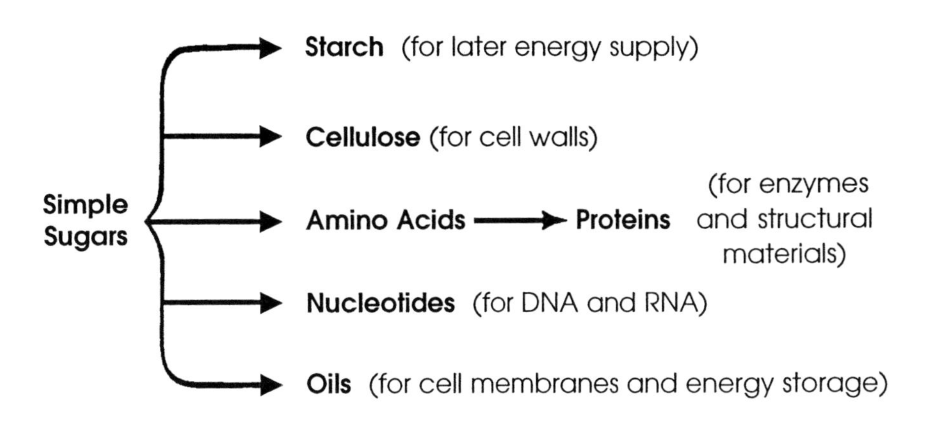

? QUESTION

1. When light shines on chlorophyll, the light energy is transformed into . . .

2. Some biologists refer to the two parts of photosynthesis as the light reactions and the synthesis reactions. Which part do you think that they call the light reactions?

3. Which part of photosynthesis do you think they call the synthesis reactions?

4. The synthesis reactions happen in the daytime, but are sometimes called "dark" reactions because light energy is not directly required. What kind of energy is required to run the synthesis reactions?

   Where does that energy come from?

Characteristics of Light

Our description of energy includes anything that is non-material, travels in waves, and having the capacity to move and change matter. There are many different kinds of energy, and each has its characteristic *wavelength*. When all of these energy waves are arranged on a scale from short waves to long waves, that scale is called the ***electromagnetic spectrum***.

Electromagnetic Spectrum

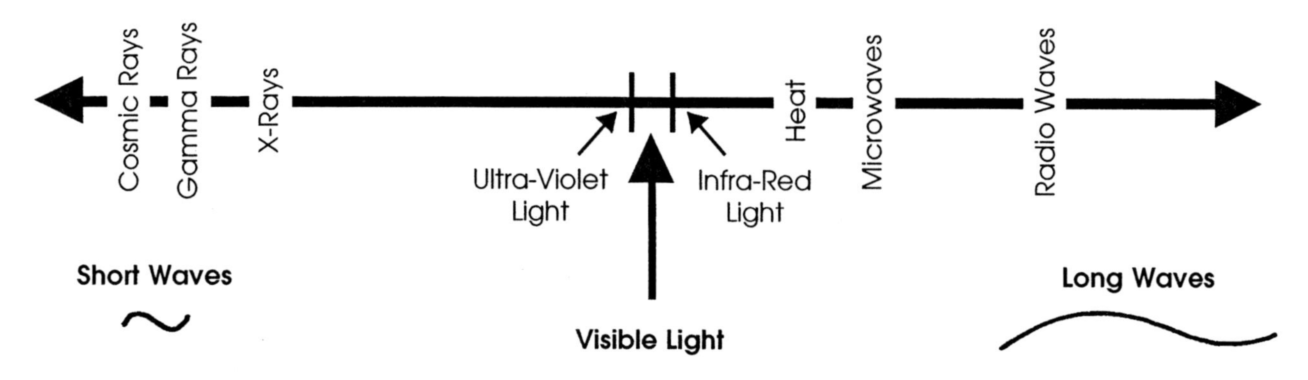

Energy waves can be too strong or too weak for life as we know it.

You can see that visible light is only a very small slice of the electromagnetic spectrum. An important feature of electromagnetic energy is that shorter waves have more energy than longer waves. This is the key to understanding why only a narrow segment of waves in the electromagnetic spectrum is ideal for biological reactions like photosynthesis and vision. Some people humorously refer to this idea as the "Goldilocks" principle—the wavelength has to be "just right."

The energy waves just shorter than violet light are called ***ultra-violet***, and the waves just longer than red light are called ***infra-red***.

Ultra-violet waves are damaging to life. These waves have too much ________________________. The molecules exposed to ultra-violet waves become over-activated and are chemically changed or destroyed.

The molecules exposed to infra-red waves are "warmed up," but these waves do not have enough ________________________ to activate electrons (essential for biochemical reactions).

White light (visible light) is made of different wavelengths of energy (different colors). You can remember where the various colors occur in the visible light spectrum by recalling the image of a rainbow in your mind.

A rainbow is the visible light spectrum.

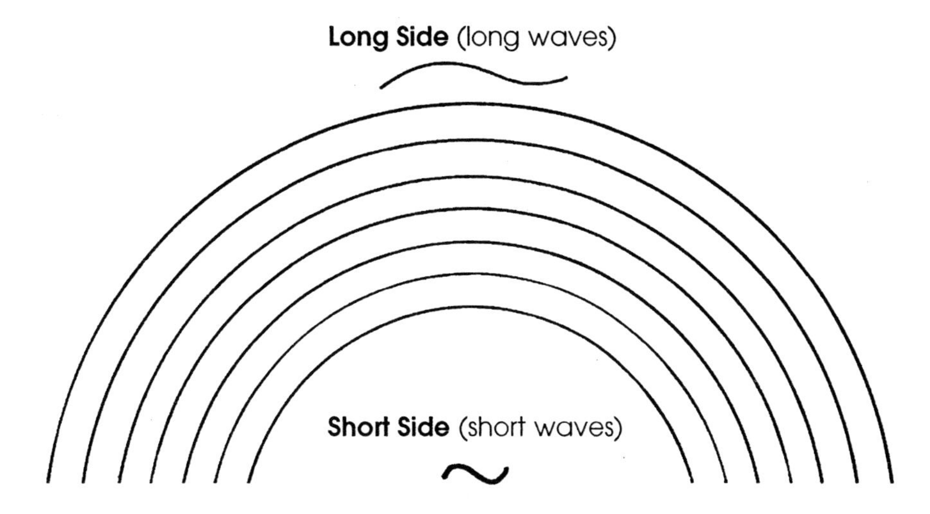

? QUESTION

1. Which color is on the short side of the rainbow?

2. Which color is on the long side?

3. Which color is about in the middle?

4. Put these colors in the rainbow diagram above. (The long side of the rainbow is the longer wavelength.)

5. Which color do you see when you look at plants? (The color that you *see* is the color that plants do *not* use in photosynthesis.)

6. Which two primary colors do you *not* see when looking at a plant? (These are the two wavelengths of energy that plants mostly use for photosynthesis, so they are *absorbed* by the plant rather than reflected back.

______________________________ ______________________________

There is much more to photosynthesis than is covered in this Exercise. You will have to get those details from lecture or your textbook.

EXERCISE #3

"Respiration"

Respiration produces ATP energy from food.

Respiration was presented from the energy perspective in Exercise #1. In this Exercise we consider the changes in substances (matter) during cellular respiration. Respiration is the chemical breakdown of food molecules, converting food energy into usable energy (ATP) for the cell.

Historical Discovery Process

Investigation of the chemical changes during respiration coincided with experimental revelations about photosynthesis. The earliest experiments on respiration were performed about 300 years ago, and involved both plants and animals.

Mouse in a sealed jar ⟶ Mouse dies in an hour or so

Mouse in a sealed jar with a large plant in the light ⟶ Mouse lives just fine

Mouse in a sealed jar with a large plant in the dark ⟶ Mouse dies in an hour or so

These early experiments revealed several facts:

- ▶ There was something in the air that animals needed to live.
- ▶ Somehow plants were able to "regenerate" the air that animals needed.
- ▶ Light was necessary for plants to "regenerate" the air.

Later experiments measured changes in the weight of substances during respiration.

? QUESTION

1. Based on the experiment illustrated on the left, where did the weight of the mouse go?

 _____________ and _____________

Mouse in a sealed jar with enough air but no food

Experimental Results:
Mouse loses weight.
Mouse produces water.
Air gets heavier.

2. The air started out light in weight. What was the substance of that air? (Refer to "light air" in Exercise #2.)

3. The air ended up heavier in weight. What was the substance of that air? (Refer to "heavy air" in Exercise #2.)

4. Write the molecular formulas for each substance in this basic equation for respiration below. (Use the same formula for mouse tissue that we used for plant tissue in Exercise #2.)

Mouse Tissue + Light Air ⟶ Water + Heavy Air

5. Radioactive isotope experiments have shown that atoms of carbon, oxygen, and hydrogen are rearranged just like in photosynthesis, only the equation is in reverse. (Refer to the photosynthesis equation on page 113.) Draw a dotted line in the respiration equation to show where the carbon atoms in the food molecule go.

6. Draw a dotted line in the respiration equation to show what happens to the oxygen atoms in the air that we breath.

7. Where do the oxygen atoms in CO_2 come from? (Show this with another dotted line in the respiration equation.)

Respiration Compared to a Burning Candle

Experimenters discovered that a burning candle is the same basic chemical process as animal respiration. The burning of any substance is called ***combustion***.

$$\text{Candle} + O_2 \longrightarrow CO_2 + H_2O$$

? QUESTION

1. Respiration is the opposite of ____________________.

2. Which must have evolved first—animals that required oxygen or the process of photosynthesis?

3. Assume that all oxygen in the atmosphere came from photosynthesis. Knowing this, what has been happening faster—respiration or photosynthesis?

4. Evidence suggests that the early atmosphere of our planet was very high in CO_2. Based on your answer to question #2, where did the CO_2 go?

5. If your answer to question #3 is true for this planet, then where is all the "extra" plant material that hasn't burned or been eaten? **Hint:** think of two major resources.

____________________ ____________________

6. If all the resources referred to in question #5 were burned, what change would happen in the atmosphere?

Does Respiration Happen in Plants?

Plant
in a sealed jar
in the dark

Results:
Plant loses weight.
Plant produces water.
Air gets heavier.

Conclusion:
Plants do respiration.

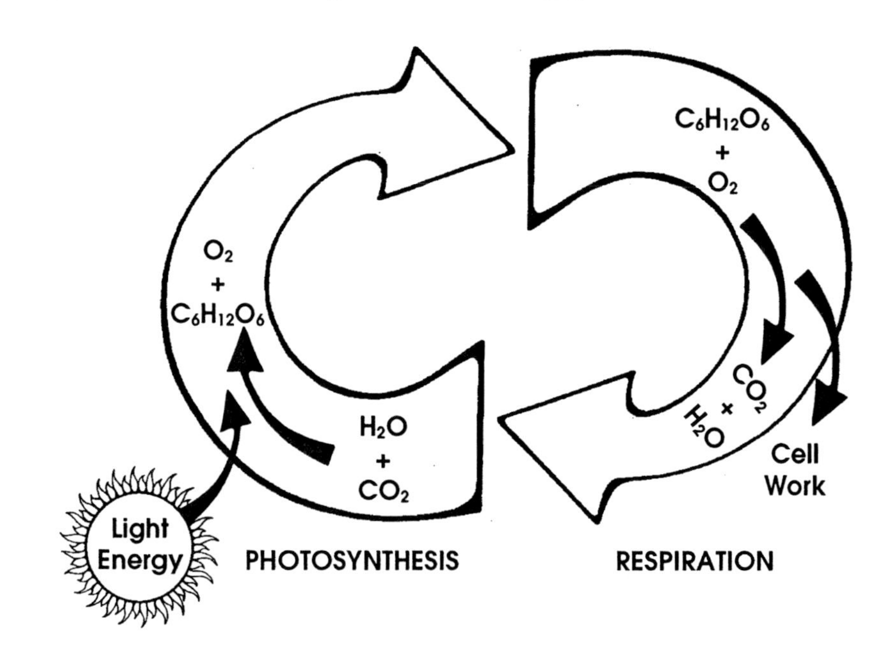

? QUESTION

A house plant was kept one month in each of three rooms in the house. Compare the intensity of photosynthesis and respiration by the plant in each of the rooms.

In room #1 the plant lost weight and started to die.

____

In room #2 the plant gained weight.

____

In room #3 the plant survived but didn't gain weight.

____

Chemical Breakdown of Food

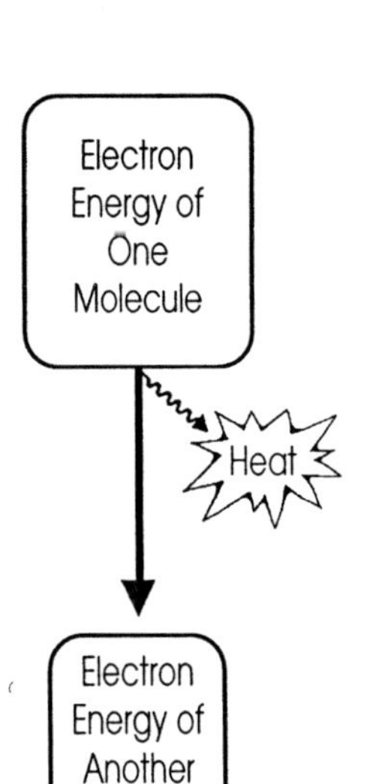

Using a few basic rules about chemical reactions, you can understand much about respiration.

Rule: Electrons have energy.

Rule: Sometimes electrons can move from one molecule to another.

Rule: Whenever hydrogen atoms are added to or removed from a molecule in a chemical reaction, assume that the number of electrons also changes (one hydrogen added = one electron added).

Rule: The Second Law of Thermodynamics applies to situations in which the electron energy of one molecule is transformed into the electron energy of another molecule.

? QUESTION

1. In the reaction shown, is the molecule gaining or losing energy?

2. Which has more electron energy? (circle your choice)

   $C_6H_{12}O_6$ or $6CO_2$

3. How many electrons are removed from sugar during respiration ($C_6H_{12}O_6 \longrightarrow 6CO_2$)?

4. Cells need a special molecule called ATP to do the work of life. Assume that the energy of one electron from food ($C_6H_{12}O_6$) can be transformed into the energy in 3 ATP molecules. How many ATP molecules are generated during the breakdown of one sugar molecule during respiration? (Refer to your answer for question #3.) ____________

ATP

ATP (adenosine triphosphate) is a special high-energy molecule in the cell. This molecule can also exist in a low-energy form called ADP (adenosine diphosphate). ATP has more high-energy electrons than ADP. That extra electron energy comes from food molecules.

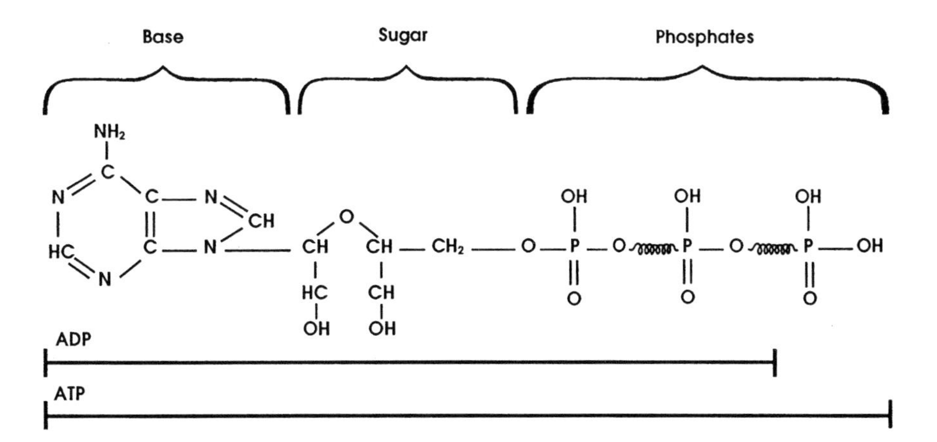

ATP carries high-energy electrons.

ATP delivers high-energy electrons to other energy-requiring processes in the cell. The two processes (ADP ⟶ ATP and ATP ⟶ ADP) create an energy exchange system in the cell.

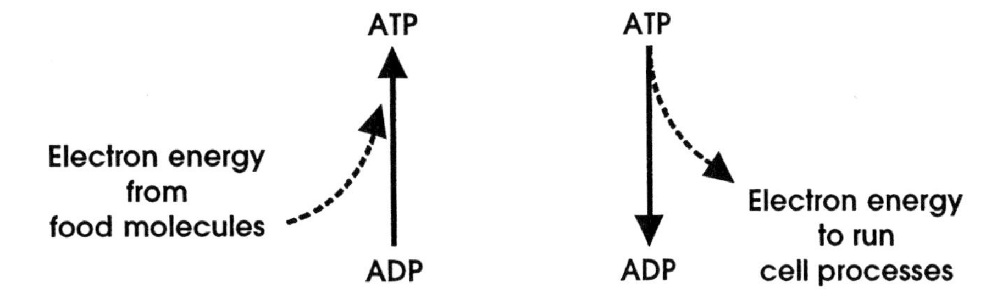

Aerobic and Anaerobic Respiration

You concluded previously that cells requiring oxygen for respiration must have evolved after photosynthesis. We know that some chemical reactions can happen without oxygen. So, could there have been a form of respiration that might have existed before photosynthesis? The answer is yes. Respiration can occur without oxygen, and there were primitive cells living by this process before photosynthesis evolved.

Respiration without oxygen present is called ***anaerobic*** (without air). Respiration with oxygen is called ***aerobic***. Aerobic respiration occurs inside a specialized organelle called the ***mitochondrion***, whereas anaerobic processes (also called *fermentation*) are associated with other membranes in the cytoplasm.

The sugar molecule is only partially broken down during anaerobic respiration. Have all high-energy electrons been removed from the sugar molecule below? ________________

Anaerobic Respiration in Human Muscles

When you shift into anaerobic respiration, you are close to the end of your capacity.

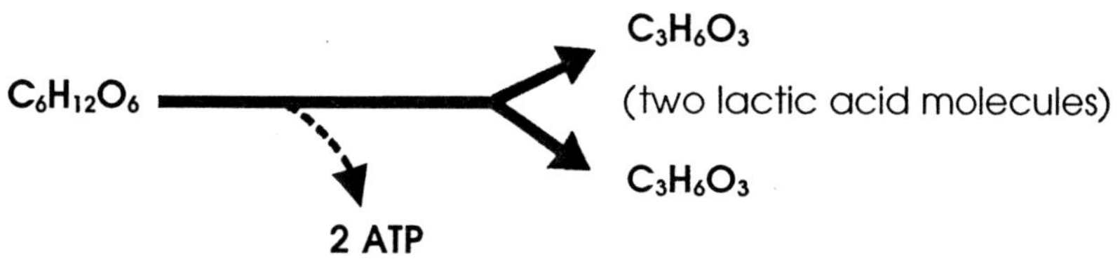

$C_6H_{12}O_6$ → 2 ATP

→ $C_3H_6O_3$
(two lactic acid molecules)
→ $C_3H_6O_3$

The energy of two ATP molecules is generated when sugar is "split" into the two lactic acid molecules. These two ATP molecules are the *only* energy captured from the food molecule during anaerobic respiration. High-energy electrons remain in the lactic acid. This means that anaerobic respiration is very *inefficient* compared to aerobic respiration. (Remember: In question #4 under "Chemical Breakdown of Food," you calculated that 36 ATP are generated during aerobic respiration.)

In humans, anaerobic respiration happens for short periods of time only in the skeletal muscles. During strenuous exercise, this maintains metabolism when oxygen is temporarily depleted. Other organs of your body are incapable of anaerobic respiration, and their cells begin to die when oxygen is used up.

? QUESTION

Complete the following table comparing aerobic and anaerobic respiration.

Comparisons	Aerobic	Anaerobic
Is oxygen necessary?		
Which came first on the planet?		
What are the end products?		
How much ATP energy is generated?		
Where in the cell does it happen?		

Mitochondria

Aerobic respiration is the process in living organisms that extracts electron energy from the chemical bonds in food (organic molecules), and converts that energy into a more useful form of energy (called ATP) to run cell activities. This cell process uses oxygen and produces carbon dioxide. The complete equation is:

Mitochondria are called the "powerhouses" of the cell because they are centers for aerobic respiration.

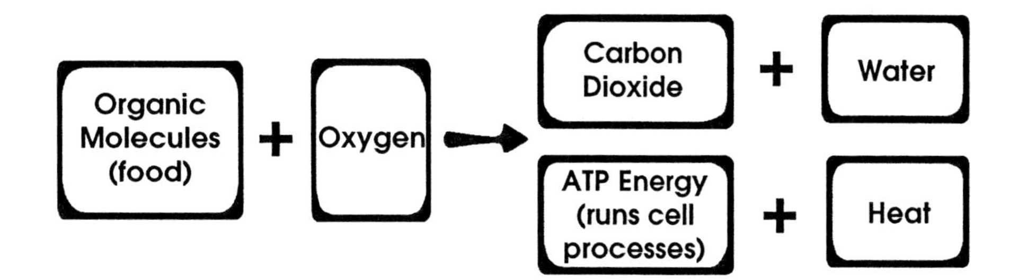

Aerobic respiration occurs inside the mitochondria, which are cellular organelles in both plant and animal cells. The mitochondria have a remarkable structure that is somewhat like a factory. The high-energy electrons are stripped off the food molecule in the fluid matrix, and then the energy of those electrons is used to generate ATP energy along the ***cristae*** membranes.

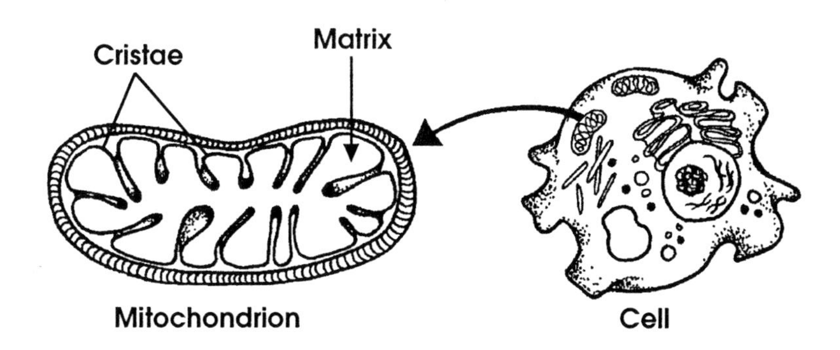

? QUESTION

1. Which cells evolved first—eukaryotic or prokaryotic?

2. Which cells have mitochondria—eukaryotic or prokaryotic?

3. Some of the cells of your body have many mitochondria and other cells have few mitochondria. Why would there be differences?

Metabolism of Nutrients

Sugar is not the only type of food molecule that can be metabolized during cell respiration. Fats, proteins, and starch are other energy sources for the generation of ATP.

Nearly everything in food can be metabolized.

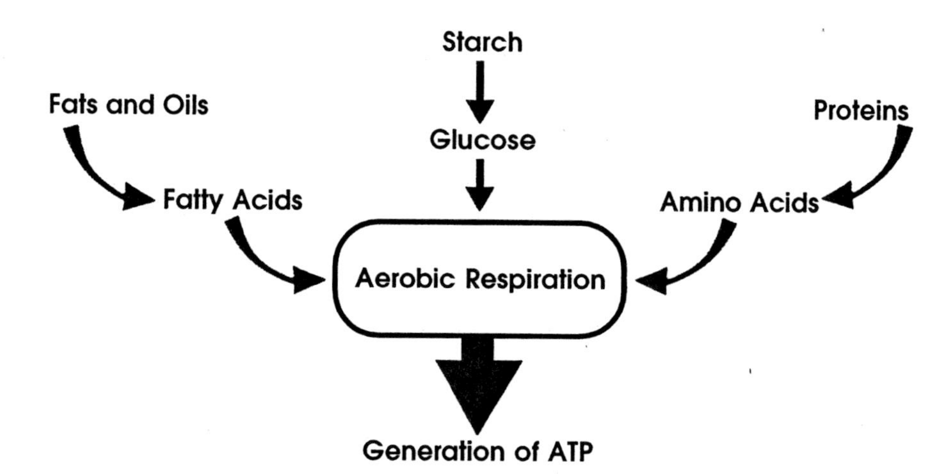

The amount of ATP generated by these nutrients depends on the size of the molecule and the number of high-energy electrons that can be stripped off. Starches are easy to metabolize because they consist of glucose sugar molecules hooked together.

Proteins must first be broken into amino acids, which are then modified. The nitrogen atoms are broken off the amino acid molecule, and ammonia is produced which is then converted into urea. Urea is dumped into the urine. The part of the amino acid remaining after nitrogen removal can be broken down by aerobic respiration. An amino acid generates about the same amount of ATP energy as does an equal weight of sugar.

Fat molecules have many more high-energy electrons than an equal weight of either sugar or protein. Protein and sugar provide about 4 Calories of energy per gram. Fat provides about 9 Calories of energy per gram. Now you can see why it's so easy for those high-energy electrons to pile up!

? QUESTION

1. What factor determines the amount of energy that can be gained from a nutrient molecule?

2. Which nutrient provides the most ATP energy per molecule metabolized?

3. Urea is one of the substances that gives urine its characteristic smell. Urea in the urine means that you have been metabolizing which nutrient?__________

This completes our discussion of cellular respiration. Your textbook will present many more details about the process, and applications related to health and nutrition.

Photosynthesis

Photosynthesis is the process by which plants use sunlight energy to make new plant tissue, and in doing so, they create the by-product of oxygen, which animals need.

All chemical reactions involve changes in the electrons of atoms. The key to ***photosynthesis*** is a special molecule called ***chlorophyll***. This large and complex molecule has electrons that become "activated" when light shines on it. This reaction is unusual because most substances only heat up when exposed to sunlight; that is, their electrons aren't "activated" by light energy. But the electrons in chlorophyll are activated by light, and this electron energy is used to make new plant tissue ($C_6H_{12}O_6$).

This week's lab on photosynthesis focuses on the association between animals and plants. This relationship is so important, that without it, all animals (including humans) would quickly die.

Plants change the energy of light into food energy.

Plants also provide us with our oxygen.
In the early history of our planet, 4 to 5 billion years ago, there was no oxygen in the atmosphere. Oxygen was released into the air only *after* photosynthesis evolved.

Water	+	Carbon Dioxide Gas	(Light Energy) → (Chlorophyll)	Organic Molecules	+	Oxygen Gas
H_2O		CO_2		$C_6H_{12}O_6$		O_2

Exercise #1 "Light Activation of Chlorophyll" 126
Exercise #2 "Leaf Pigments" 127
Exercise #3 "CO_2 Uptake by Plants" 128
Exercise #4 "O_2 Production by Plants" 129
Exercise #5 "Oxygen Demand for Humans" 130
Exercise #6 "How Big of a Plant Does it Take to Keep You Alive?" 131

Special Note: Exercises #3 and #4 will require about an hour of your time. You should set up these two experiments *early* in the lab so that you won't run out of lab time to finish them.

EXERCISE #1

"Light Activation of Chlorophyll"

Botanists tell us that the electrons of the chlorophyll molecule are "charged up" by light energy, and those electrons release that energy immediately to make organic molecules (food) during photosynthesis.

In this Exercise, you will see for yourself whether chlorophyll can be "charged up" by light.

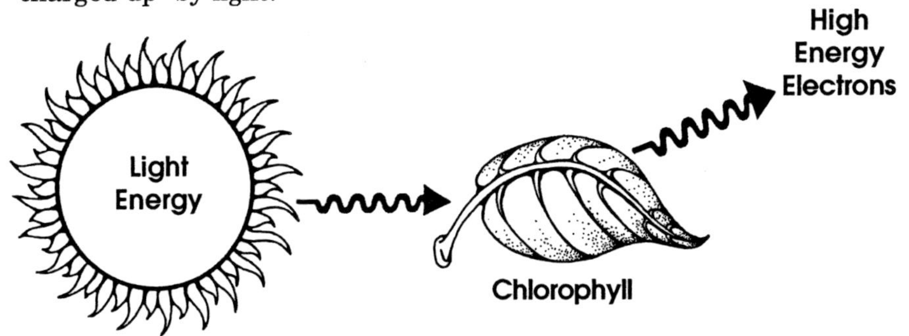

Procedure

[1] Your instructor will take you into a dark room and shine a blue light on a *pure chlorophyll solution*. (Blue light contains no other light colors in it.)

[2] Your group is to observe. Then, go out of the room and discuss what you saw. (Your instructor may shine the blue light on green food coloring as a control experiment.)

Hint: Pure chlorophyll cannot pass any energy onto the rest of the photosynthesis process (to make food) unless the chlorophyll is contained in the chloroplast. That does not mean that the chlorophyll can't react. It only means that it can't make food.

? QUESTION

1. When light activates the electrons of chlorophyll, then those electrons have . . . (circle your choice)

Less energy or More energy

2. Physics tells us that if a substance absorbs energy, then eventually it will lose that energy in one form or another. What did you observe about the chlorophyll solution when the light was shined on it?

3. Based on the results of this experiment, fill in the empty box.

4. Plant cells, *under normal conditions*, convert activated electron energy into what?

EXERCISE #2

"Leaf Pigments"

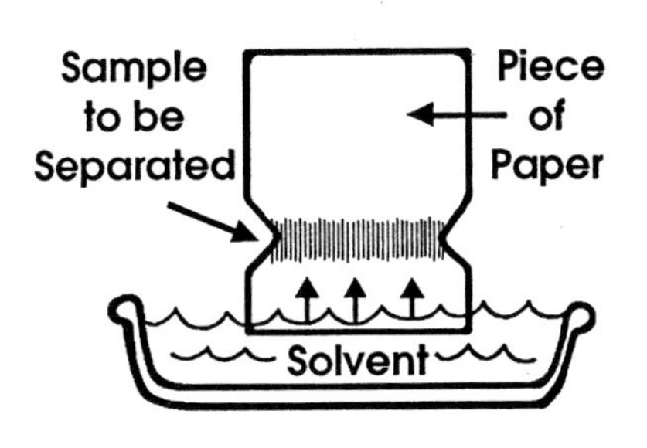

There are many thousands of different kinds of organic molecules (food) on this planet, and sometimes they can be all mixed together in something that we want to analyze. A sample may appear to be *one* substance, but it often is a mixture of *many* different substances.

Chromatography is a very basic chemical process used to separate organic molecules from each other. During this process a solvent passes through a sample that has been impregnated on a piece of paper.

As the solvent travels up the paper, *heavier* or more chemically charged molecules will be left near the *bottom*. The other molecules (lighter or less chemically charged) will be carried *up* the paper.

The secret to understanding the results that you see here is that each *different kind* of organic molecule in the sample will be picked up by the solvent at different rates. (This depends on the individual characteristics of each substance.) Therefore, the organic molecules will be spread out along the paper according to their individual qualities.

Materials

- A chromatography jar and cork.
- A chromatography paper and scissors.
- A spinach leaf and a penny.

Procedure

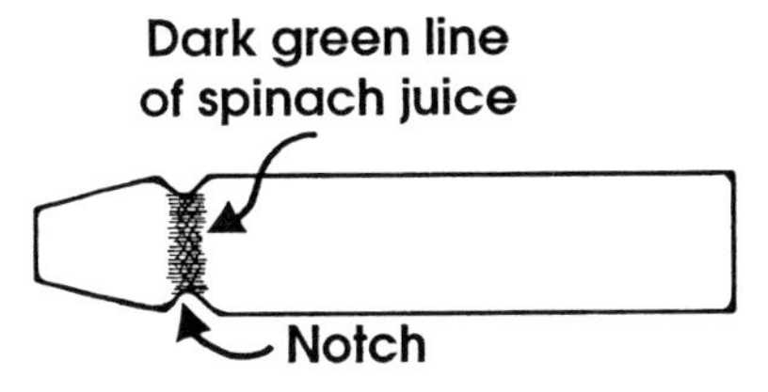

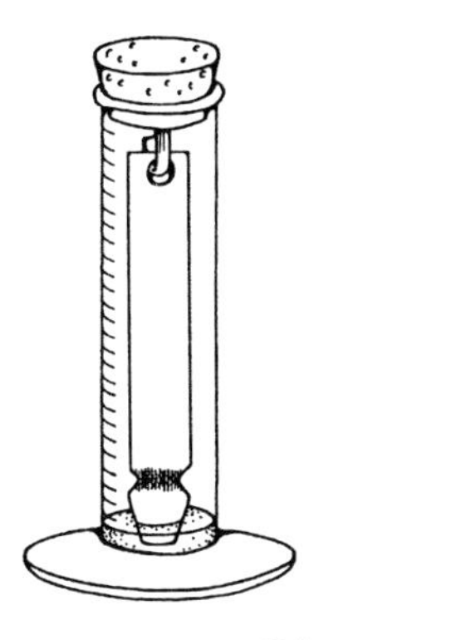

1. Wash your hands with soap so that the substances normally on your hands (French fry grease and hamburger relish) don't become part of the chromatography separation.
2. Cut a point on the end of the chromatography paper. Cut two small notches in the sides about 1.5 cm up from the point. These notches force the solvent to go through the spinach juice.
3. Roll a penny across a spinach leaf to squash a line of juice between the two notches. Make sure this line is dark green. Go over it several times. (The ridges of a quarter will work even better than a penny.)
4. Set up the chromatography jar. Place the notched paper so that when it hangs from the cork, the point *just touches* the bottom of the jar.
5. You must do the rest of the experiment *under a fume hood* or outside in the open air. ***Be careful! The solvent you are using is highly flammable!***
6. Pour the solvent into the chromatography jar to a depth of about 0.5 cm. Plug the cork with the hanging paper into the jar. Leave the jar under the fume hood and *don't move it.*

Observe

During the next 10–30 minutes, the spinach juice will be separated into its individual pigments. Determine how many pigments are present. Each may be a slightly different shade, or might be the same color, but at a different location on the chromatography paper. *Present your answer, and show the evidence to your instructor.*

In Conclusion

When you have finished the chromatography separation, pour the solvent into the waste jar in the fume chamber, and return the chromatography setup to the lab classroom. ***Do not wash out the setup!*** Solvents collect in the air spaces of city drain systems, and can be deadly to sewer workers.

EXERCISE #3

"CO_2 Uptake by Plants"

The photosynthesis equation: $H_2O + CO_2 \xrightarrow[\text{(Chlorophyll)}]{\text{(Light Energy)}} C_6H_{12}O_6 + O_2$ says that carbon dioxide is used to make part of the organic molecule product (food) during the process of the reaction. If this is true, then we should be able to observe that happening.

There is a very simple way to show changes in CO_2 level. Phenol red is a substance that turns yellow when CO_2 is added, and then it turns back to red when CO_2 is removed.

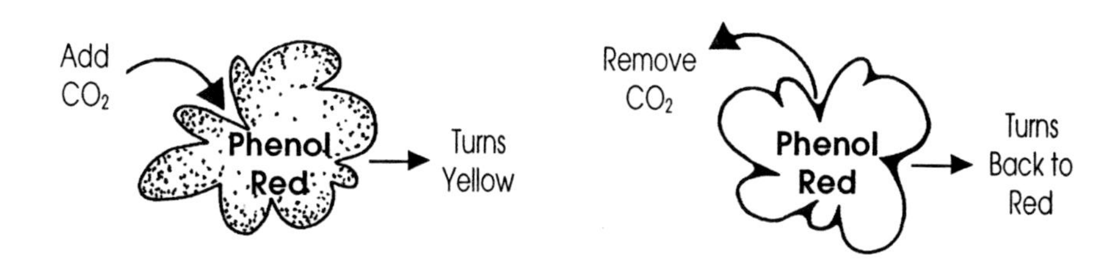

Experimental Question

We can use phenol red as an experimental tool to answer the question:

"Do plants use CO_2 during photosynthesis?"

Experimental Setup

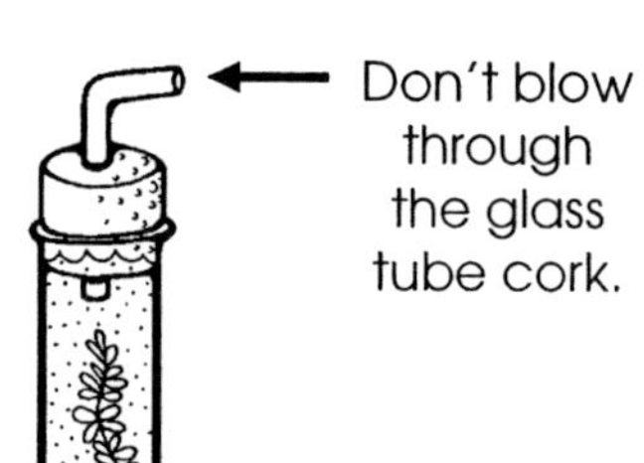

1. Put a small piece of *Elodea* plant (about 10 cm) into a test tube two-thirds filled with a dilute phenol red solution.
2. Using a straw, or your cupped hand, blow very slowly into the top of the phenol until there is a color change through the test tube. Why did it change? Add enough water to fill the test tube close to the top. (Blow into the phenol if it turns back to red.)
3. *Carefully* put the bent glass tube cork into the test tube, leaving *no air bubbles.*
4. Put the experimental setup in front of a light source for 30 minutes. What happens?

Procedure

[1] Your group is to design a simple experiment that will test question #2 on the next page. Be sure to include a control.

[2] Check with your instructor when you think you have a good design for the experiment.

[3] Now, do it!

[4] *Please put the used* Elodea *plants into the special container!* They have some phenol red on them that will contaminate the rest of the *Elodea* and kill it.

? QUESTION

1. If CO_2 is removed from the phenol red solution, then what process is going on in the *Elodea*?

2. Is light required by the plant during photosynthesis?

3. Describe your controls. (There are two.)
   a.
   b.

4. What is the purpose of having a control?

EXERCISE #4

"O_2 Production by Plants"

The photosynthesis equation $H_2O + CO_2 \xrightarrow[\text{(Chlorophyll)}]{\text{(Light Energy)}} C_6H_{12}O_6 + O_2$ says that oxygen is produced. If this is true, then we should be able to observe it.

Experimental Question

How much oxygen is produced by the Elodea plants in one hour?

Experimental Setup

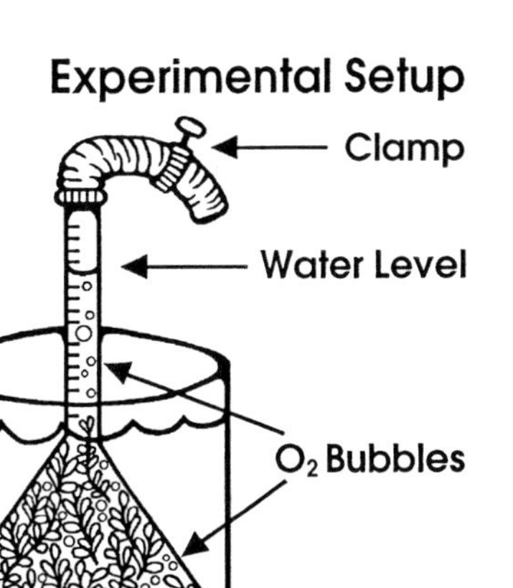

1. You need ten 2" pieces of healthy *Elodea* plants. Trim $\frac{1}{8}$" off each stem. A fresh cut will allow oxygen to bubble out of the plant.
2. You have to suck water up the funnel and up the tube, and then *clamp* the hose at the top to keep the water level from falling.
3. If the experiment is working, the oxygen bubbles will collect at the top of the tube and push the water level down. This drop in water level is what you are to measure during the experiment.
4. Put the light source close to the *Elodea*, but make sure to put a large beaker of clear water *between* the light and the *Elodea* container. (This prevents the *Elodea* from overheating by absorbing heat from the bulb, yet still allows the light to pass through to the container.)

Procedure

[1] We will do only one of these experimental setups for the whole class to observe. Select one person in your group to work with the instructor to set up the apparatus.

[2] Have your group's representative record the O_2 production every 15 minutes for one hour.

[3] Record the total milliliters (ml) of oxygen produced during one hour. You will use this production value during Exercise #6.

ml of O_2 produced by the plant in one hour = ____________

EXERCISE #5

"Oxygen Demand for Humans"

Problem

How much oxygen does a human need to survive one hour of biology lab class?

Next week we will actually measure the O_2 consumption of a mouse under different temperature conditions and compare different animals and plants. However, this week we can borrow an estimate of human oxygen demand from experimental research.

The O_2 used by a human in one hour can range from $\frac{1}{4}$ of a liter of O_2 per kg of body weight to as high as 8 liters. (Although that high rate of metabolism could be maintained for only about 2 minutes without total exhaustion.)

A person in biology lab class uses about 0.4 liters of O_2 per kg of body weight in one hour as long as they aren't walking around all the time.

Procedure

1. Assume that the O_2 used by a person during one hour of biology lab is about 0.4 liters (400 ml) per kg of body weight.
2. Assume that the average human weighs 60 kg.

? QUESTION

What is the oxygen demand for an average person during one hour of biology lab?

__________ = ml of O_2 used by a human in one hour

You will use this calculation again in Exercise #6.

EXERCISE #6

"How Big of a Plant Does it Take to Keep You Alive?"

You have an estimate of the amount of O_2 (in ml) produced during one hour by the *Elodea* Plant (see Exercise #4), and you have an estimate of the amount of O_2 used by a human in one hour (see Exercise #5).

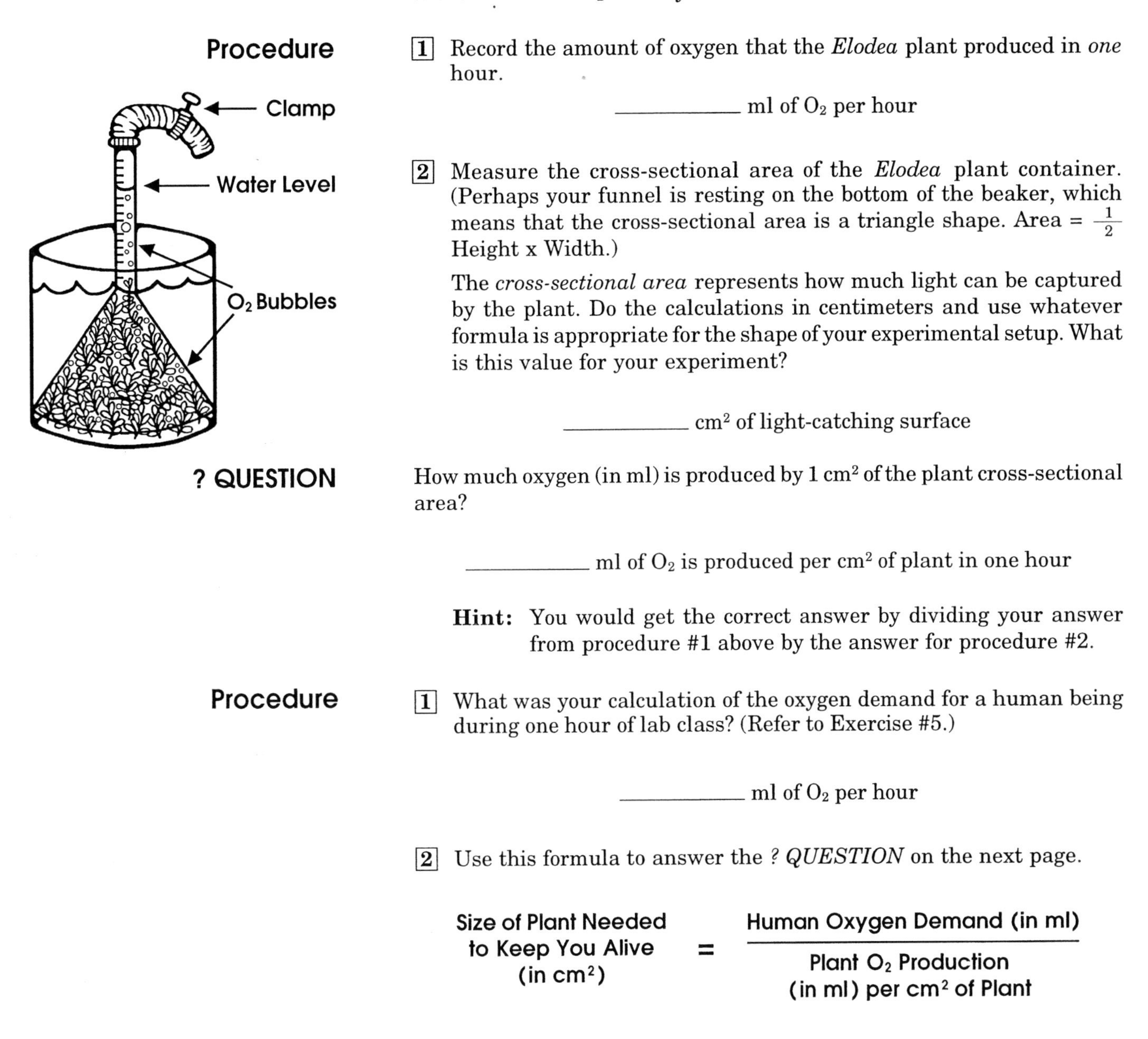

Procedure

[1] Record the amount of oxygen that the *Elodea* plant produced in *one* hour.

__________ ml of O_2 per hour

[2] Measure the cross-sectional area of the *Elodea* plant container. (Perhaps your funnel is resting on the bottom of the beaker, which means that the cross-sectional area is a triangle shape. Area = $\frac{1}{2}$ Height x Width.)

The *cross-sectional area* represents how much light can be captured by the plant. Do the calculations in centimeters and use whatever formula is appropriate for the shape of your experimental setup. What is this value for your experiment?

__________ cm^2 of light-catching surface

? QUESTION

How much oxygen (in ml) is produced by 1 cm^2 of the plant cross-sectional area?

__________ ml of O_2 is produced per cm^2 of plant in one hour

Hint: You would get the correct answer by dividing your answer from procedure #1 above by the answer for procedure #2.

Procedure

[1] What was your calculation of the oxygen demand for a human being during one hour of lab class? (Refer to Exercise #5.)

__________ ml of O_2 per hour

[2] Use this formula to answer the *? QUESTION* on the next page.

$$\text{Size of Plant Needed to Keep You Alive (in cm}^2\text{)} = \frac{\text{Human Oxygen Demand (in ml)}}{\text{Plant } O_2 \text{ Production (in ml) per cm}^2 \text{ of Plant}}$$

? QUESTION

1. What size of plant is required to keep you alive?

   ____________ cm^2

2. Change this plant size to m^2 by dividing your answer above by 10,000 (there are 10,000 cm^2 in one m^2). What is your answer in m^2?

   Size of plant required = ____________ m^2

3. If you determine the square root of the plant area above, then you will have calculated the *side measurement of a square shape* representing the plant area.

   Side = __________ m

4. In the above calculations, you have determined how big a plant is required to keep you alive during daylight hours, but what will keep you alive at night? **Remember:** Plants don't photosynthesize at night.

5. Does this change your estimate of how big a plant it takes to keep you alive both day and night?

In Conclusion

Go outside and mark off on the ground how big of a plant is required to keep you alive.

Respiration

Cellular ***respiration*** is the process in living organisms that extracts electron energy from the chemical bonds in *food* (organic molecules), and converts that energy into a more useful form of energy (called **ATP**) to run cell activities. This cell process uses oxygen and produces carbon dioxide. The complete equation is:

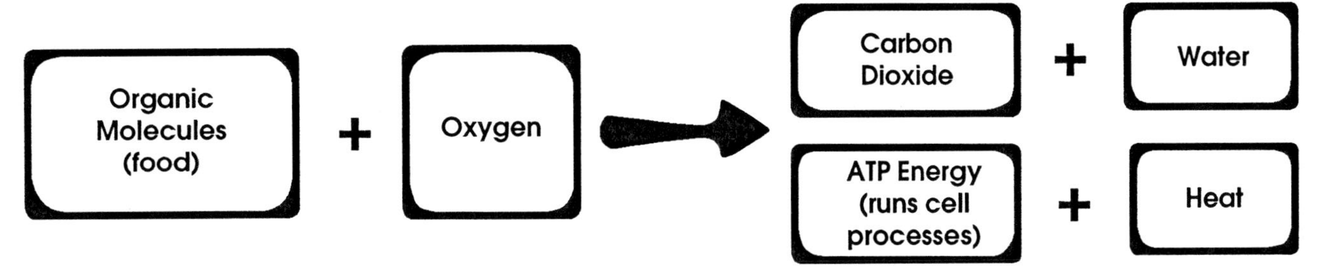

Respiration occurs inside the ***mitochondria***, which are cellular organelles in both plant and animal cells. Refer to your textbook for the structural and functional description of this organelle.

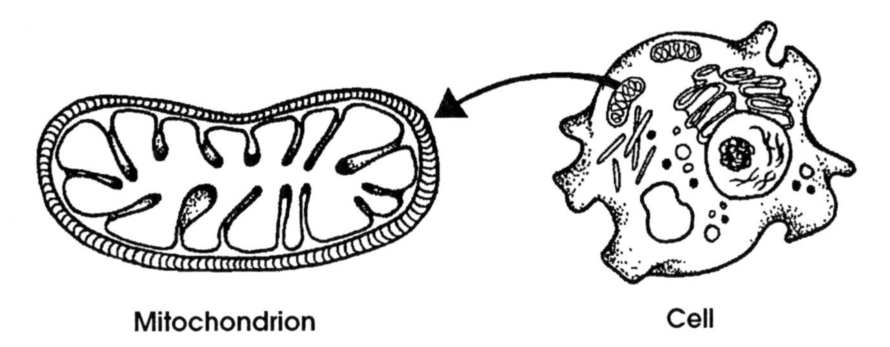

During this lab we will investigate some aspects of cellular respiration including the effects of environmental temperature on the rate of respiration in ***endotherms*** (internally heated animals) and ***ectotherms*** (externally heated animals).

Exercise #1 "Heat Production During Respiration" 133
Exercise #2 "Respiration in an Endotherm" 134
Exercise #3 "Comparison of Endotherm and Ectotherm" 138
Exercise #4 "Food Demand for Humans" 139
Exercise #5 "Respiration in Plants" 140

EXERCISE #1

"Heat Production During Respiration"

The Second Law of Thermodynamics states that heat is released whenever any form of energy is transformed into another form.

Since *respiration* is described as the conversion of food energy into usable energy for the cell, we should be able to observe heat being given off during the process.

Observe

Live Seeds

Dead Seeds

Two experimental containers were set up yesterday. One of the containers was filled with *dead seeds* killed by boiling, and the other container was filled with *live seeds.* These seeds demonstrate the basic respiration process that is going on in all living organisms.

Record the temperature of each container.

Temperature of live seeds = ____________

Temperature of dead seeds = ____________

? QUESTION

1. What does the equation for respiration say about *heat*?

2. What does this experiment suggest is occurring in live seeds and not in dead seeds?

3. What would happen to the respiration process in the container of live seeds if we pumped the oxygen out?

What would happen to the temperature in that container?

EXERCISE #2

"Respiration in an Endotherm"

Mice are ***endotherms***. That is, they get most of their heat from *inside* their own body (*endo* means inside). Cellular respiration generates the heat that keeps these animals warm. (Refer to the Equation for Respiration on the first page.)

During this Exercise you will monitor the ***rate of respiration*** (also called ***metabolic rate***) in a mouse. In addition, you will investigate the influence of environmental temperature on the mouse's rate of respiration by comparing a mouse in a *cold environment* with a mouse in a *warm environment.*

Later, in Exercise #3, you will compare the differences between an endotherm (mouse) and an ectotherm (frog).

Right!

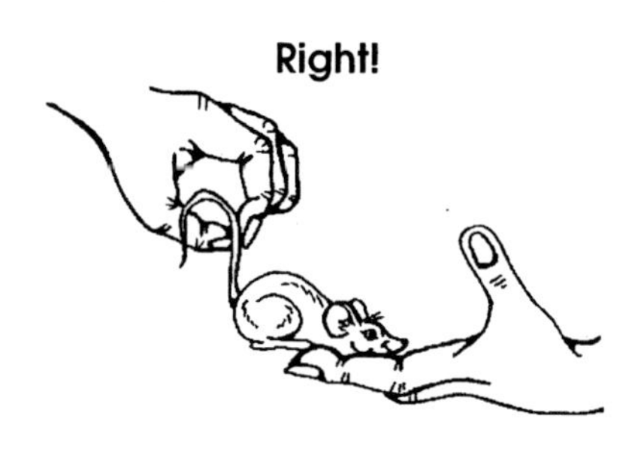

How to Handle Mice

Mice should be *picked up by their tail* and immediately *rested on your hand,* and then marched into the Metabolic Cage.

Do not grab them. Grabbing scares the hell out of them, and they may bite you or pee on you because of that fear.

Also, *don't play with the mice* (on table tops, etc.) because there is a possibility of them getting loose on the floor. These are professional mice. They work several years for us, and we treat them very well. So, please be careful.

Wrong!

Experimental Apparatus

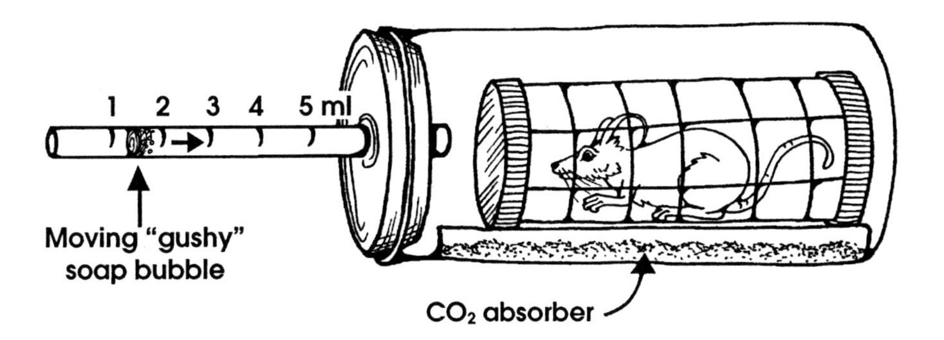

Experimental Design

The basic question is: ***What effect does environmental temperature have on the metabolic rate of an endotherm (mouse)?***

Do this experiment at two temperatures: Room Temperature and Packed in Ice.

Room Temperature

Procedure

[1] Weigh the wire cage part of the chamber: __________ grams.

[2] Go get your mouse, and put it into the wire cage. Then weigh the cage with the mouse in it.

Cage + Mouse		Cage		Weight of Mouse
__________ g	-	__________ g	=	__________ g

[3] Put one tablespoon of CO_2 absorber (soda lime) into the trough at the bottom of the Metabolic Rate Chamber.

[4] Wet the inside of the glass tube with soapy water. This will help prevent the "gushy" bubble from "popping" during the experiment.

[5] Put the caged mouse into the chamber and seal the cork tightly. *Don't worry! The mouse won't suffocate.* Leave the chamber alone for 10 minutes *(sealed up—cork on—no soap bubble)* to equalize the temperatures inside and outside of the chamber.

[6] Use your finger to make a "gushy" soap bubble on the open end of the glass tube. Then, measure the time it takes (in seconds) for the bubble to move between the marks on the tube until 5 ml of O_2 have been consumed by the mouse.

Perform three trials.

__________ seconds	__________ seconds	__________ seconds
Trial 1	**Trial 2**	**Trial 3**

? QUESTION

1. Food + O_2 ⟶ CO_2 + H_2O. During respiration a mouse will consume O_2, and CO_2 will be produced in its place. If no CO_2 absorber had been used in your experiment, would you have seen a change in air *volume*?

2. If you use a CO_2 absorbing substance in the Metabolic Rate Chamber, then what happens to the CO_2 that is produced during respiration?

3. Now, with the absorbing substance in the chamber, what happens to the *air volume* during your experiment as the O_2 is consumed during respiration?

Packed In Ice

Procedure

[1] If ice is packed around a Metabolic Rate Chamber like the type we are using, the temperature inside will stabilize at 5°C. This cold air temperature will *not harm* the mouse as long as the mouse is removed before 45 minutes. Our experiment will take less than 20 minutes.

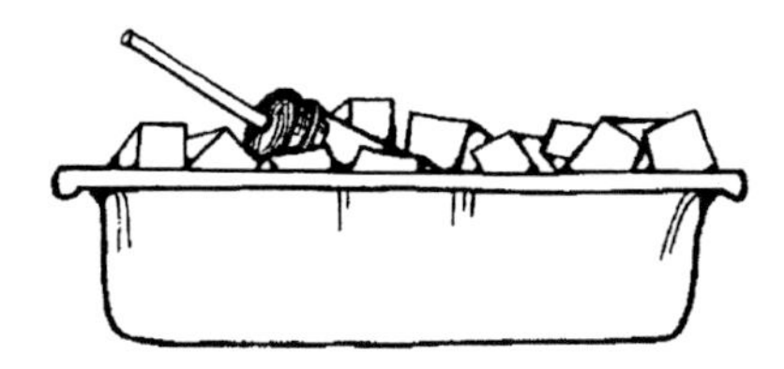

[2] Now perform the "Packed in Ice" experiment. Let the chamber equalize the temperatures inside and out for 10 minutes before applying the "gushy" soap bubble.

[3] After 10 minutes, apply a "gushy" soap bubble and perform the three separate measurements of the rate of respiration.

________ seconds	________ seconds	________ seconds
Trial 1	**Trial 2**	**Trial 3**

[4] Disassemble the chamber, carefully returning your mouse to its home, and dump all CO_2 absorber and feces into the special waste jar. Don't wash the apparatus unless you are told to do so. The chamber must be dry for the next lab class. ***Wash your hands!***

Respiration Calculations

You must convert the mouse's O_2 consumption to an hourly metabolic rate. This is accomplished by dividing the bubble time (in seconds) into 3,600 (the number of seconds in one hour). That number is to be multiplied by 5 (5 ml of O_2 used in each trial).

Calculation 1

Calculate the *average* time of the three trials at room temperature.

5 ml O_2 consumed in ________ seconds (average time)

Calculation 2

Based on Calculation 1, how much O_2 would your mouse consume in one hour?

$$\frac{3{,}600}{\text{Calculation 1}} \times 5 = \_\_\_\_\_\_\_\_\_\_ \text{ ml } O_2 \text{ consumed in one hour}$$

Calculation 3

In order to have a metabolic rate that can be compared with an animal of different weight, we must correct the calculations considering the mouse's weight.

$$\frac{\text{Calculation 2}}{\text{Weight of Mouse}} = \_\_\_\_\_\_\_\_\_\_ \text{ ml } O_2 \text{ per hour per gram of weight}$$

Procedure

[1] You have finished the calculations for room temperature. Record your answer below. Repeat the three calculations for "Packed in Ice," and record your answer below.

Metabolic rate of your mouse at room temperature (20°C) = __________ ml O_2 per hour per gram of weight

Metabolic rate of your mouse packed in ice (5°C) = __________ ml O_2 per hour per gram of weight

[2] Put a dot on the graph for each of the metabolic rate values in your experiment.

[3] Draw a line between those two dots, and write the word ***endotherm*** on the line.

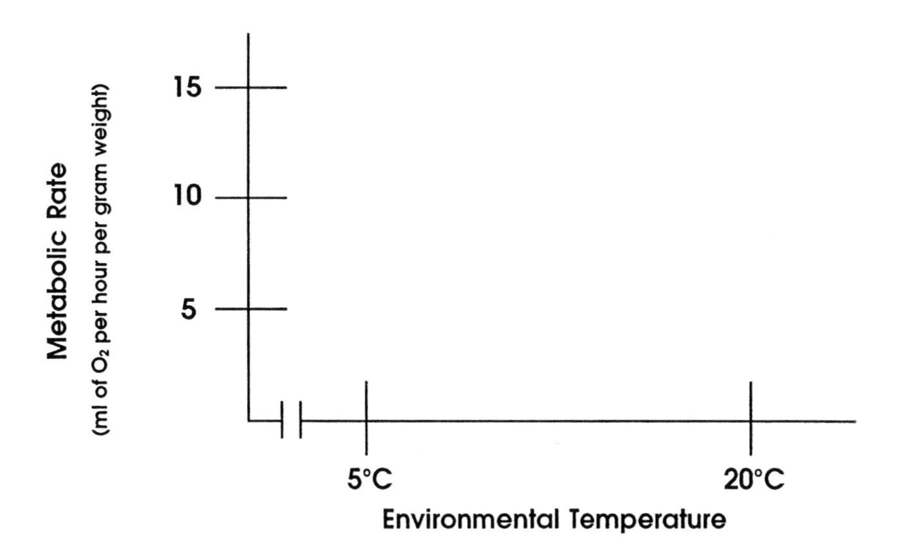

[4] Check with other lab groups to see how your calculations compare with theirs.

EXERCISE #3

"Comparison of Endotherm and Ectotherm"

An ***ectotherm*** gets its heat from the environment (*ecto* means outside). The body temperature of an ectotherm is warm when the environment is warm, and the body is cooler when the environment is cold.

Information

The following results are taken from some experiments that measured the metabolic rate in a frog (*ectotherm*) of about the same size as your mouse.

	Metabolic Rate Packed in Ice (5° C)	Metabolic Rate at Room Temperature (20° C)
Frog #1	0.05	0.30 ml O_2 per hour per gram of weight
Frog #2	0.03	0.28
Frog #3	0.04	0.25

Procedure

1. Calculate the *average* metabolic rate for the three frogs at each of the two temperatures.
2. Put a dot on the graph for each of the average values.
3. Draw a line between those two dots, and write the word ***ectotherm*** on the line.

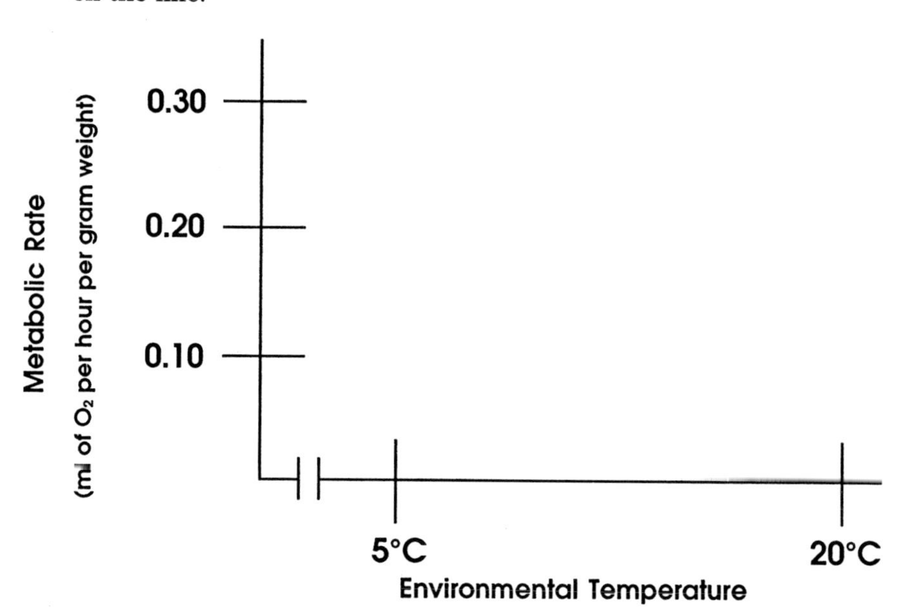

? QUESTION

1. Which organism has the slowest rate of respiration? (circle your choice)

   Endotherm or Ectotherm

2. Which organism needs less food to survive? (circle your choice)

   Endotherm or Ectotherm

   Explain why.

3. How much food does the *ectotherm* need compared to the *endotherm*?

__________ %

4. Which organism would do better if the amount of food is very limited, but the environment is fairly warm? (circle your choice)

Endotherm or Ectotherm

5. In what areas of the world would you expect to find ectotherms?

6. Which organism would do better in cooler environments where the food is plentiful? (circle your choice)

Endotherm or Ectotherm

7. Will the organism in question #6 do fine in warmer environments if the food is plentiful?

Why or why not?

EXERCISE #4

"Food Demand for Humans"

How much food does a human need to survive one hour of biology lab class?

We can borrow data from experimental research to help us estimate the amount of food that is required to support a human. Our calculations will be based on grams of sugar as the nutrient. Also, notice that the word *Calorie* is capitalized. When capitalized, this term represents 1000 times the value of a single calorie.

The Caloric demand for food varies greatly for a human depending on activity and environmental conditions. The energy demand might be as slow as 50 Cal per hour during sleep to as fast as 2,000 Cal per hour during extreme exercise. (Although that high rate of metabolism could be maintained for only about 2 minutes without total exhaustion.)

An average student in biology lab class uses about 100 Calories per hour as long as they aren't walking around all of the time.

Information

1. Assume a food demand of 100 Cal/hour for students.
2. A human gets about 3.85 Cal of energy from 1 gram of sugar.

? QUESTION

How many grams of sugar are required to “fuel” an average student during one hour of biology lab class?

__________ grams of sugar used in one hour

Procedure

Weigh out that much sugar and show it to your lab instructor.

EXERCISE #5

“Respiration in Plants”

The Respiration Equation states that CO_2 is produced as O_2 is used. If that is true, then we should be able to use CO_2 production as an indicator that respiration is occurring.

Phenol Red Test for CO_2

There is a very simple way to show changes in CO_2 level. Phenol red is a substance that turns yellow when CO_2 is added, and then turns back to red when CO_2 is removed.

Add CO_2

Phenol Red

Turns Yellow

Remove CO_2

Phenol Red

Turns Back to Red

Experimental Setup

Yesterday we put a small piece of *Elodea* plant into a test tube filled with dilute phenol red solution. The tube was red because the water had very little CO_2 in it. We put this experimental setup into a closed cabinet until today.

Ask your instructor where the plant is, and make your observations.

? QUESTION

What do you conclude about plants in the dark?

Genes and Protein Synthesis

(Tutorial)

Ngoma kicked at the ground as he sat outside the math lecture hall. He was waiting for his best friend Uta. She was a graduate student at the college, and understood science better than anyone else he knew. Ngoma wasn't a science student, and he wanted her opinion right now.

In his morning biology class, Ngoma had learned that small pieces of DNA, called *genes,* are responsible for producing traits in people. The genes control how amino acids assemble into proteins. These proteins are the fabric of life, and they produce all the structures and functions found in living beings. This new explanation of life destroyed some of the imagined mystery he had before hearing the lecture.

The science explanation was very different from the creation stories Ngoma listened to and loved as a boy. He lived many adventures in his daydreams because of those stories. It was these boyhood fantasies that first inspired Ngoma to draw and paint. Eventually that led him to study art in college.

Uta burst from the classroom, walking briskly in the direction of the Chemistry Building.

"Uta!" he shouted. Ngoma knew that it was up to him to catch up with her.

As their paths paralleled each other, Uta glanced quickly at her friend and said, "Ngoma, you look upset. Tell me, what's the matter."

"Uta, how do you keep the magic in your heart after you discover the way things really happen in nature?"

Uta stopped abruptly, spinning to face him, and exclaimed in her usual determined voice, "Ahh, Ngoma, I learned that from *you*! You are the one who notices every little thing. You are the one who shows me the details that I often miss."

Ngoma blushed with surprise. His face filled with a warm smile. "Uta, I'd like to take you to lunch. How about it?"

She smiled back. "Sure—after chemistry and physics. But, what was bothering you?"

"Oh, it was nothing," Ngoma said touching her hand. "For a moment, I just forgot the details."

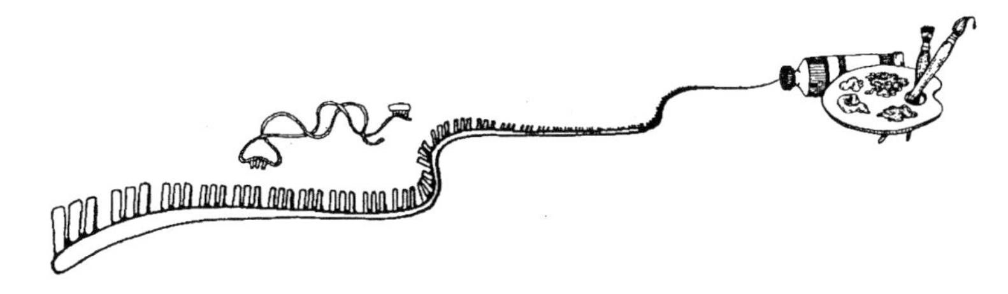

Exercise #1 "The Molecular Blueprint" 141

Exercise #2 "Genes and Enzymes" 144

Exercise #3 "Genes and Traits" 148

EXERCISE #1

"The Molecular Blueprint"

Proteins make up the fabric of cell structure, and also form the enzymes that control biochemical reactions. The *kinds* of protein are what make one cell *different* from another cell.

DNA is the blueprint of life

So, what makes proteins? The nucleic acids DNA and RNA are the biochemical blueprint molecules for making proteins in the cell. These molecules are similar in structure, and both are examples of the earliest molecules that could be considered "alive." Today, RNA and DNA determine the biochemical processes in living cells.

DNA is the master blueprint molecule that is passed from generation to generation. ***RNA*** is a copy of the DNA blueprint, and it does the actual mechanics of building proteins.

DNA Structure

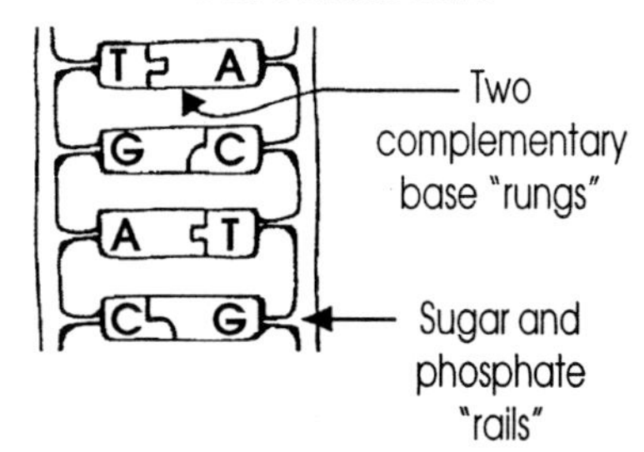

DNA is an extremely thin, long, ladder-like molecule that has two "rails" made of sugar and phosphate, and many "rungs" made of special complementary bases.

The DNA bases (rungs) are molecular units that combine only in two kinds of pairs.

The base ***adenine*** (A) is always paired with ***thymine*** (T), and ***cytosine*** (C) is always paired with ***guanine*** (G). Therefore, if you know *one* of the complementary bases, you can easily figure out the other.

The term ***nucleotide*** is the name for these repeating subunits in a DNA molecule. A nucleotide is actually a phosphate, a sugar, and a base hooked together as a basic building unit.

? QUESTION

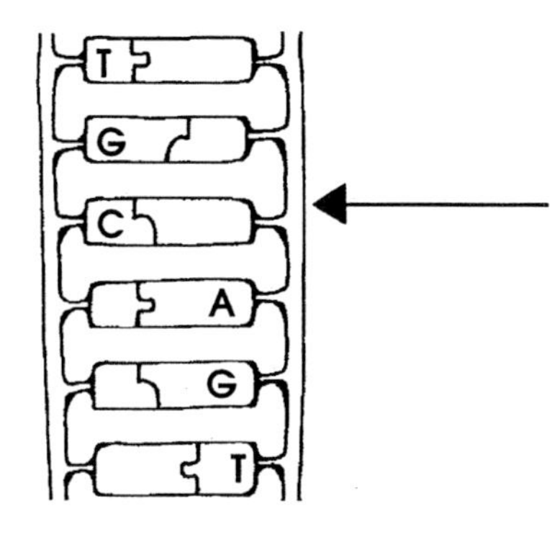

1. Guanine always pairs with ________________.

2. Thymine always pairs with ________________.

3. Fill in the complementary nucleotides on this DNA ladder.

RNA Structure

RNA is a single-stranded molecule made up of nucleotides linked together end to end. Its structure is similar to DNA except for a slightly different sugar in the four nucleotides. In addition, RNA's single strand has one different nucleotide: RNA has ***uracil*** instead of thymine. RNA has no thymine.

Laboratory experiments have shown that it is easy for DNA to replicate itself. If a DNA molecule is unzipped (by slightly heating or using an enzyme), and immersed into a tube containing a mixture of DNA nucleotides and special enzymes, it will make copies of itself until all the nucleotides are used up. This happens because of the specific nucleotide pairing (A with T, and G with C).

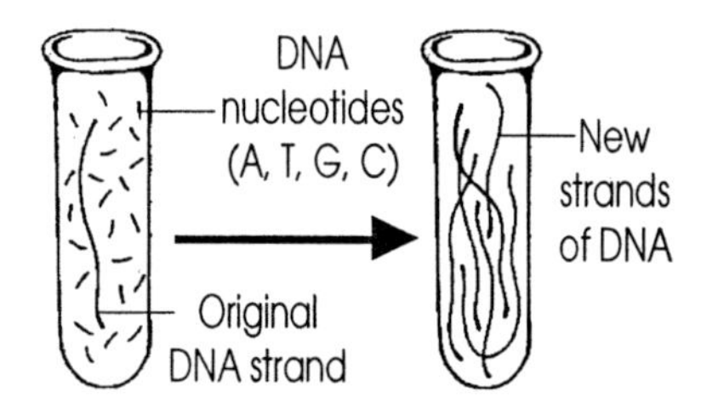

Other experiments demonstrate that DNA can also make RNA. If a DNA strand is immersed into a tube of RNA nucleotides, it will make RNA copies until all the nucleotides are used up. This happens because of the specific nucleotide pairing. The difference is that the U (uracil) of RNA pairs up with the A (adenine) in DNA. There is no T (thymine) nucleotide in RNA. Because uracil has a similar shape to thymine, it pairs up with the adenine nucleotide of the DNA molecule.

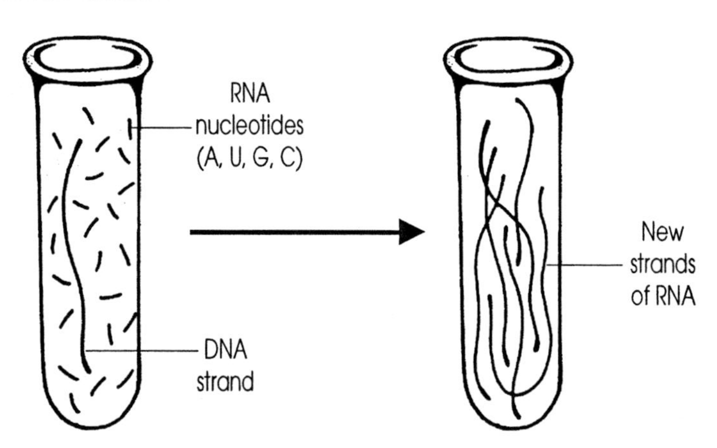

Draw arrows to show where each of the RNA nucleotides will line up along the DNA strand.

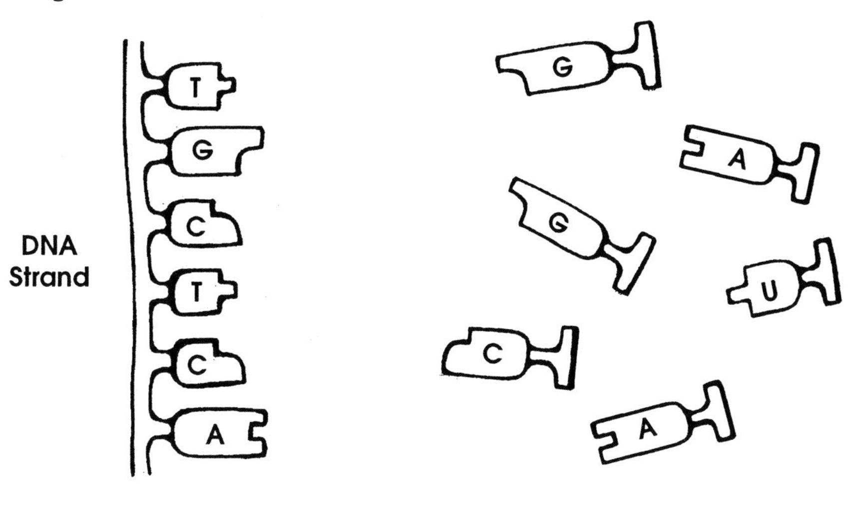

? QUESTION

1. List the four nucleotides of DNA:

    __________ __________ __________ __________

2. List the four nucleotides of RNA:

    __________ __________ __________ __________

3. What are the subunits of a protein?

4. Which molecule (RNA or DNA) is passed on from generation to generation?

5. Which molecule (RNA or DNA) does the actual mechanics of building proteins?

6. Which is a single-stranded molecule (RNA or DNA)?

7. An original strand of DNA has the following sequence of nucleotides:

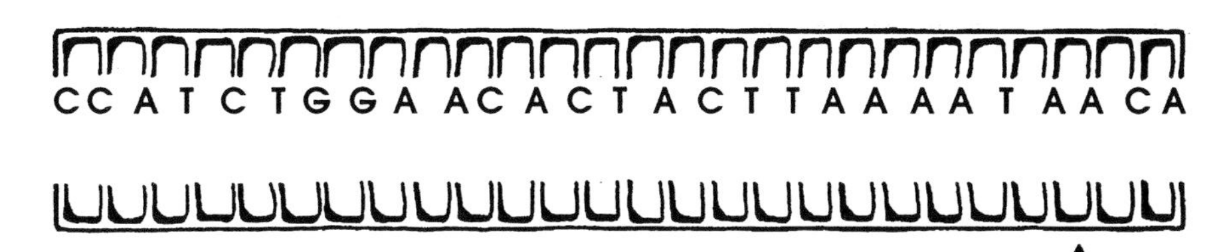

Fill in the corresponding nucleotides for the RNA strand.

EXERCISE #2

"Genes and Enzymes"

The "one-gene-one-enzyme" hypothesis was the simplest explanation to start with.

By the 1950s, it was well established that a gene is a portion of the DNA, and that each gene contains a "message" for making a particular protein. The first proteins investigated were enzymes involved in the biochemical pathways of simple organisms. The idea was termed: ***one-gene-one-enzyme hypothesis***.

Further investigation revealed that each gene in the DNA makes a different enzyme for the organism's biochemistry. Actually, the one-gene-one-enzyme hypothesis is an oversimplification. Some enzymes are synthesized under the direction of several genes, with each gene being responsible for a *different piece* of the enzyme. In other cases, certain genes produce structural proteins needed by the cell.

The diagram below shows how a number of genes control the production of different enzymes involved in the synthesis of a single product—melanin.

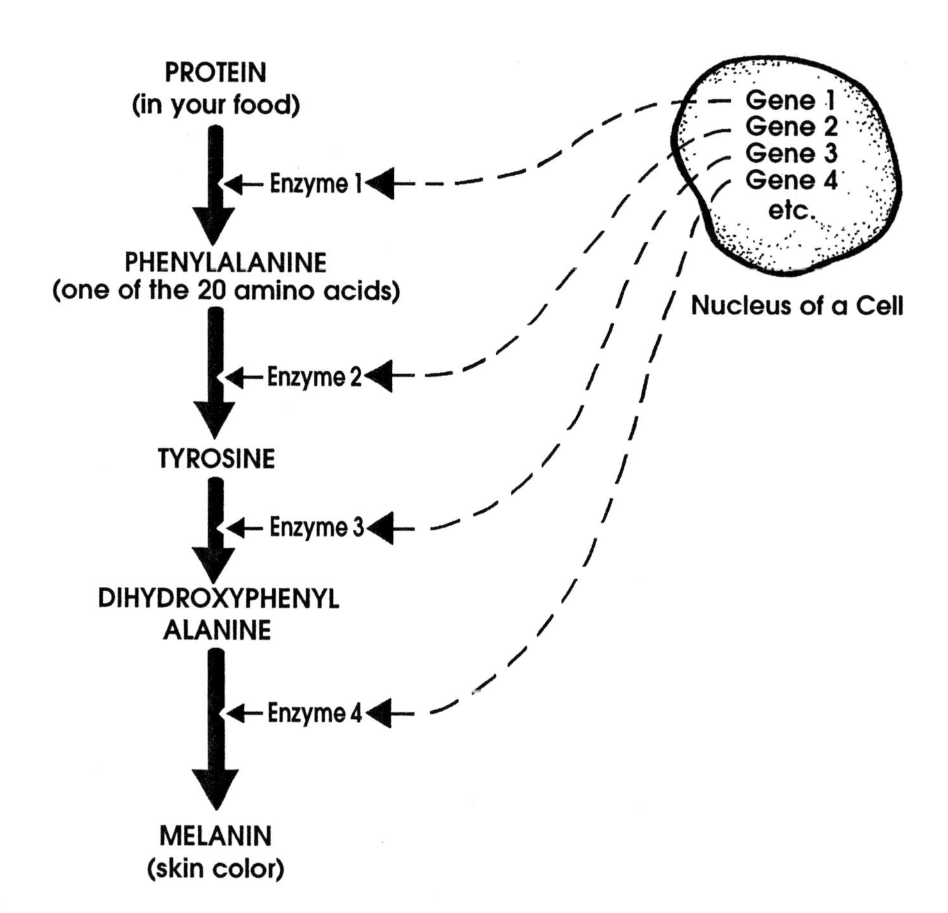

Gene Copied by mRNA

The message is transcribed from DNA language into RNA language.

If you wanted to make many copies of a valuable sculpture piece, one method of manufacture would be to use the "original" each time you make a copy. You would hesitate doing this because the original could be damaged and would soon wear out. A reasonable solution would be to make a mold of the original, and send that mold to the production line. When the mold is worn out, you make a new one!

This analogy helps us to appreciate the cellular process whereby RNA copies genes and assembles proteins. RNA saves the wear and tear on the DNA original. Since a gene contains the "message" for building a particular protein, the RNA that copies the gene is called ***messenger RNA***, or simply ***mRNA***.

The process of mRNA copying the gene is called ***transcription***. The message is "transcribed" from DNA language into RNA language.

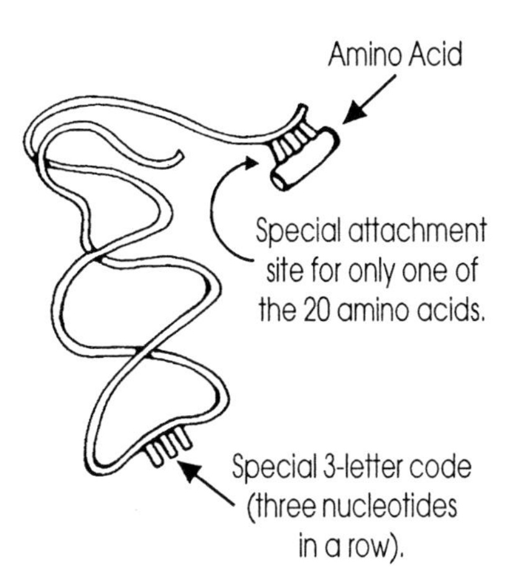

tRNA

Some of the genes in DNA are responsible for making another kind of RNA, called ***transfer RNA***, or simply ***tRNA***. The tRNA is a twisted strand with two important reaction sites. One site attaches to a particular amino acid. Amino acids are the building blocks for all of life's proteins. There is a special tRNA for each of the 20 different amino acids. Each of the 20 different amino acids is "carried by" or "transferred" around the cell by its own special tRNA.

The second important reaction site is on the other end of the tRNA; it is a special 3-letter code (three nucleotides in a row) that is critical in protein synthesis. Your textbook calls this 3-letter code an *anticodon*.

Codons

Each mRNA molecule is a copy of a different gene. Some genes are as small as 30 nucleotides long, and other genes have a few hundred or even a few thousand nucleotides. The mRNA has a "start" end and a "stop" end. Between the two ends, there is a sequence of special 3-letter codes (three nucleotides in a row), called ***codons***.

Each codon of the mRNA is responsible for lining up a particular amino acid that is to be assembled into the protein. This is accomplished because a codon binds with only one kind of tRNA which carries only one kind of amino acid.

Each codon is responsible for lining up a particular amino acid.

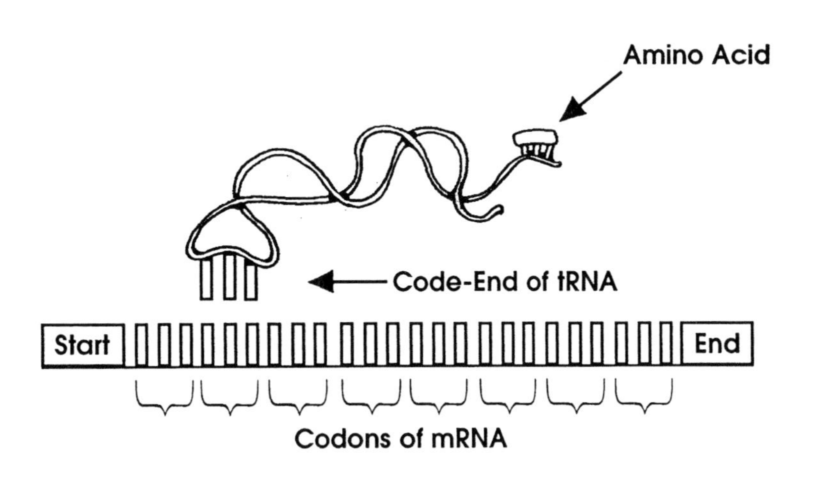

mRNA CODONS RESPONSIBLE FOR LINING UP EACH OF THE 20 AMINO ACIDS

Amino Acid	mRNA Codons*	Code-End of the tRNA
Alanine	GCU	
Arginine	AGA	
Asparagine	AAU	
Aspartic Acid	GAU	
Cysteine	UGU	
Glutamic Acid	GAA	
Glutamine	CAA	
Glycine	GGU	
Histidine	CAU	
Isoleucine	AUU	
Leucine	CUU	
Lysine	AAA	
Methionine	AUG	
Phenylalanine	UUU	
Proline	CCU	
Serine	UCU	
Threonine	ACU	
Tryptophan	UGG	
Tyrosine	UAU	
Valine	GUA	

* There are 64 codons. Some amino acids have several mRNA codons. There is, however, no overlap of codes.

? QUESTION

1. You should be able to fill in the 3-letter "code-end" of the tRNA molecules in the table above. Remember, in RNA ***A*** pairs with ***U***, and ***G*** pairs with ***C***. There is no thymine. Fill in the table.

2. What is the one-gene-one-enzyme hypothesis?

3. Which molecule copies the DNA message for building a particular protein? ____________________ What is this process called?

4. Where are tRNA and mRNA made?

5. How many different amino acids are there in living organisms?

6. Describe two special reaction sites on the tRNA.

7. How many nucleotides in a row make one codon?

Events at the Ribosome

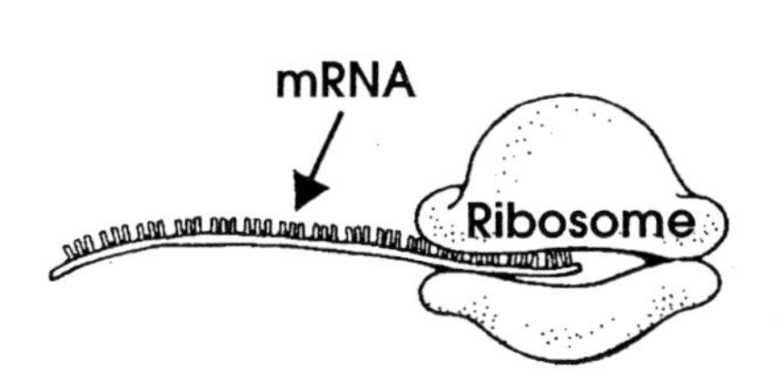

The protein synthesis reactions at the ribosome are sometimes called ***translation***. The "message" in the mRNA is "translated" into the sequence of amino acids in a particular protein. Remember, when there is control over the line up of amino acids, then an exact protein of any kind can be synthesized.

The mRNA attaches to the ribosome and passes through it one codon at a time.

As each of the codons passes through the ribosome, the appropriate tRNA carries in its amino acid. The 20 different tRNA molecules carry the 20 different amino acids.

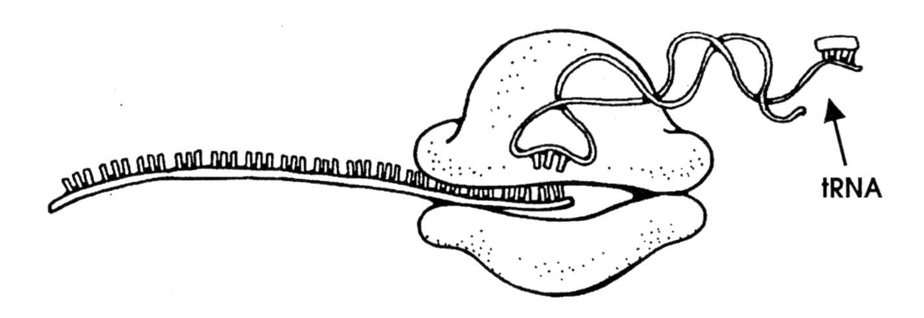

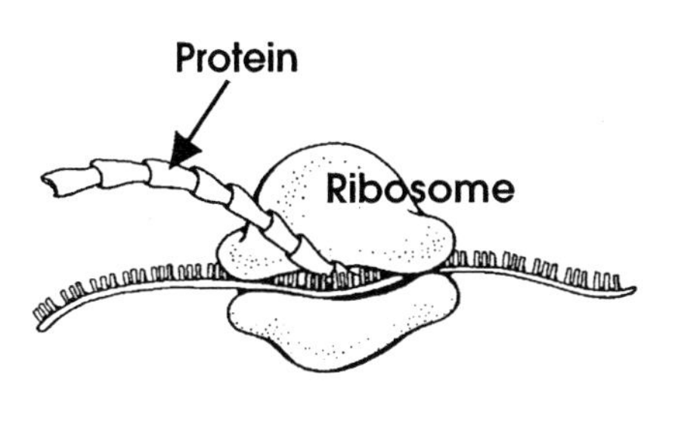

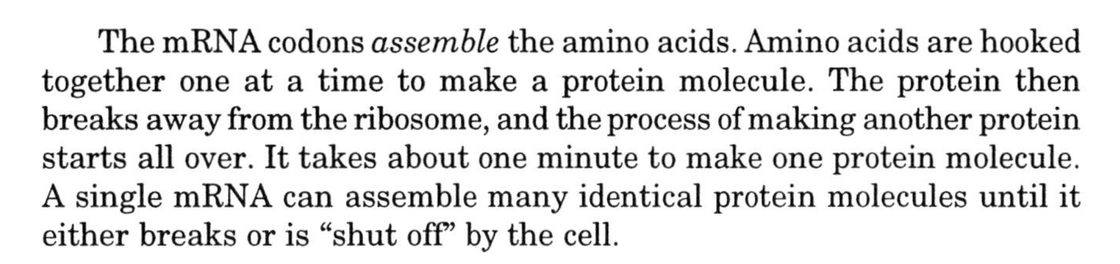

The mRNA codons *assemble* the amino acids. Amino acids are hooked together one at a time to make a protein molecule. The protein then breaks away from the ribosome, and the process of making another protein starts all over. It takes about one minute to make one protein molecule. A single mRNA can assemble many identical protein molecules until it either breaks or is "shut off" by the cell.

? QUESTION

1. Define *translation*. Where does it happen?

2. Genes are changed by exposure to radiation. Radiation can cause one of the nucleotides in DNA to be replaced by another. When nucleotides are changed in the DNA, what happens to the codons in mRNA, and what effects result during protein synthesis? (Be specific.)

3. Practice your understanding of the events at the ribosome by writing a paragraph summarizing the important highlights.

EXERCISE #3

"Genes and Traits"

Genetic traits are features (either structural or biochemical) of the organism that are produced by the action of one or more genes. If you have the gene, then you develop the trait. If you don't have the gene, then you don't develop the trait.

Turning Genes On and Off

All of the cells of your body have the same genes. However, in each cell certain genes are turned "on" and other genes are turned "off." How the cell functions is determined by which genes are *on* and which genes are *off*.

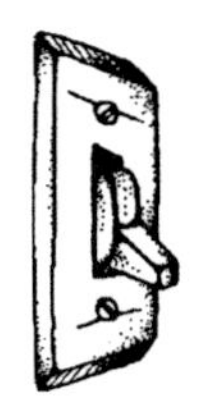

Type of Gene	Skin Cell	Liver Cell	Intestine Cell
Basic Metabolism Genes	On	On	On
Sugar Releasing Gene	Off	On	Off
Melanin Gene	On	Off	Off
Digestive Enzyme Gene	Off	Off	On

Science is not yet able to decipher all the processes controlling which genes are on or off in a particular cell. However, certain general mechanisms have been discovered. For example, in some cases, a substance in the environment around the cell can activate a gene. Sometimes the proteins produced by one gene will activate or inhibit another gene. There are also master control genes that activate clusters of other genes.

During embryonic development, the genes of many cells are turned on or off in a particular sequence. This causes a cell to become specialized so it can function in part of the body (as bone or muscle or brain cells). Some of the genes that are turned on during embryonic development are essential only at that stage. These same genes are turned off in the adult person, and would cause serious diseases like cancer if they were to accidentally turn on again.

? QUESTION

1. List the different ways that genes are controlled in the cell.

2. What is the relationship of the gene to the trait?

Traits

The complexity of genetic processes responsible for the different kinds of traits in an organism is beyond the scope of this exercise. Your textbook and lecture class will expand on this discussion. This brief review of gene-controlled traits will introduce you to the topic.

Single-Gene Traits

The simplest example of genetic traits is the ***single-gene trait***. Gregor Mendel, who pioneered the science of modern genetics, discovered that pea plants with a "purple-making" gene produced purple flowers. This gene controls the synthesis of an enzyme that activates the biochemical pathway leading to the purple color. Other peas plants possessed a *defective* gene that would not synthesize the purple-making enzyme. These plants had white flowers. They could not make the purple color.

Multiple-Gene Traits

The height of a person and the color of the skin are examples of ***multiple-gene traits*** because there are several duplicate genes contributing to the same trait. If skin color were a single gene trait, then there would be two colors of skin—brown and albino. Refer to the melanin pathway example in Exercise #2. If you have the gene for *enzyme 3* in this pathway, then you make melanin (brown pigment). A defective gene produces *non-functional* enzyme 3, so no melanin is made (no pigment).

Now, consider the situation in which there may be six duplicate skin color genes in each person. There are two possible choices for each color gene (melanin or no-melanin). People show a distribution of skin color from light to dark, depending on how many melanin or no-melanin genes they have.

The size of a person depends on the amount of growth hormone they produce. This is another multiple-gene trait similar to skin color.

Environment

A yellow pigment in certain plants can collect in the fat of rabbits eating those plants. It results in yellow-colored fat. In some rabbits, there is a metabolic enzyme that breaks down the yellow pigment into a white by-product. A gene controls the synthesis of this enzyme.

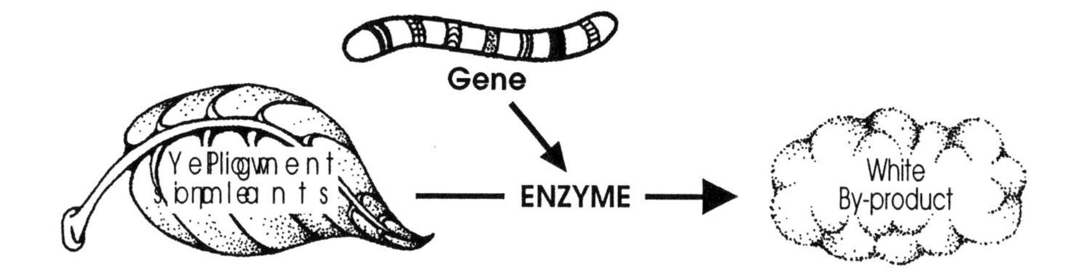

Rabbits that have the active form of this gene produce white-colored fat. Rabbits who have the defective form of this gene produce yellow-colored fat. However, there is an environmental complication of this trait. (See the following questions.)

? QUESTION

1. If a rabbit that has the active form of the gene is fed plants with the pigment, what color is its fat?

2. If a rabbit with the defective gene is fed plants with the pigment, what color is its fat?

3. If a rabbit with the defective gene is *not* fed plants with the pigment, what color is its fat?

Can you see how the environment influences the expression of a trait?

4. What is different about a gene that creates albino color?

5. Describe the variation in the appearance of individuals in a population when a trait is single-gene controlled compared to a trait that is multiple-gene controlled.

Mutations

Mutations are the source of all change.

A ***mutation*** is a change in the sequence of nucleotides in the gene (DNA). A gene can be mutated (changed) by radiation, chemicals in the environment, or other spontaneous events that are surprisingly common on this planet.

? QUESTION

1. Which amino acid is supposed to be lined up at codon 6? (**Hint:** Refer to the table in Exercise #2.)

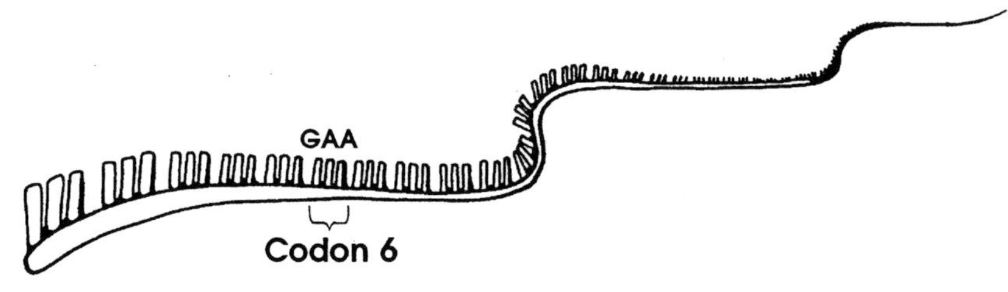

2. What would happen if a mutation in DNA changed codon 6 in the mRNA to GUA? (This one nucleotide substitution creates sickle-cell anemia.)

3. How does a mutation change the events at the ribosome?

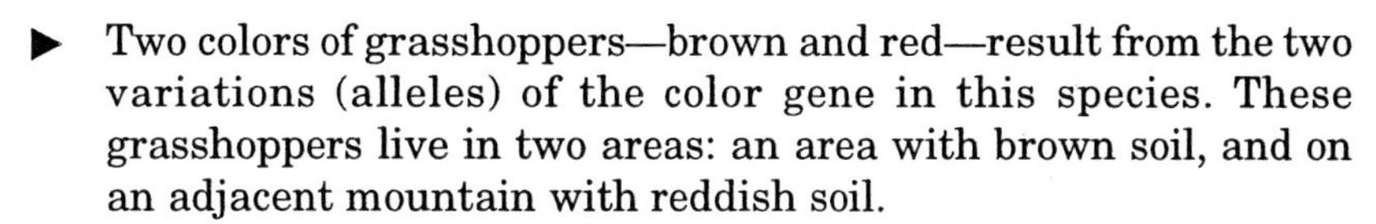

Evolution

Biologists define ***evolution*** as the change in the frequency of ***alleles*** (variations of a gene) in a population. Consider the following situation:

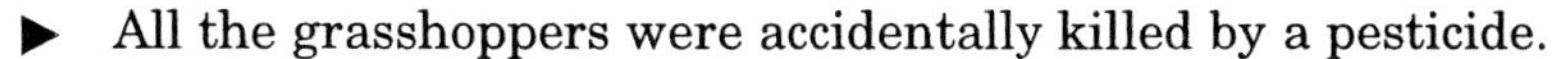

- Two colors of grasshoppers—brown and red—result from the two variations (alleles) of the color gene in this species. These grasshoppers live in two areas: an area with brown soil, and on an adjacent mountain with reddish soil.
- All the grasshoppers were accidentally killed by a pesticide.
- Grasshoppers are a great food source for any bird that can catch them. Ecologists re-introduced 1000 red and 1000 brown grasshoppers into each of the two soil areas.

? QUESTION

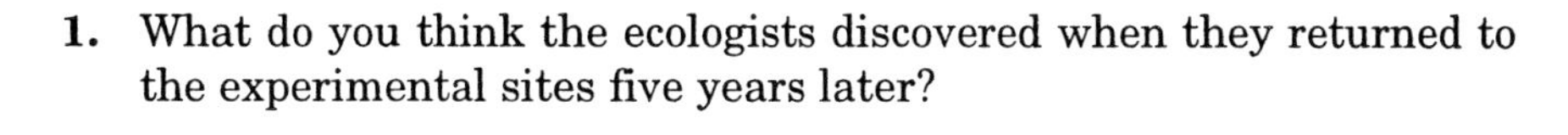

1. What do you think the ecologists discovered when they returned to the experimental sites five years later?

2. What factor controls the actual color of a grasshopper?

3. What determines the color frequency of grasshoppers in the two soil areas?

4. Is there only one direction of change during evolution? Explain.

Sameness and Variety

Is it better to be the same as everyone else?
or
Is it better to be different?

We struggle with these questions in our personal lives. Would it surprise you to learn that all life, in terms of its reproductive strategy, has struggled with the same basic questions? It might also surprise you to learn that there is no *one* answer to reproduction, but *two* answers.

The strategy of producing offspring that are genetically the same as the parent is called ***asexual reproduction,*** and is accomplished through a cell division process termed ***mitosis***. Asexual reproduction is the simplest and oldest form of reproduction, and it relies on a single parent.

The strategy of producing offspring that express genetic variety is termed ***sexual reproduction***. It is accomplished through a cell division process called ***meiosis*** and a fusion process called ***fertilization***. Sexual reproduction is complex and it usually relies on two parents.

Many organisms have lost the means to reproduce asexually except for cell replacement or growth. But some less specialized species use both modes, depending on the time of year. The one certainty is that organisms exist today only because they have incorporated both sameness and variety in their struggle to live and reproduce.

Exercise #1 "DNA Replication" 153
Exercise #2 "Chromosome Sets" 155
Exercise #3 "Asexual Reproduction of Cells—Mitosis" 159
Exercise #4 "Chromosome Movement During Mitosis" 160
Exercise #5 "Onion Root Tip" 163
Exercise #6 "Sexual Reproduction" 164
Exercise #7 "The Events That Create Variety" 165

EXERCISE #1

"DNA Replication"

You don't reproduce —your DNA does!

During the last 40 years, the study of biochemistry revealed a fact that stunned the scientific establishment and transformed our approach to biology.The fact is this: ***A chemical called DNA reproduces—not the individual.***

Also, it was discovered that the sorting of DNA by the cell is what controls the *sameness* or *variety* of the next generation. This does not mean that the organism isn't important. The organism houses the cells that contain the DNA molecules. Without the organism, DNA would destabilize and fall apart. But the unit that actually gets passed on to the next generation is the DNA. That's why the discussion of reproduction is centered around this unique molecule.

So, first let's take a look at the structure of DNA and how it reproduces itself. Then, we will see how DNA uses an organism to achieve sameness by asexual reproduction, or achieves variety by sexual reproduction.

DNA Structure

DNA Molecule

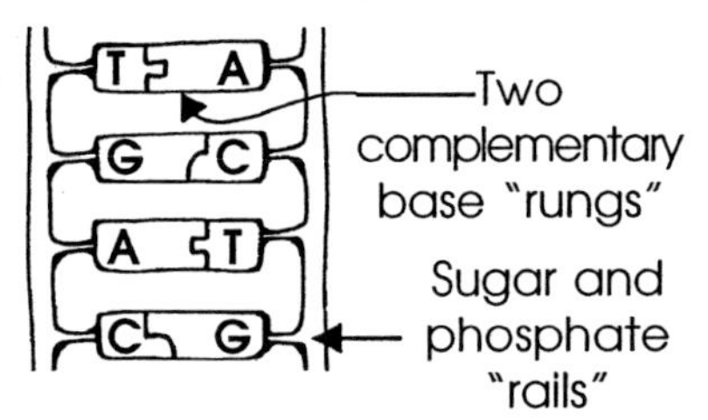

DNA is an extremely thin, long, ladder-like molecule that has two "rails" made of sugar and phosphate, and many "rungs" made of special complementary bases. The DNA bases (rungs) are molecular units that combine only in two kinds of pairs.

The base ***adenine*** (A) is always paired with ***thymine*** (T), and ***cytosine*** (C) is always paired with ***guanine*** (G). Therefore, if you know *one* of the complementary bases, you can easily figure out the other.

The term ***nucleotide*** is used as a name for the repeating subunits in a DNA molecule. A nucleotide is actually a phosphate, a sugar, and a base hooked together as a basic building unit.

? QUESTION

1. Guanine always pairs with ________________.
2. Thymine always pairs with ________________.
3. Fill in the complementary nucleotides in the DNA ladder to the left.

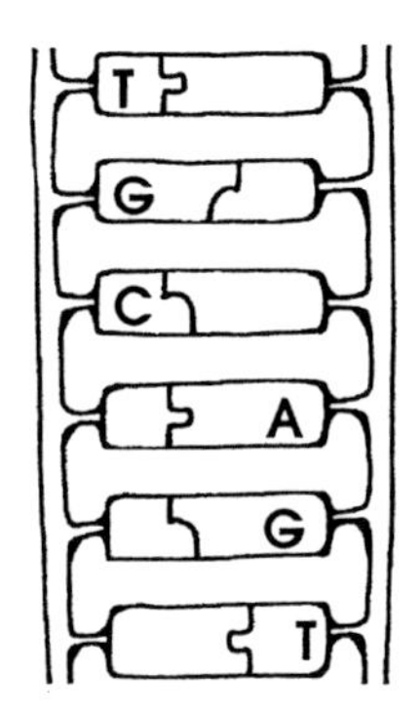

Reproduction of DNA

Before DNA can copy itself the cell must make lots of extra A, T, G, and C. Then the DNA unzips between the two bases and adds nucleotides to each side of the unzipped DNA molecule.

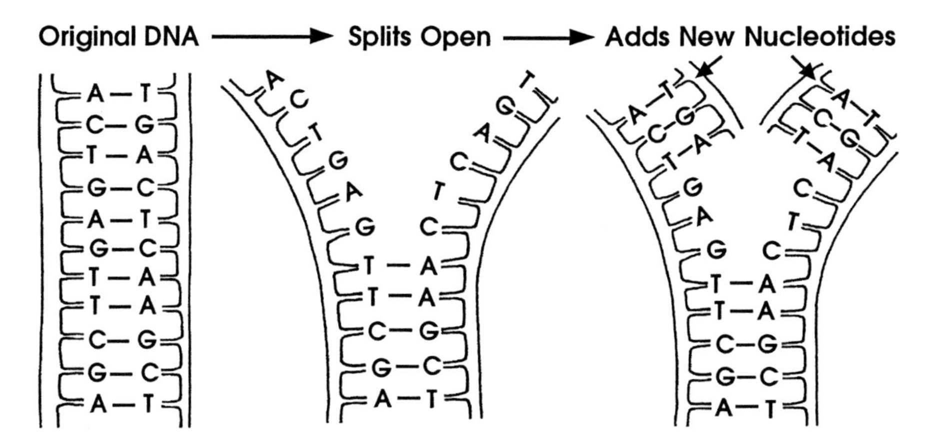

Now, two DNA molecules exist where there was only one before.

? QUESTION

1. Finish unzipping the DNA molecule pictured above, and draw the completed picture of the two "new" DNA molecules in the left margin. Use the same DNA sequences as the example above.
2. What are the two "new" DNA molecules built from? Explain.
3. Are the two "new" DNA molecules absolutely identical to the "original" DNA molecule?

Example

Expressed using apples, DNA replication looks like this.

Observe

Go look at the model of DNA on the demonstration table.

We have been presenting DNA as a straight ladder, but actually it is twisted on itself like a spiral staircase. This shape is called a ***helix***.

The Chromosome

Coiling

Normal DNA

Coiled DNA (chromosomes)

Normally DNA exists as loose strands (***chromatin***) in the nucleus of a cell. This nuclear DNA sends a message (RNA) to the ribosomes where protein and enzymes are synthesized.

When stretched out, the length of one DNA molecule in a human cell is almost 4 cm. However, the cell itself is but a tiny fraction of that size. During cell reproduction the DNA must be able to move around. So it shortens its length by tightly coiling up. In doing so, the DNA strands become wider and are visible under a microscope. Visible DNA is called a ***chromosome***.

? QUESTION

1. A chromosome is made up of tightly ______________________ DNA.
2. Explain the reason for this shape.

3. When not reproducing, DNA is found in the ______________________ of a cell and exists in the form called ______________________.
4. Visible DNA is called a ______________________.

EXERCISE #2

"Chromosome Sets"

The number of chromosomes in a cell varies from species to species, but it is exactly the same among individual members of the same species.

The One-Set Concept of Chromosomes: What is Haploid?

All species have one or more sets of chromosomes. This means that chromosomes come in *sets*, and the *number* of chromosomes in a set depends on the particular species.

A set of chromosomes includes *one copy* of all of the genes necessary to control the biochemical activities of a species. Most species have either one or two chromosome sets.

In genetics a single set of chromosomes is symbolized by the letter "**n**." Any cell that has only one set of chromosomes is termed ***haploid***. Haploid means that the cell has *one* of each *kind* of chromosome.

The set concept will be used throughout the rest of this lab, so the following questions were designed to aid you in understanding and recognizing sets. Remember that a set is a group of objects related in function and generally used together.

? QUESTION

1. Pretend that the *fingers* of one hand represent chromosomes. (Count your thumb as a finger.) Hold up your hand.

a. How many fingers (chromosomes) do you have on one hand?

____

b. Do you have different kinds of fingers on one hand?

____

c. Do you have more than one of each kind of finger on one hand? __________

d. Judging by the definition of a set, you have __________ set(s) of fingers on one hand.

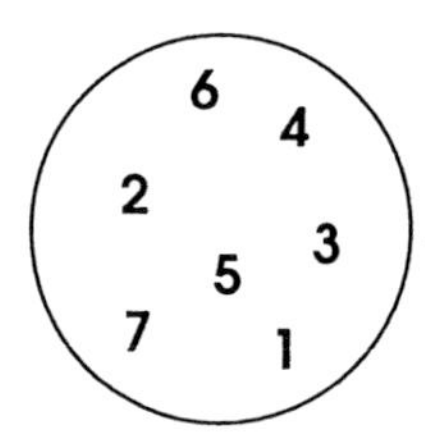

2. Pretend that all the numbers within the circle represent chromosomes.

a. How many numbers are there? __________

b. Are there different kinds of numbers? __________

c. Is there more than one of each kind of number? __________

d. Judging by the definition, you have __________ set(s) of numbers.

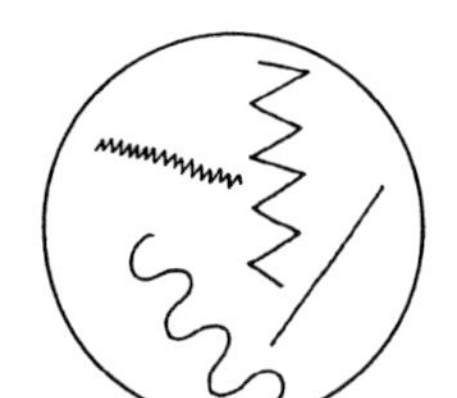

3. Pretend these lines represent chromosomes.

a. How many lines are there? __________

b. Are there different kinds of lines? __________

c. Is there more than one of each kind of line? __________

d. Judging by the set definition, you have __________ set(s) of lines.

Answers:

#1	#2	#3
a=5	a=7	a=4
b=yes	b=yes	b=yes
c=no	c=no	c=no
d=1	d=1	d=1

4. Explain the one-set concept of chromosomes.

The Two-Set Concept of Chromosomes: What is Diploid?

"Diploid" means two.

Simple organisms and the gametes of complex organisms are haploid. That is, they have a single set of chromosomes. Complex organisms require *two sets* of chromosomes to survive. We will discuss the details of this two-set requirement in a later lab.

In genetics, the two-set condition is symbolized as "**2n**," and is called ***diploid***. Diploid means that the cell has *two* of each *kind* of chromosome.

? QUESTION

1. Pretend that the fingers of *both* of your hands represent chromosomes. Hold up your hands.

a. How many fingers (chromosomes) do you have? ____________

b. Regarding both hands, do you have a *duplication* of each of the *kinds* of fingers? ____________

c. How many sets of fingers do you have? ____________

d. How many fingers are in a single set? ____________

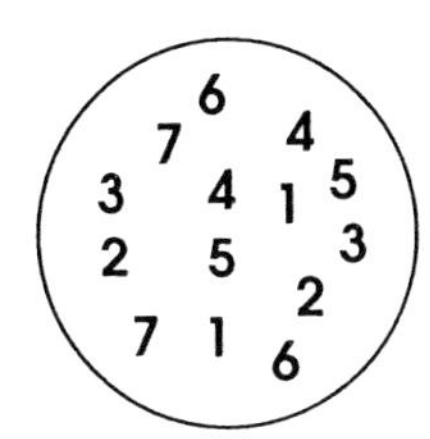

2. Pretend that all the numbers within the circle represent chromosomes.

a. How many total numbers are there? ____________

b. Is there a duplication of each kind of number? ____________

c. How many sets of numbers are there? ____________

d. How many numbers are in each set? ____________

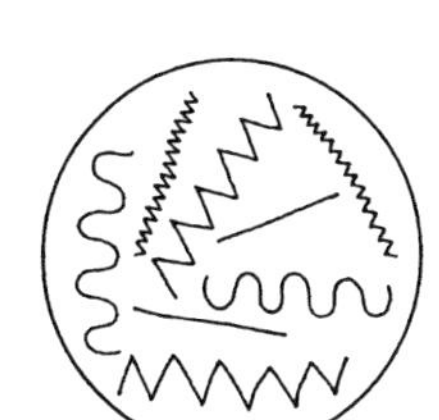

3. Pretend that the lines within the circle represent chromosomes.

a. How many lines are there? ____________

b. Is there a duplication of each kind of line? ____________

c. How many sets of lines are there? ____________

d. How many lines are there in each set? ____________

Answers:

#1	#2	#3
a=10	a=14	a=8
b=yes	b=yes	b=yes
c=2	c=2	c=2
d=5	d=7	d=4

4. Explain the two-set concept of chromosomes.

Homologous Chromosomes

Humans have 23 pairs of homologous chromsomes.

The word *homologous* means "the same," and the term ***homologous chromosomes*** refers to the pairs of chromosomes in a diploid cell that carry genes for the same traits. We will discuss more details about cell division in the following Exercises, but for now, keep this note in mind: ***Each chromosome of a homologous pair comes from a different parent.*** Humans are diploid and have 46 chromosomes (two sets of 23). This means that we have 23 homologous pairs.

Although both chromosomes of a particular homologous pair carry the *same genes*, these genes may be slightly different *forms*. For example, one might be the "blue eye" form, and the other might be the "brown eye" form. (More about that later.)

? QUESTION

1. Pretend that the fingers of both of your hands are chromosomes. Hold up both your hands.

a. How many individual fingers (chromosomes) do you have? ___________

b. How many homologous pairs are there? ___________

c. How many sets of fingers do you have? ___________

d. How many homologous pairs are in one set of fingers? ___________

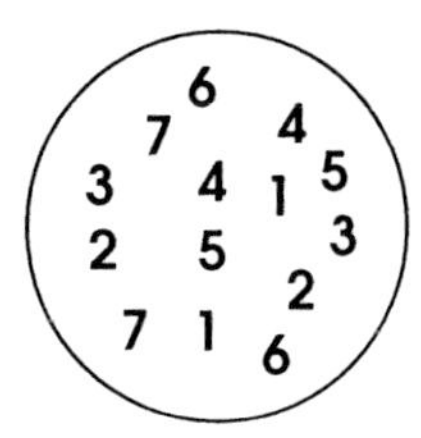

2. Pretend that the numbers in the circle are chromosomes.

a. How many individual numbers are there? ___________

b. How many homologous pairs of numbers are there? ___________

c. How many sets of numbers are there? ___________

d. How many homologous pairs are in one set of numbers? ___________

3. Pretend that all the lines within the circle are chromosomes.

a. How many individual lines are there? ___________

b. How many homologous pairs of lines are there? ___________

c. How many sets of lines are there? ___________

d. How many homologous pairs of lines are in one set? ___________

Answers:

#1	#2	#3
a=10	a=14	a=8
b=5	b=7	b=4
c=2	c=2	c=2
d=0	d=0	d=0

4. Explain in simple terms what homologous chromosomes are.

EXERCISE #3

"Asexual Reproduction of Cells—Mitosis"

Mitosis produces more of the same.

Asexual reproduction of cells is called ***mitosis***. Immediately before this cell division process begins, the DNA of a cell (either haploid or diploid) duplicates itself creating two identical copies of every DNA molecule (and chromosome). The DNA copies move to opposite ends of the cell. Then the cell partitions itself into two cells (each with *exactly* the same DNA as the original cell). The purpose of asexual reproduction by mitosis is to create new cells that are genetically *identical* to the original cell.

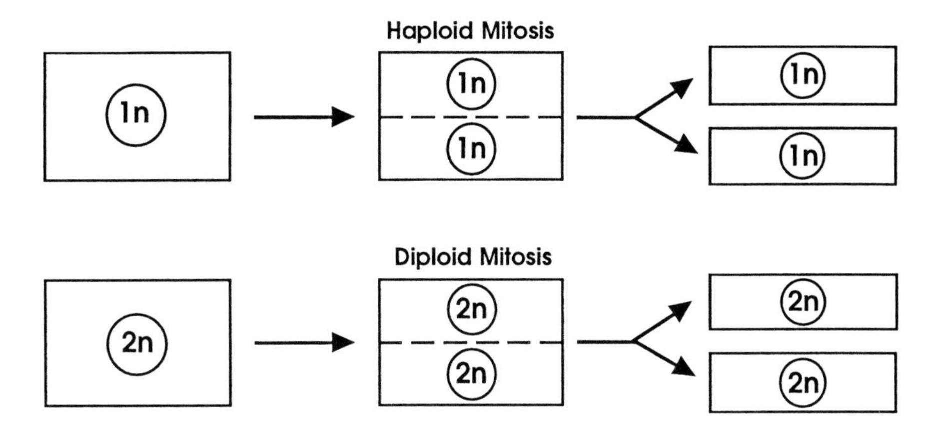

Problem

1. Let's imagine what would happen to the amount of DNA material in a cell if, when it reproduced, it was *halved* instead of duplicated. Complete this cell box.

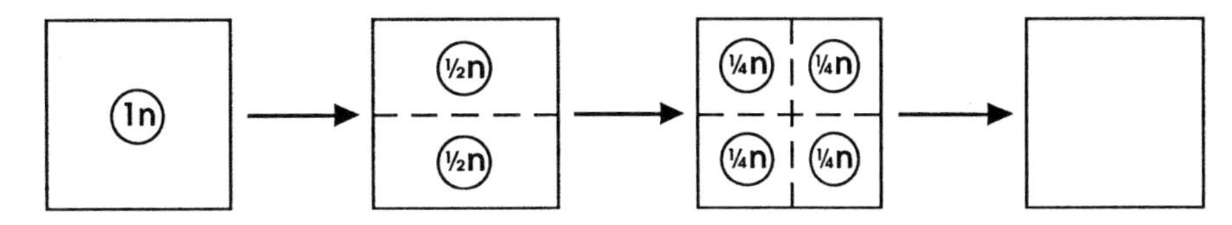

2. Explain what this mistake creates in the new cells.

In Conclusion

As stated above, some cells start with *one set* of DNA molecules (**1n**), and other cells start with *two sets* of DNA molecules (**2n**). For a dividing cell to maintain its original set # in the new cells, it must duplicate its genetic material prior to beginning mitosis. In addition, the genetic material must be divided in such a way that no new cell is missing any DNA, or has more DNA than the original cell.

Whatever amount of DNA the original cell has prior to mitosis, its offspring cells must end up with the same amount after the process is complete.

? QUESTION

1. When the process of mitosis is used for organism reproduction, are the new organisms exact genetic duplicates of the parent organism?

2. Are the organisms that use mitosis to reproduce, exhibiting *sameness* or *variety* in their offspring?

3. If asexual reproduction produces identical offspring, then how does such an organism "change" over time?

4. What would be the *advantage* of reproducing asexually?

5. What would be the *disadvantage* of reproducing asexually?

6. Starting with the *diploid* cell below, draw the next *three* generations of that cell as it reproduces.

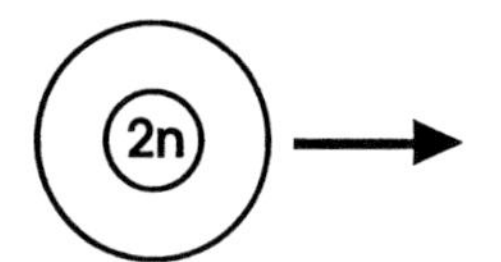

EXERCISE #4

"Chromosome Movement During Mitosis"

Chromosomes move during mitosis. They replicate themselves (see Exercise #1), and the copies separate, allowing two cells to be created from one. This movement of DNA material in the form of chromosomes (coiled DNA) has several "phases," which are described in this Exercise.

Materials

- One red and one yellow crayon.
- A package of chromosome beads. Each package should contain 8 chromosomes.

Procedure

[1] The human has 46 chromosomes (23 homologous pairs). We will follow the movements of 4 chromosomes (2 homologous pairs) as an example of what all the chromosomes are doing during mitosis. Start with 4 of the chromosomes from your package. This is how a cell would look *prior* to duplicating its genetic material and undergoing the process of mitosis:

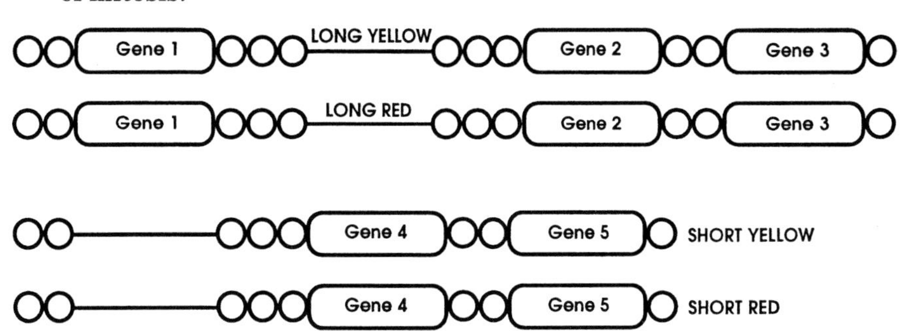

[2] The two short chromosomes represent *one* homologous pair, and the two long chromosomes represent *another* homologous pair. The *red* set represents the chromosomes from your mother, and the *yellow* set is the chromosomes from your father.

[3] Color the chromosome beads above with your crayons, and as you go through this Exercise, use the crayons to help you keep track of the chromosomes that came from your father and from your mother.

[4] The bead chromosomes are labeled A^1, A^2, B^1, and B^2 in the next section.

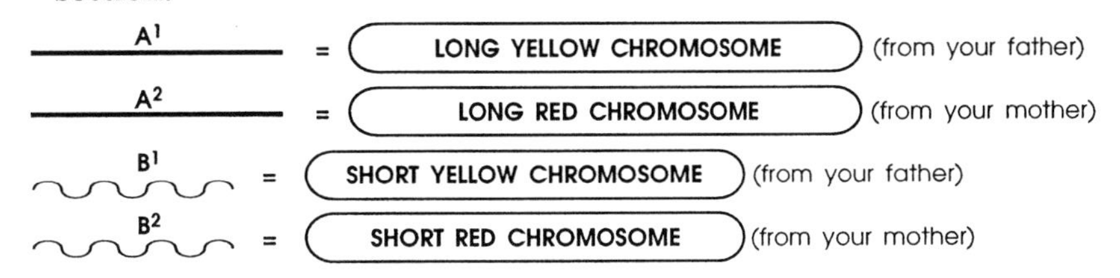

Phases of Mitosis

Biologists sometimes describe mitosis as having several phases. Ask your instructor if you are required to memorize the names of the phases. If so, remember the phrase: **P**ay **M**e **A**ny **T**ime. This will help you to remember the sequence of phases in mitosis.

Pay attention to the different events as they occur in each phase, and mimic the phases by using your bead chromosomes. Interphase is usually considered to be the stage *before* mitosis actually begins. We include it as part of the mitosis discussion, but your textbook will say that mitosis begins with prophase.

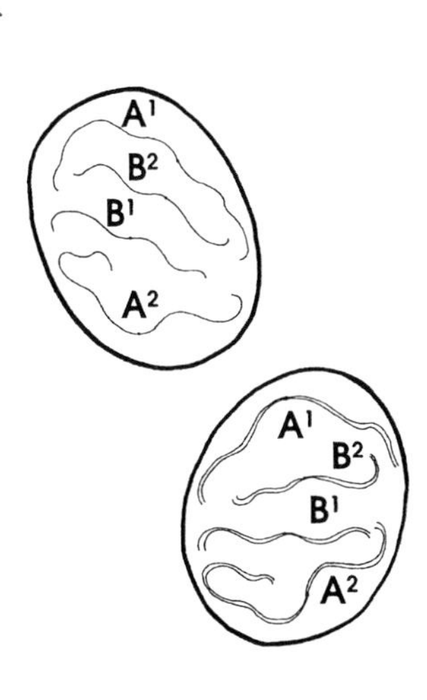

1. **Early Interphase:** Although we have diagramed the DNA as long, thin chromosomes, in reality it is not coiled up yet, and is *not* visible as chromosomes until ***prophase***. However, it is best to label the DNA at this stage so that we can remember what the parent cell starts with.

2. **Later Interphase:** Each DNA molecule has duplicated itself. We have diagramed these "doubled" DNA molecules as though we could see them. Actually, DNA is still in the long, thread-like form.

   Duplicate your beads now, using the other four bead chromosomes from the package.

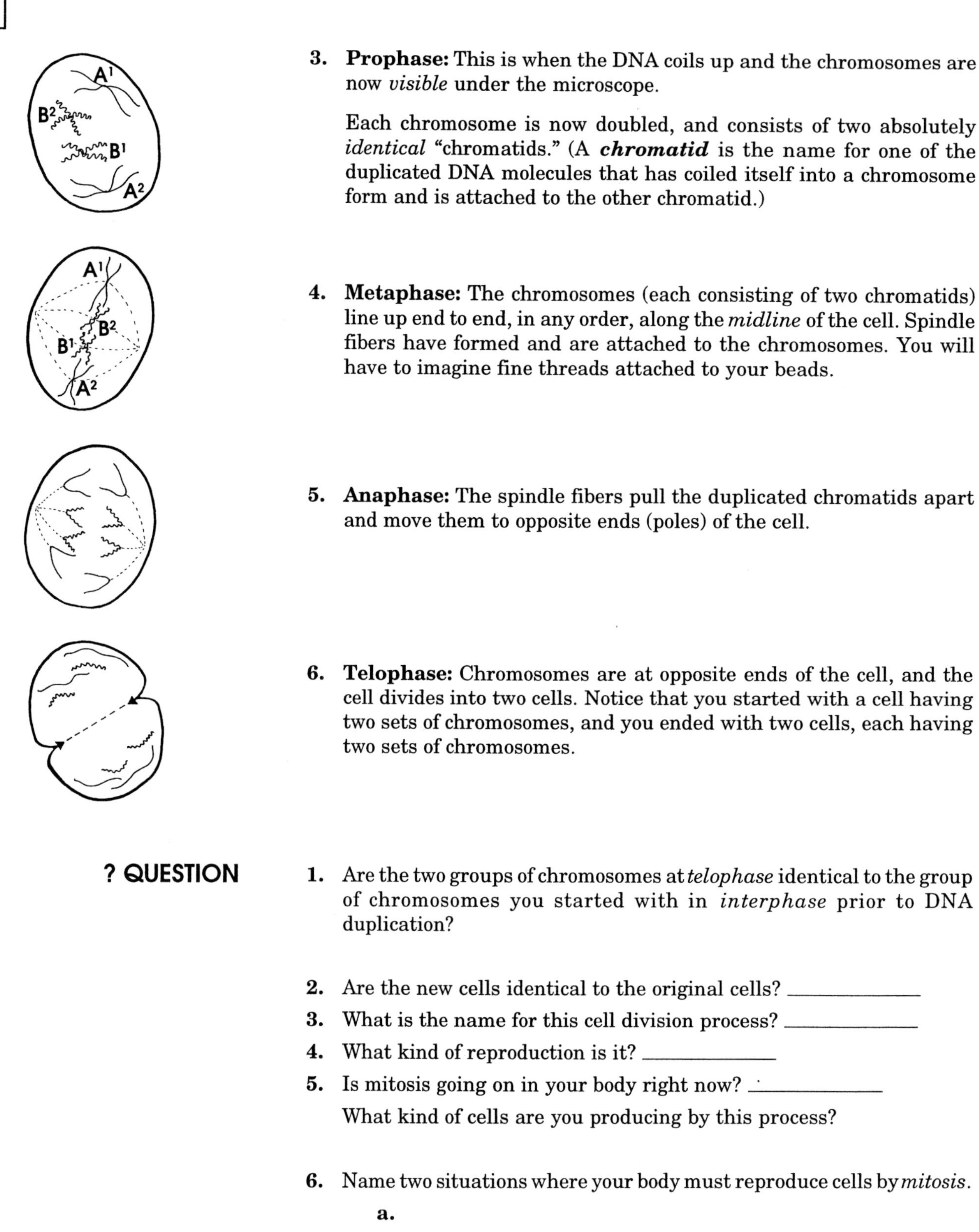

3. **Prophase:** This is when the DNA coils up and the chromosomes are now *visible* under the microscope.

   Each chromosome is now doubled, and consists of two absolutely *identical* "chromatids." (A ***chromatid*** is the name for one of the duplicated DNA molecules that has coiled itself into a chromosome form and is attached to the other chromatid.)

4. **Metaphase:** The chromosomes (each consisting of two chromatids) line up end to end, in any order, along the *midline* of the cell. Spindle fibers have formed and are attached to the chromosomes. You will have to imagine fine threads attached to your beads.

5. **Anaphase:** The spindle fibers pull the duplicated chromatids apart and move them to opposite ends (poles) of the cell.

6. **Telophase:** Chromosomes are at opposite ends of the cell, and the cell divides into two cells. Notice that you started with a cell having two sets of chromosomes, and you ended with two cells, each having two sets of chromosomes.

? QUESTION

1. Are the two groups of chromosomes at *telophase* identical to the group of chromosomes you started with in *interphase* prior to DNA duplication?

2. Are the new cells identical to the original cells? ____________
3. What is the name for this cell division process? ____________
4. What kind of reproduction is it? ____________
5. Is mitosis going on in your body right now? ____________

   What kind of cells are you producing by this process?

6. Name two situations where your body must reproduce cells by *mitosis*.

   a.

   b.

7. What do you think might be going on as you age (get wrinkles, grey hair, loose your hair, etc.)?

EXERCISE #5

"Onion Root Tip"

It's time to review what you learned in Exercise #4. You will use a microscope to find the various phases of mitosis (cell division) in the root tip of an onion.

Materials

- A compound microscope.
- A prepared slide of an onion root tip.

Procedure

Onion Root Tip

Zone of Elongation

Zone of Cell Division

Root Cap

1. Under low power, note that the root tip is covered by a root cap (like a thimble over your finger). Behind the root cap is an area of square-shaped cells that are undergoing cell division.
2. Look at this area under the high power (430x). If the cells are *rectangular,* then you are in the wrong place.
3. *Find every stage of mitosis.*
4. Draw a simple sketch of what you see at each phase of mitosis.

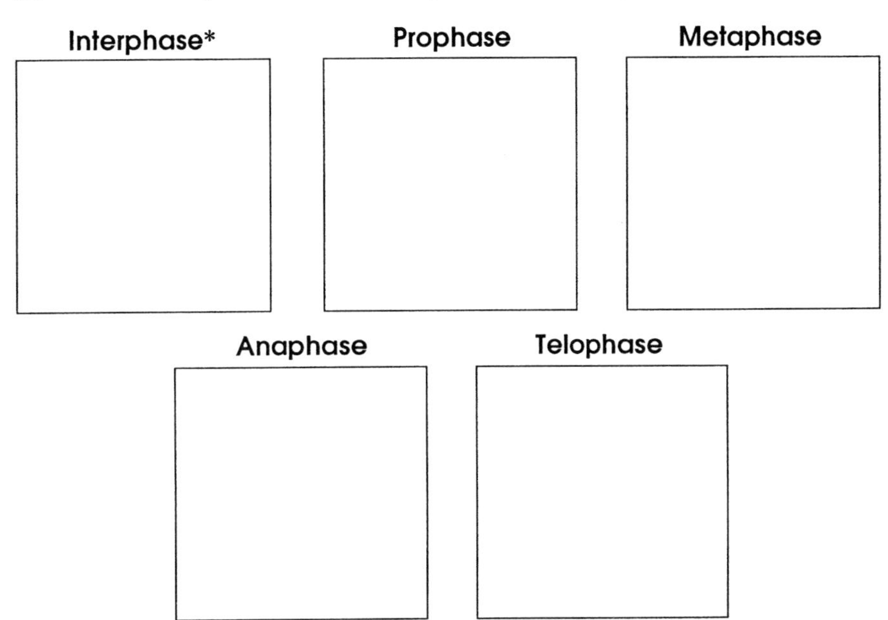

* Interphase will look like a stained nucleus. You won't be able to see the DNA threads.

5. Optional: *(Ask your instructor if you are to do this experiment.)* You can estimate the relative amount of time that a cell spends in each phase of the cell cycle by counting all of the cells in the zone of cell division and recording how many of them are in each phase. Then calculate what percent each stage is of the total. This is an indication of the relative time a cell spends in each stage of cell division. Does this make sense to you?

Phase	# of Cells in Phase	% of Total Cells

EXERCISE #6

"Sexual Reproduction"

Sexual reproduction is the process of creating *variety* in the offspring of a species. It consists of two parts: ***meiosis*** and ***fertilization***.

Meiosis and fertilization produce variety.

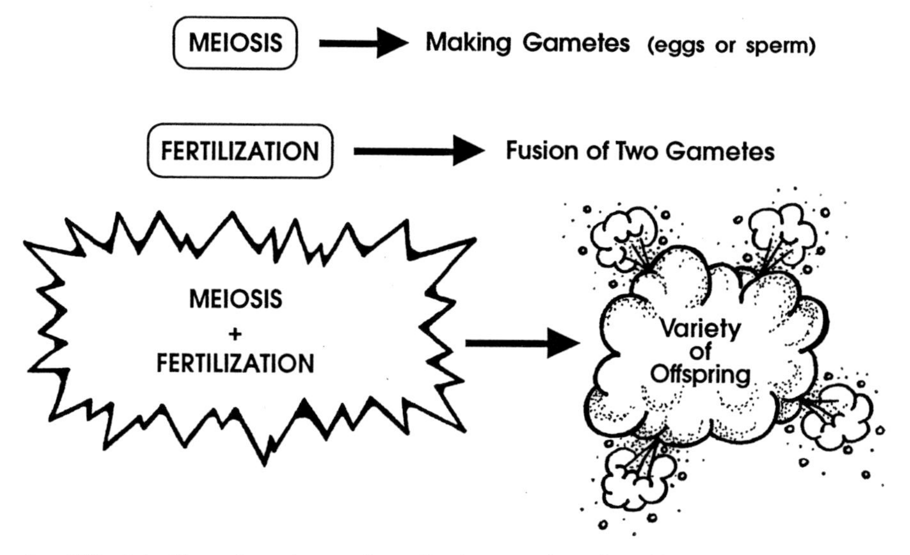

? QUESTION

1. What is the advantage of producing variety in offspring?

2. What is the disadvantage of producing variety in offspring?

Set Changes During Meiosis and Fertilization

Meiosis

Meiosis starts with a single *diploid* cell and ends with four *haploid* cells. The original cell has two sets (2n) of DNA molecules which are then duplicated. After DNA duplication, the original cell divides *twice,* producing *four* cells—each with a *single* set (1n) of chromosomes.

These four cells produced by meiosis are haploid and their chromosomes have been mixed to produce *genetic variety*. Some details of that process are presented in Exercise #7.

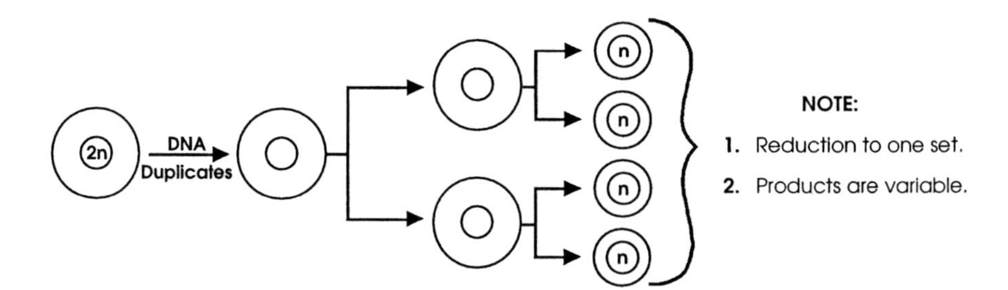

The product cells of meiosis are called ***gametes***, and these can fuse with gametes from another organism of the same species to begin the next generation.

Fertilization

Fertilization is the fusion of *two haploid gametes*. It results in the chromosome set number returning to 2n. This allows the next generation of the species to have the same "set" number of chromosomes as the parent generation.

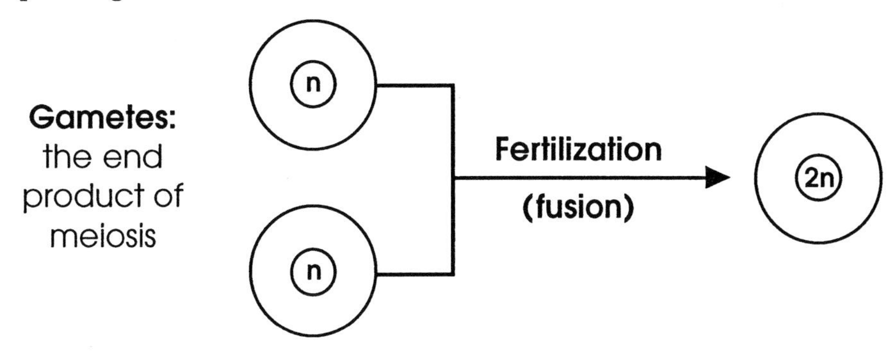

? QUESTION

1. What type of cell division reduces the set number? ______________
2. What process takes this reduced set number and returns it to match the original set number of the parent cell? ______________
3. How many *sets* of chromosomes are in a human sperm? ______________
4. How many *chromosomes* are in a human sperm? ______________
5. How many *sets* of chromosomes are in a human egg? ______________
6. How many *chromosomes* are in a human egg? ______________
7. When a human egg and sperm fuse, how many *sets* of chromosomes are there? __________
8. When a human egg and sperm fuse, how many *chromosomes* are there? __________

EXERCISE #7

"The Events that Create Variety"

Three specific events during meiosis and fertilization produce variety.

The essence of meiosis is the production of haploid cells from diploid cells. The essence of fertilization is the recombination of two haploid gametes to produce the next diploid generation. During meiosis and fertilization there are *three* events that create *genetic variety* in the next generation: ***crossing-over***, ***independent assortment***, and ***random fusion of gametes***.

None of these genetic variety events would be possible without a very special process during meiosis called ***synapsis***. This is the single most critical event that makes meiosis so different from mitosis.

Synapsis

Synapsis is defined as the "pairing up" process of *homologous pairs* early in meiosis. Let's illustrate the differences between mitosis and meiosis using the same chromosomes from our earlier mitosis diagrams.

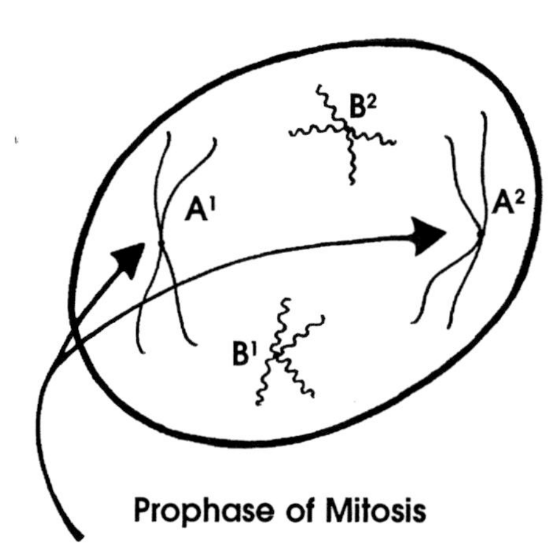

Prophase of Mitosis

Notice that the two members of the "A" homologous pair have duplicated themselves, and they will be moved around the cell *separately* from each other.

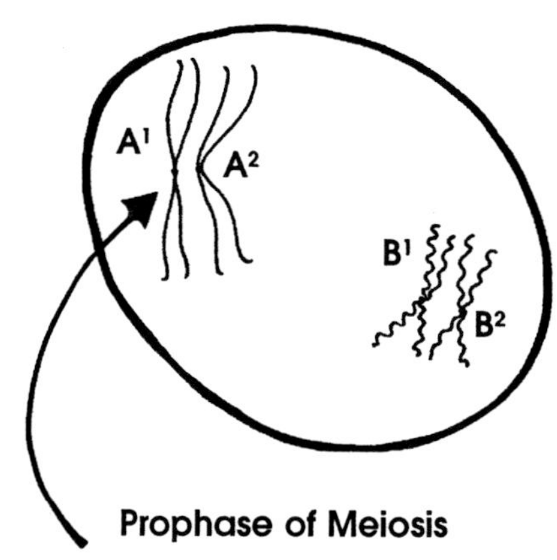

Prophase of Meiosis

Notice that the two members of the "A" homologous pair have duplicated and they have "paired up" *(synapsis)*. This paired grouping is called a ***tetrad*** (meaning four chromatids) and they will be moved around the cell *together*.

? QUESTION

1. The two "A" chromosomes are concerned with the same traits and are called ____________________ pairs.
2. Are the chromatids of the "A^1" chromosomes identical or different? ____________________
3. Are the chromosomes "A^1" and "A^2" absolutely identical? ____________
4. Are the "A" and "B" chromosomes homologous pairs? ____________
5. In meiosis, do the "A" and "B" chromosomes pair up with each other (synapse)? ____________

Crossing-Over

Crossing-over is the exchange of DNA between the four chromosomes (chromatids) of a *tetrad*.

Sexual reproduction creates variety at the expense of quality.

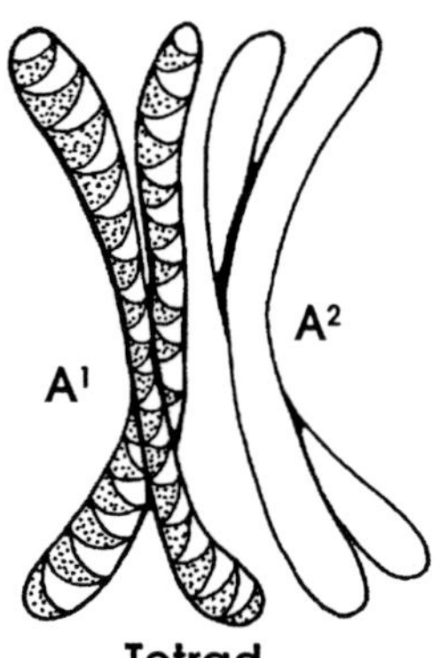

Tetrad

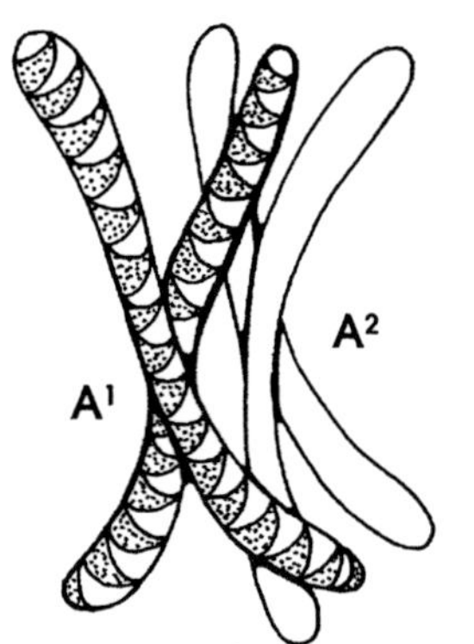

Tetrad Crossing-Over

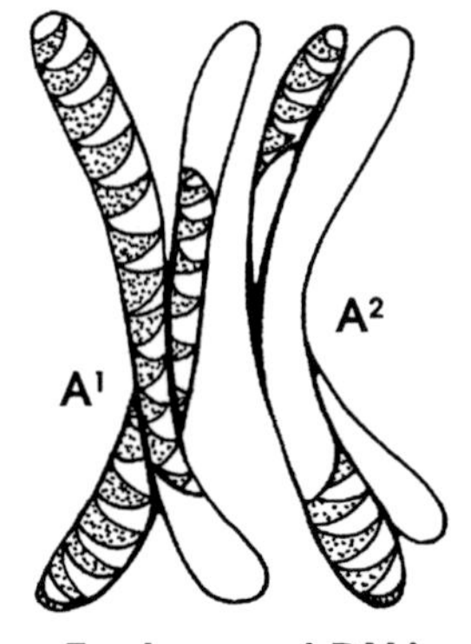

Exchanged DNA from Crossing-Over

If you consider that a single chromosome may carry a thousand or more genes, then these small cross-over exchanges are capable of creating hundreds of mixtures of chromosomes. Remember, you received one of the homologous chromosomes of a pair (A^2) from your mother and the other (A^1) from your father. The process of crossing-over makes "*new mixes*" of those chromosomes.

? QUESTION

1. What event during meiosis prophase (as opposed to mitosis prophase) makes it possible for crossing-over to occur?

2. As a result of crossing-over, will the "A" chromosome that you pass on to your children be your mother's, your father's, or will it be a mixture of your mother's and father's?

Independent Assortment

During meiosis the tetrads formed by synapsis move around the cell and divide in different ways than we saw in mitosis. The simplest description of the difference is:

- ▶ Tetrads line up in the middle of the cell (during metaphase).
- ▶ Cell divides twice, separating the tetrads first into pairs of chromatids and then into single chromosomes, resulting in *four separate cells.*

Each of these cells contains only *one of each kind of chromosome.* They are haploid.

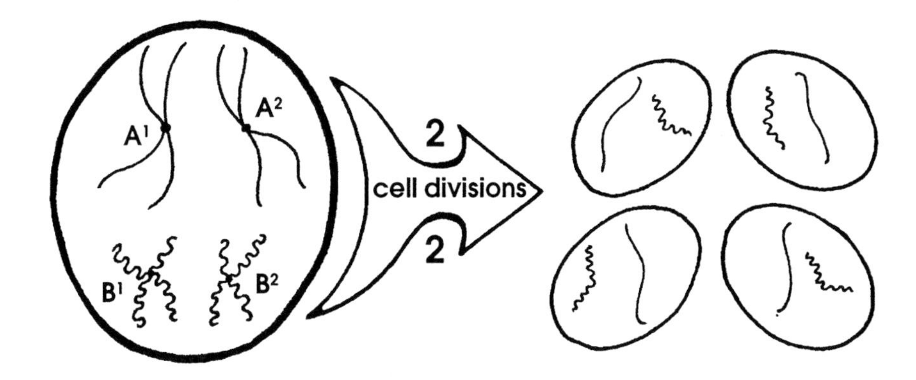

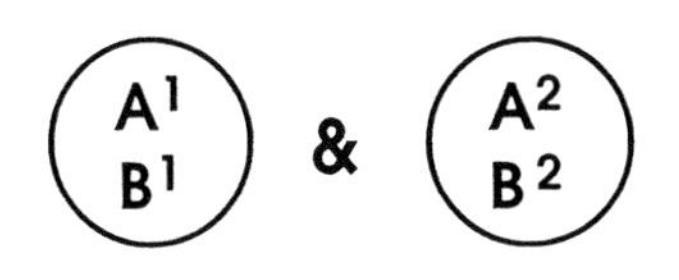

However, there is a very important detail called *independent assortment* that occurs during the chromosome separations of meiosis. Notice that the tetrads are drawn so that the A^1 and B^1 chromosomes are placed on the left side of the cell, and the A^2 and B^2 chromosomes are lined up on the right. If the tetrads separated equally as drawn above, you would see only two kinds of gametes from this process (as shown on the upper left).

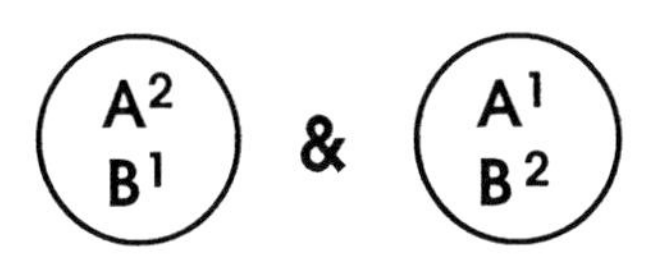

But, if the top tetrad had originally lined up during metaphase with the A^1 chromosomes on the right, then A^2 would have moved with B^1 into a gamete, and A^1 would have moved with B^2 into a gamete as pictured on the lower left.

This production of gametes containing different combinations of chromosomes is called ***independent assortment*** because one pair of homologous chromosomes is separated (segregated) into individual gametes independently of how another pair is separated. (A^1 and A^2 have been separated independently of how B^1 and B^2 have been separated.)

Remember: Every gamete gets a complete set of chromosomes with only one A chromosome and only one B chromosome.

? QUESTION

1. How many *genetically different* gametes were produced in the independent assortment of these two homologous pairs?

2. The human contains 23 pairs of homologous chromosomes, all of which are independently assorted. What do you think the chance would be that one of your gametes would contain either all your mother's or all your father's chromosomes?

Random Fusion of Gametes

Someone out there is more genetically like you than either of your parents.

Two mating individuals have the same kinds and number of chromosomes, but those chromosomes are *not exactly* identical. Because individuals possess different variations of genes, the ***random fusion of gametes*** (fertilization) from any two individuals will result in more genetic variety in the offspring. Through fertilization the diploid set number is recreated with the offspring receiving one chromosome of a homologous pair from one parent, and the other chromosome of the pair from the other parent.

? QUESTION

To keep the example relatively simple, let's consider only one of the homologous pairs of the human—the A chromosome. (Actually, human chromosomes are referred to by numbers from 1 to 23.)

A^1
A^2
You

A^1 came from your father

A^3
A^4
Your Mate

A^3 came from your mate's father

1. Where did chromosome A^2 come from?
2. Where did chromosome A^4 come from?
3. What are your possible gametes?
4. What are your mate's possible gametes?
5. Determine the four possible combinations of your gametes with your mate's gametes. In other words, what genetic variety can we expect in your offspring?

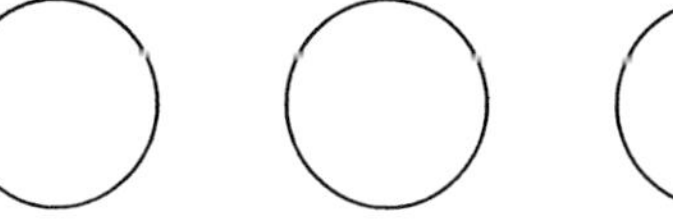

6. Are any of the offspring identical to either parent?

7. What are the three events during *sexual reproduction* that prevent identical children?

Genetics

More than a hundred years ago Gregor Mendel discovered that hereditary particles are passed from parent to offspring during the reproductive process. These particles were later named ***genes,*** and the science of studying inheritance was called ***genetics.***

Genes can be traced backwards to the very origin of life about 4 billion years ago. When we do so, we find that all traits were new at some point in time, and that a gene's success is determined by natural selection. However, genes do not last forever. And most have already gone extinct.

The investigation into the mechanics of inheritance—the mixing, the passing on, and the function of genes—is one of the greatest scientific puzzles of the 20th century. Understanding genetics has led to the prevention and curing of numerous hereditary diseases. It has substantiated the principle of evolution by natural selection, and has helped human beings to realize their place in Nature's Family Tree. As a contemporary student, you should note that your individuality is not the result of possessing a trait that no other individual has, but is a result of a particular *combination* of genes. These genes came from your parents and their ancestors before them.

This lab will explore some of the basic principles of genetics, introduce you to basic terminology, and help you apply genetic rules to some hypothetical problems.

Exercise #1 "Basic Terminology" 169
Exercise #2 "Some Rules of Genetics" 173
Exercise #3 "How to Solve Genetic Problems" 175
Exercise #4 "Genetic Problems" 177
Exercise #5 "Sex-Linked Traits" 179

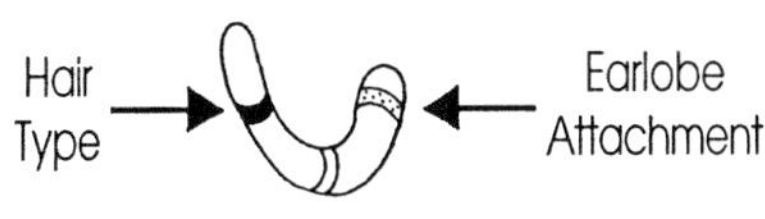

One Chromosome

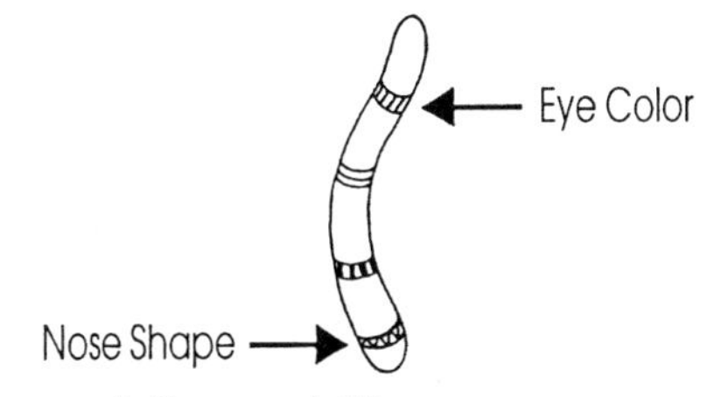

A Second Chromosome

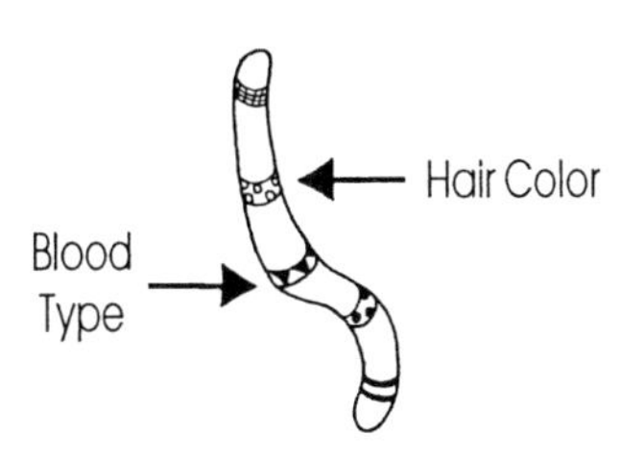

Another Chromosome

EXERCISE #1

"Basic Terminology"

In order to understand the mechanics of inheritance, you must understand the terminology used to describe this very complex process.

Genes and Chromosomes

A ***gene*** is a segment of the DNA molecule and is responsible for manufacturing a protein that either becomes part of the organism's structure or becomes an enzyme that controls biochemical events.

Every organism has a certain number of chromosomes—the exact number depends on the species—and each chromosome is made of many genes. DNA coils up into the form of a chromosome during cell division, and a gene becomes a distinct particle on that chromosome. **Remember:** Chromosomes and DNA molecules are basically the same thing.

Genes can be described by their exact location on a chromosome. The process of locating genes is called *mapping.* The location of a gene is its ***locus,*** and geneticists go through great efforts to pinpoint these locations. Knowing where a gene is found on the chromosome is what allows scientists to do genetic research.

? QUESTION

1. The particles that control inherited traits are called ______________.

2. These particles are segments of ____________, and are responsible for manufacturing a ______________ that becomes either . . .

   ______________ or ______________.

3. Every living thing on the planet has the same number of chromosomes. ________ (T or F) Explain your answer.

4. Every chromosome has the same identical genes as every other chromosome. ________ (T or F) Explain your answer.

5. The place where you find a particular gene on a chromosome is called the ______________.

Homologous Chromosomes

We have discussed homologous chromosomes before. This idea is essential to the understanding of genetics, so we will review it again.

Information

- ▶ Very simple organisms have only one set of chromosomes and they are *haploid.*
- ▶ More complex organisms have two sets of chromosomes and are *diploid.*
- ▶ Haploid organisms have one of each kind of chromosome and one of every kind of gene.
- ▶ Diploid organisms have two of each kind of chromosome and two of every kind of gene.
- ▶ The two chromosomes of each kind in a diploid organism are called ***homologous chromosomes*** because they are carrying the same kind of traits (genes). *Homo* means "same."
- ▶ A human has 23 different kinds of chromosomes that are given numbers from 1 to 23. Because we are diploid organisms we have two of each of the different kinds. So, we have 46 chromosomes in all, made up of 23 *homologous pairs*.

? QUESTION

1. How many sets of DNA molecules or chromosomes does a diploid organism have? __________
2. How many sets of DNA molecules or chromosomes does a haploid organism have? __________
3. Humans are ______________________. (haploid or diploid)
4. How many homologous pairs of chromosomes does a human have? __________

5. Because chromosomes occur in pairs in a diploid organism, how many genes for one trait would a diploid organism possess? ____________

6. How many genes for a trait would a haploid organism possess?
____________ Why?

Alleles: The Various Forms of a Gene

Humans are diploid, and they have two copies of every kind of gene. One of the purposes of genetics is to figure out which form (variation) of these two genes you have, and what expression of those genes you can expect.The *alternate forms* of a particular gene are called ***alleles***. For example, there are three alternate forms—three alleles—for blood type: A, B, and O.

The reason all species have various alleles (forms of genes) is that ***mutation*** events change the structure of genes. A gene can be mutated (changed) by radiation, by chemicals in the environment, or by other spontaneous events that are surprisingly common on this planet. There may have been a time when all the genes for eye color were identical and resulted in brown eyes. But over time, mutations occurred and changed the DNA of this eye color gene, creating a new "allele" (variation) for the eye color trait. Perhaps this new allele was for blue eyes.

Alleles are always for the same trait, and are located at the exact *same* spot on homologous chromosomes. (This is how we know that they are truly alleles of each other, and not different genes.) **Remember:** Alleles are variations of the same gene!

? QUESTION

1. What is an allele?

2. Where are alleles located?

3. What process creates the various alleles in a species? Explain how.

4. Which of the following genes (1 through 9) are alleles?

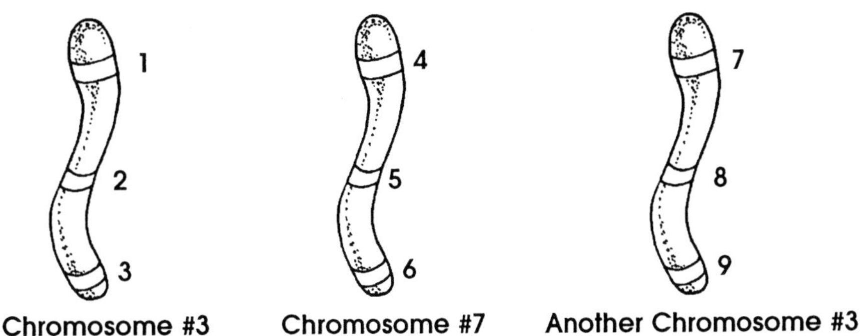

Homologous Chromosomes

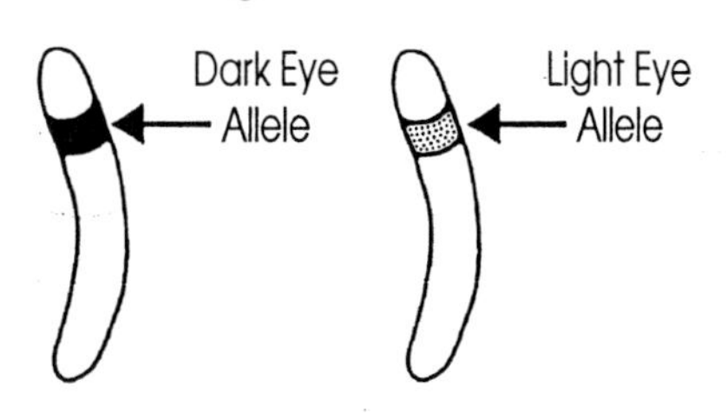

Genotype

A ***genotype*** is the description of the alleles an individual possesses for a particular trait.

Observe the following situation where there are two different alleles for a particular trait. **Note:** Even though the two genes look different, they are alleles because they are at the *same locus* on homologous chromosomes.

Procedure

Genotype = your genes.

[1] Study the chromosomes above. Draw and label the three combinations of eye color alleles that are possible in individuals of the same species.

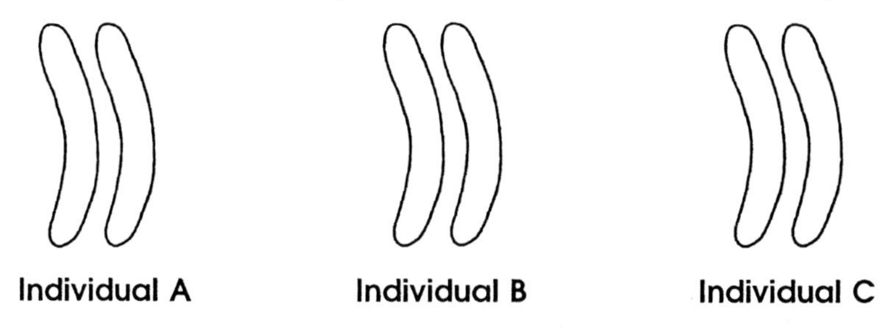

Individual A	Individual B	Individual C
________________	________________	________________

[2] If an organism has ***two identical alleles,*** we say that it is ***homozygous*** for that trait (meaning the "same" two alleles).

[3] If an organism has ***two different alleles,*** we say that it is ***heterozygous*** for that trait (meaning "different" alleles).

[4] Go back to the diagram of the three individuals above and label each as to whether it is *homozygous* or *heterozygous.*

Phenotype

The physical expression of the alleles—what an organism looks like—is termed the ***phenotype.***

Because there are different possible combinations of alleles (genotypes), there are alternative possible phenotypes for a trait that can be expressed in a population.

Procedure

Draw this survey chart on the chalkboard, and record your phenotype for each of the six traits. After everyone in the class has recorded their phenotypes, write the class totals on the chart below. Your instructor will tell you the genotypes of these traits at the end of the next Exercise.

Phenotype = what you look like.

Trait	Class Phenotype Totals		
Eye Color	*Dark # ________	Light # ________	
Earlobes	Attached # ________	Unattached # ________	
PTC Paper	Can taste # ________	Cannot taste # ________	
Hairline	Widow's peak # ________	Straight across forehead # ________	
Hair Type	Straight # ________	Wavy # ________	Curly # ________
Fingers	Five # ________	Six # ________	

* Dark is considered to be black, brown, hazel, green, or grey.

Information

- ▶ The phenotype is the description of the physical expression of a trait (brown eyes), whereas the genotype is the description of the exact combination of alleles (for example, 1 allele for brown eyes + 1 allele for blue eyes).
- ▶ The genotype results from the combination of genes you inherited from your parents.
- ▶ The phenotype results from the expression of the genes in the genotype, and also may be influenced by the organism's environment. In some cases there may be only two phenotypes for a trait, and in other cases there are *more than two* phenotypes for a trait.

EXERCISE #2

"Some Rules of Genetics"

Nine times out of ten, in the arts as in life, there is actually nothing to be discovered; there is only error to be exposed.

—H. L. Mencken
American editor and critic (1880–1956)

Rule of the Gene

The parent must possess the gene in order to pass it on.

The source of all genes in the offspring is the parents. Always look to the parents to figure out what genes the sperm or egg can possibly carry, and remember that a parent does not possess all of the genes found within a reproducing population of a species.

? QUESTION

1. How many different *alleles* for a single trait can a homozygous parent pass on? ___________
2. How many different *alleles* for a single trait can a heterozygous parent pass on? ___________

Rule of Segregation

Only one gene of the two alleles that you have is put into each gamete that you make.

Alleles are located on homologous chromosomes, and since homologous chromosomes are segregated during meiosis, the genes are also segregated.

Numerous gametes are formed during gamete production, and if the alleles are different (heterozygous), 50% of the gametes will carry one gene and 50% of the gametes will carry the other. When alleles are the same (homozygous), 100% of the gametes will carry the same allele.

? QUESTION

1. A parent possesses two copies of each gene. When this parent passes on its alleles for a gene, how many does it contribute to each of the offspring? _______________
2. How many copies of a gene does the other parent contribute to each offspring? _______________
3. How many copies of each gene for the trait does each offspring receive? _______________

Rule of Dominant and Recessive Alleles

Some alleles control the phenotype even if they are paired with a different allele.

If two different alleles are together in an organism, and only one phenotype is expressed, then the allele that is expressed is called ***dominant***. The other allele that is "hidden" is called ***recessive***. One example of a dominant allele is the dark-eye allele that will create the dark-eye phenotype in an individual even if the allele for light eyes is present.

? QUESTION

1. Can the individual carry an allele that is not expressed? Explain.

2. What word is used to describe the *genotype* condition in which two different alleles occur together in the same organism? ______________

3. What word is used to describe the *genotype* condition in which two of the same alleles occur together in the same organism? ___________

Information

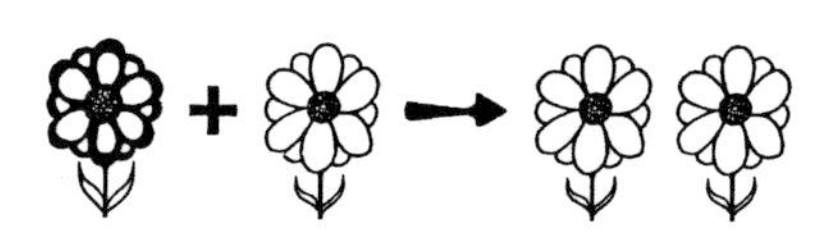

Since dominance and recessiveness have intricate biochemical explanations, the only way of determining dominance is to cross two individuals that are homozygous (pure) for the two different phenotypes. This produces the heterozygous condition. Whichever phenotype is exclusively expressed is said to be the *dominant phenotype.*

? QUESTION

1. A homozygous blue-eyed mouse with short whiskers mates with a homozygous brown-eyed mouse with long whiskers. All of their offspring have brown eyes and short whiskers. Which traits are the dominant traits?

2. A homozygous five-clawed cat is crossed with a homozygous six-clawed cat and all of the kittens have six claws. Which trait is dominant?

Dominant does not mean frequent.

3. In humans, the five-fingered condition is *recessive* to the six-fingered condition. Yet, most people have five fingers. Explain how this can happen.

Ask your instructor which alleles are dominant in the class Phenotype Chart. It is a common mistake to assume that the trait found most frequently is always the dominant trait. *Natural selection* determines the success of a trait.

Rule of Incomplete Dominance

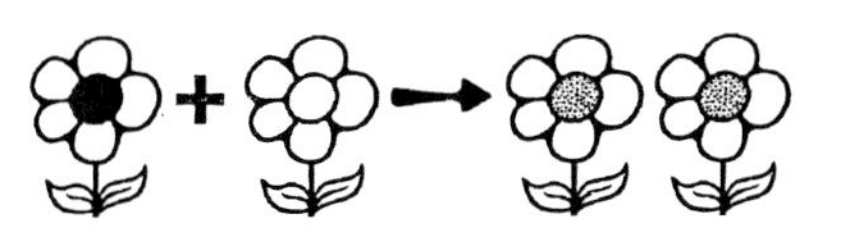

When two different pure-breeding strains are crossed, and their offspring show a blending of phenotypes, then neither allele is dominant.

This is easily recognized when the phenotype is somewhere between two extremes. Counting the parents, there are *three* phenotypes (*black, white, grey*) being expressed in these flowers instead of only two, and that third phenotype is *intermediate* between the other two. This heterozygous condition is called ***incomplete dominance***.

? QUESTION

1. On the chart you did earlier, which of the three hair types (wavy, curly, or straight) represents incomplete dominance—the *blended* heterozygous condition? ____________________
2. You cross a herd of red cattle with white cattle and all of the calves appear to be roan (reddish white). Is this an example of incomplete dominance? ____________ How do you know?
3. You cross a blue flowering pea plant with a white flowering pea plant and all of the offspring are blue flowered. Is this an example of incomplete dominance? ____________ How do you know?

EXERCISE #3

"How to Solve Genetic Problems"

Using Letters for Alleles

For convenience, the genes of an allele pair are usually symbolized by a letter from the alphabet. A *large* letter is used for the dominant trait and a *small* letter for the recessive trait. When we want to describe the genotype of an organism, we use both letters to represent the alleles inherited from the parents.

For example, free earlobes is a dominant trait and attached earlobes is recessive. You would use a capital "**F**" to indicate the dominant allele and a small "**f**" to indicate the recessive allele in describing an individual.

? QUESTION

1. Write the three genotypes for earlobe attachment as it applies to the following individuals.
   - **a.** Heterozygous ________ ________
   - **b.** Homozygous Dominant ________ ________
   - **c.** Homozygous Recessive ________ ________
2. When it comes to symbolizing incomplete dominance with letters, it is best to use the letter "**C**" for one allele and "**C'**" for the other allele.

   List the three possible genotypes for hair type.
   - **a.** Curly ________ ________
   - **b.** Wavy ________ ________
   - **c.** Straight ________ ________

   Why not use a small letter "**c**" for the heterozygous genotype?

Using the Punnett Square

The ***Punnett Square*** is a method of predicting the probable outcome of genetic crosses.

Step 1

Draw a square like this:

Put the gametes of one parent here.

Put the gametes of the other parent here.

Step 2

Determine what kinds of gametes are made by each parent in the cross, and put those gametes into the boxes of the Punnett Square.

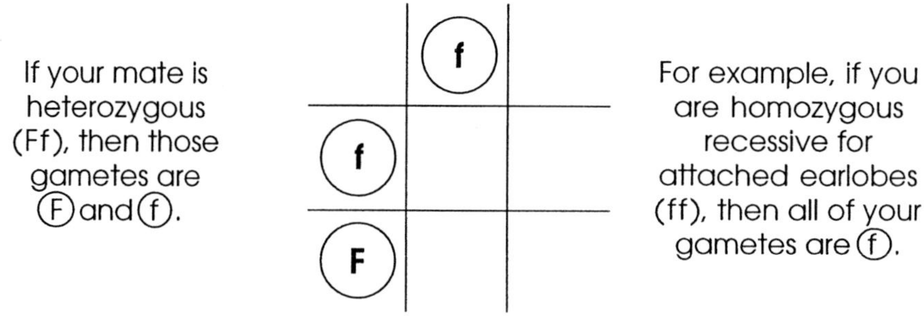

Step 3

Fill in the offspring boxes of the Punnett Square.

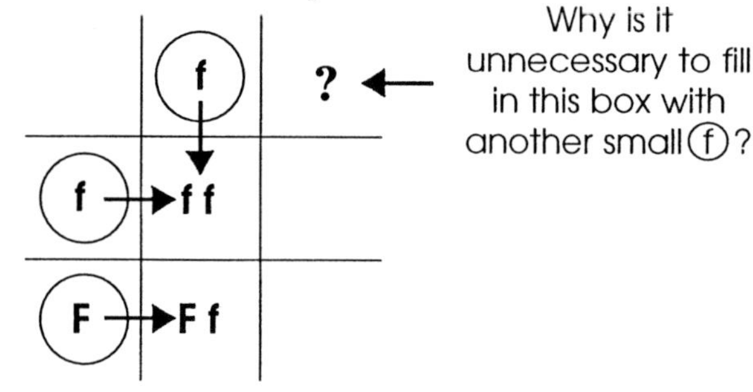

In this example there are only two possible offspring genotypes. The Punnett Square tells us to expect about 50% ff and 50% Ff.

Sometimes the Punnett Square is more complex than this and you must figure out more than one trait at a time. Nevertheless, you use the same basic method.

Procedure

Make up your own genotype example and work out the crosses.

1. Traits:
2. Symbols:
3. Male Genotype:
4. Female Genotype:
5. Offspring Genotypes:

EXERCISE #4

"Genetic Problems"

Cases of Complete Dominance

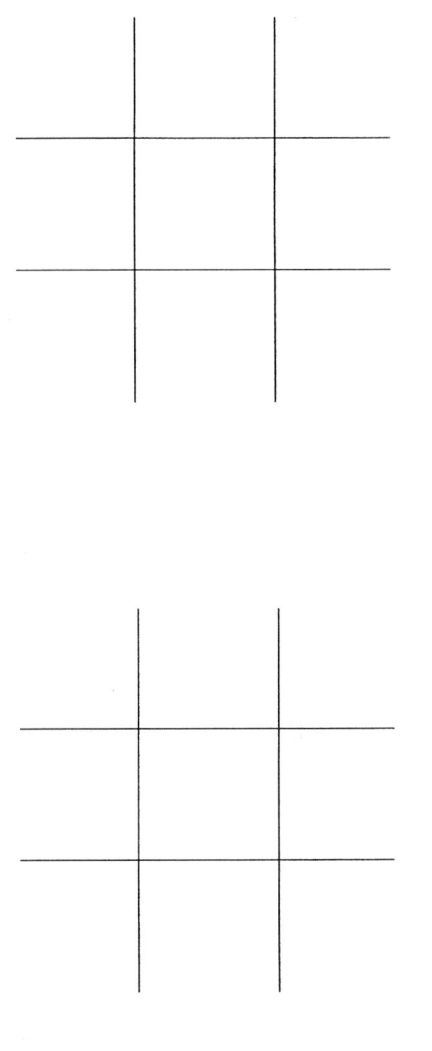

1. Gregor Mendel grew different varieties of pea plants in his garden. When he crossed yellow-seed plants with green-seed plants, he always got yellow pea seeds.
   a. What is the dominant trait?
   b. What is the genotype of all green-seed plants?
   c. Use the Punnett Square to show Mendel's cross.
   d. Do the parent yellow-seed plants have the same genotype as the offspring yellow-seed pea plant?
      Parent: __________ Offspring: __________
   e. What genetic fact do you know about any yellow-seed pea plant?
   f. If yellow-seed pea plants are dominant to green-seed pea plants, why are there mostly green pea seeds in nature?

2. A dark-eyed man mates with a light-eyed woman and they have ten dark-eyed children.
   a. What is the dominant trait?
   b. What is the genotype of all light-eyed people?
   c. What are the genotypes of the two parents?
      __________ and __________
   d. What is the genotype difference between the dark-eyed parent and the dark-eyed offspring?
      Parent: __________ Offspring: __________
   e. When two heterozygous dark-eyed people (Dd) are crossed, what is the *phenotype* ratio of dark-eyed offspring to light-eyed offspring? (Use the Punnett Square to get your answer.)

A Special Note on Eye Color

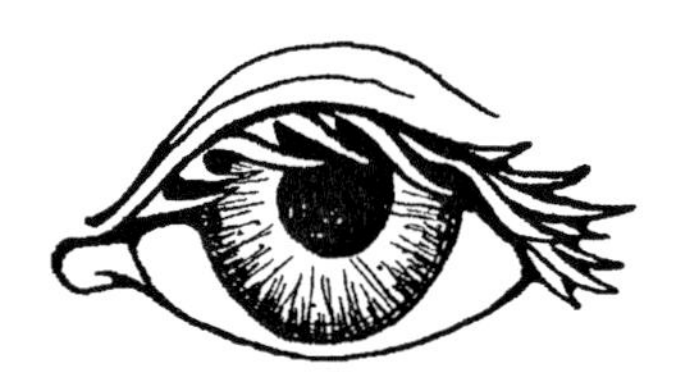

Eye color is probably due to multiple alleles and more than one gene pair. The numerous phenotypes are determined by genes that control both the *amount* and the *distribution* of a dark pigment called ***melanin***. Except for albinos, everyone has some eye pigmentation.

Eye color is determined mainly by the location of melanin in the iris of the eye. Concentrated melanin particles appear as brown; dilute melanin particles appear as yellow or yellow-brown.

Various Eye Colors

Blue: No melanin in the front part of the iris. The color is due to minimal amounts of melanin in the rear of the iris with the clear front portion scattering the light reflected off the melanin. This scattering is greatest in the blue spectrum giving the iris its blue color.

Grey: The same as blue, but with a slight amount of melanin in the front of the iris which tones down, or greys, the blue reflected from behind.

Green: A bit more melanin particles scattered in the front part of the iris create yellow. Blended with the light blue from the rear of the iris, it produces an overall green color.

Hazel: Even more melanin particles in the front of the iris give a slight brown color, and dilute melanin particles scattered throughout the iris add some yellow.

Brown: Melanin particles in the front part of the iris and throughout the iris. The amount of melanin varies, leading to gradations of brown color in the eye.

Black: Large amounts of melanin in front and throughout the iris.

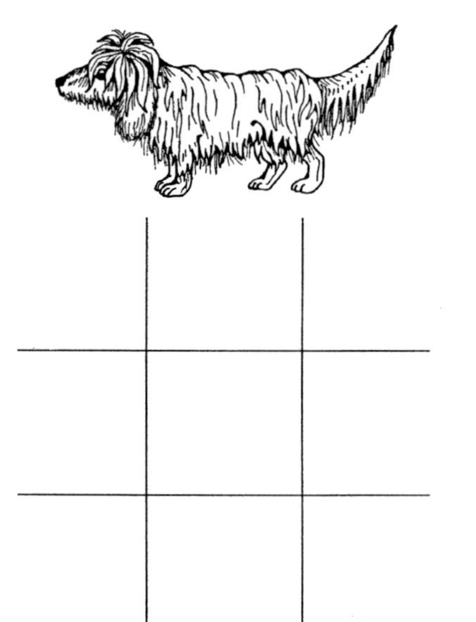

Test Cross to Check Genotype

If an organism shows the dominant phenotype, then one of its genes has to be the dominant allele, but you cannot be sure of the identity of the other allele unless you do a ***test cross*** to see if the dominant parent breeds pure.

Let's pretend that you are in the dog-breeding business. You know that long hair on a "pooch hound" is a dominant trait and short hair is recessive. You purchase a male long-haired "pooch hound." How do you figure out if your male "pooch hound" is homozygous or heterozygous for long hair? Which genotype of female would you use?

Complete the Punnett Square to show the test cross that would convince someone that your "pooch hound" is homozygous for long hair.

Male Genotype: ____________ Female Genotype: ____________

Cases of Incomplete Dominance

1. When a straight-haired mouse is crossed with a curly-haired mouse, the result is always wavy hair. Two wavy-haired mice cross.

   a. What are the genotypes of the two wavy-haired mice?

   ____

   b. Draw the Punnett Square of a cross between two wavy-haired mice, and show the probable genotypes of their offspring.

   c. What is the expected *phenotype* ratio of the offspring?

   ________ % ________ % ________ %

   d. What is the expected *genotype* ratio of the offspring?

2. Red orchids with straight petals are crossed with white orchids with curly petals. The results are pink orchids with wavy petals.

   a. What are the genotypes of the two parent orchid plants? **Remember:** You are dealing with *two different* traits.

   First parent: _____ _____ _____ _____
   (color) (shape)

   Second parent: _____ _____ _____ _____

   b. What is the genotype of the offspring orchids?

   Offspring: _____ _____ _____ _____

EXERCISE #5

"Sex-Linked Traits"

Gender Determination

Y is male.

Humans have 23 homologous pairs of chromosomes. Twenty-two of these pairs are named using the numbers 1 through 22. The 23rd pair is individually labeled with the letters "***X***" and "***Y***" for males, and "***X***" and "***X***" for females. These labels distinguish them as the ***gender chromosomes***.

During meiosis in the male two types of sperm are produced: those carrying the X and those carrying the Y chromosome. Females produce eggs carrying only the X chromosome.

If a Y chromosome is present in the cells of an embryo, then the child becomes a male. If the Y is not present, the child becomes a female. It is the presence or absence of the Y chromosome that determines the gender of a child! This means that a male child receives a Y chromosome from his father and an X chromosome from his mother. A female child receives an X chromosome from her father and the other X chromosome from her mother.

? QUESTION

Draw a Punnett Square to show a cross of X and Y chromosomes in the fertilization of male and female gametes. The offspring boxes should reveal why we have about a 50% male to 50% female ratio within the human population.

Sex-Linkage

The X and Y chromosomes are not exactly identical, and we should expect that there would be differences in how each of them carries traits (genes). These differences are expressed in the unequal frequencies of traits in the male and female offspring.

If any trait is distributed *unequally* between male and female offspring, and those differences are due to X and Y chromosome differences, then we call those traits ***sex-linked***. Actually, "sex-linked" means that the gene is carried on the X chromosome and not on the Y chromosome. We would call these genes ***X-linked***. It is easier to understand sex-linkage by looking at the sex chromosomes.

Two genetic situations are illustrated below.

The 23rd Pair of Chromosomes

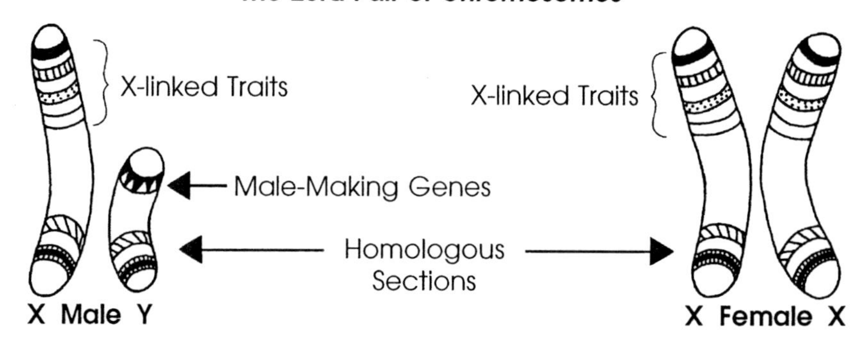

"Sex-linked" means carried on the X chromosome.

First: There is a homologous section of the X and Y chromosomes that is the same, and there will be no differences in phenotype between male and female children.

Second: Notice that the Y chromosome is very short. We would expect it to lack some of the genes that are carried on the X chromosome. There is an X-linked section on the X chromosome that carries genes that are *missing* from the Y chromosome.

? QUESTION

1. How many copies of an X-linked gene does a male have? ___________

2. Will a male be able to give his X-linked genes to his daughter? ___________ Explain.

3. Will a male be able to give X-linked genes to his sons? ___________ Why or Why not?

4. How many copies of an X-linked gene does a female have?

5. A male child gets X-linked genes from which of his parents?

6. A female child gets X-linked genes from which of her parents?

7. If a father is carrying an X-linked allele, then how many of his sons will get that allele? ___________

   How many of his daughters will get that allele? ___________

8. If a mother has a defective X-linked allele on one of her chromosomes and the other chromosome is normal, then how many of her sons will get that defective allele? ________

   Will any of her daughters get the defective allele? ________ How many? ________

9. If we found that none of the daughters actually showed the defective phenotype, then how could we explain it?

Tips for Solving Sex-Linked Genetic Problems

There is a sex-linked gene on the X chromosome that causes a disorder called *hemophilia,* where the blood fails to clot properly when a person is injured. This disorder is recessive and can be symbolized by the small letter "n." Normal blood clotting is dominant and can be symbolized by the capital letter "N."

In sex-linked cases we not only use letters to symbolize the genes, but also include the X or Y chromosome to indicate gender and to follow the sex chromosomes into the next generation.

- ▶ Using these symbols we would indicate a female who is heterozygous for clotting as $X^N X^n$.
- ▶ A homozygous female for normal clotting would be $X^N X^N$.
- ▶ A hemophilic male would be $X^n Y$.

We would diagram a Punnett Square showing the cross between a heterozygous female and a normal clotting male like this:

Follow the X and follow the Y.

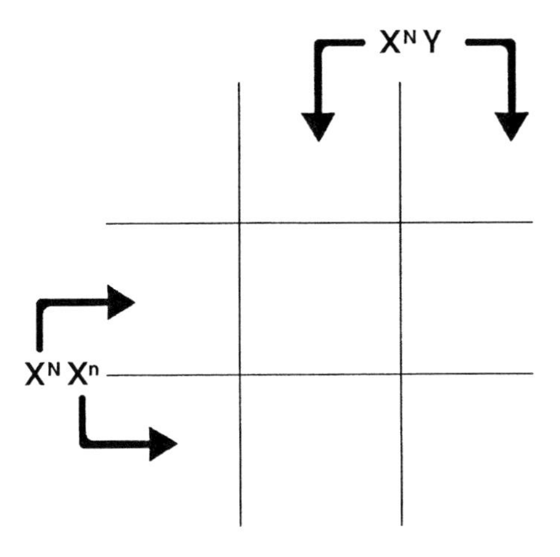

Complete the Punnett Square above showing the offspring.

? QUESTION

1. Looking at the Punnett Square you just completed, answer the following questions.
   - **a.** What is the genotype for the female parent? ____________
   - **b.** What is the genotype for the male parent? ____________
   - **c.** What are the genotypes for their offspring? ____________
   - **d.** What are the chances that any child will be a hemophiliac?
   - **e.** Is it the father or the mother that passes the hemophilia gene to the male child?

2. Failure to distinguish between red and green colors is a recessive allele and is a sex-linked gene carried on the X chromosome.

   A red-green color-blind male mates with a normal female. Of their six children (four boys and two girls), all have normal vision.
   - **a.** What is the most probable genotype of the mother? ________
   - **b.** Will any of their male children pass this disorder on? ____________ Explain.
   - **c.** Draw a Punnett Square of this cross to prove your answers.

3. A normal-visioned female gave birth to a color-blind daughter. Her husband has normal vision. He claims that the child is not his. Does the genetic information indicate that someone else is the child's father?

   Explain and prove your answer using a Punnett Square.

Evolution

(Tutorial)

Sceptical scrutiny is the means, in both science and religion, by which deep thoughts can be winnowed from deep nonsense.
—Carl Sagan, American scientist (1934–1997)

A topic like evolution seems to push everybody's button. Mention it and you're likely to start an argument. Most people think that the central issue of this controversy is about the descent of humans from other animals. It is not! The argument is not about human ancestry, it is about *change.* In its simplest definition, ***evolution is change***. Disagreements arise because humans both crave and fear change. The social argument about evolution becomes a projection of the dilemma we've created inside our own minds. Religions and philosophies meditate on change, whereas science investigates it. Both ways of thinking are natural, expected by-products of human awareness and curiosity. Both ways of thinking can provide value to people.

Your textbook and lecture class will do a thorough job of presenting the traditional scientific evidence and implications of evolution. The purpose of the following Exercises is to start you on the path of understanding evolution, and to offer you some different ways of thinking about it.

Exercise #1 "Genetic Change" 183
Exercise #2 "How Fast Can Evolution Happen?" 188
Exercise #3 "Evolution of One Species Into Two Species" 192
Exercise #4 "The Role of Mass Extinctions" 195

EXERCISE #1

"Genetic Change"

Biologists define ***evolution*** as genetic change in a species over time. Based on what you have learned about DNA, protein synthesis, sexual reproduction, and genetics, you should be able to answer one of the most profound questions about life on this planet:

Evolution is genetic change.

Is it possible for life to stay the same, or is all life destined to change?

DNA and Traits

In the chapter on "Genes and Protein Synthesis" you learned that small pieces of DNA, called genes, produce ***traits*** in organisms. These genes contain messages that make amino acids assemble into an exact sequence to produce protein. Proteins are used in cellular structure and as enzymes in an organism's biochemistry.

If it has been more than two weeks since you covered DNA and protein synthesis in lecture, then review Exercise #2 and #3 in "Genes and Protein Synthesis" now.

? QUESTION

1. A gene is a long chain of nucleotides. Each group of three nucleotides in a row along the gene is called a ________________, and is responsible for the insertion of an ______________________ into a protein.

2. Define evolution in the simplest terms from a biological perspective.

3. Does anything stay exactly the same over time? Explain.

Mutations

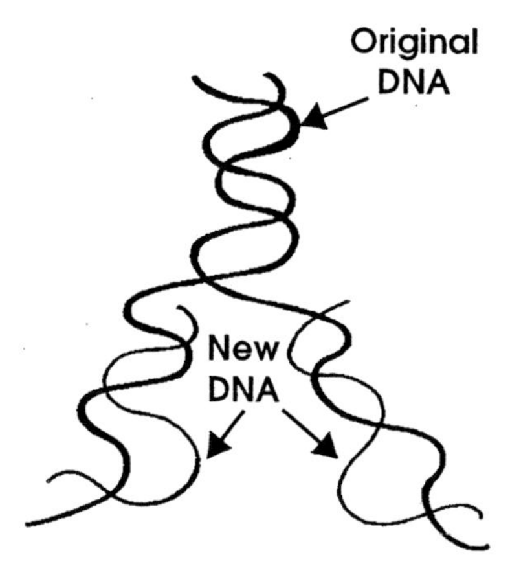

If a "mistake" is incorporated into this new DNA, then that error will be recopied the next time the DNA replicates.

A ***mutation*** is a change in the sequence of nucleotides in the gene (DNA). These changes are caused by a variety of natural events. You might think that your DNA is exactly replicated every time a cell divides or an egg or sperm is produced, but this is not always true. Small amounts of certain chemicals, produced by your own metabolism and in the foods you eat, can chemically alter the sequence of nucleotides. That new sequence will be the model for the *next* replication, and so on.

Nucleotides can also be changed when they are struck by cosmic radiation from outer space or UV rays in sunlight. If these abnormal nucleotides become incorporated into the DNA of an egg or sperm, then a mutation will be passed on to the next generation.

Another mutation event occurs when a nucleotide is accidentally skipped during DNA replication, or when an extra nucleotide is inserted in the DNA. (Accidents do happen!) Whatever the cause, a change in the nucleotide sequence will continue to be copied by all future generations of cells.

Mutations can have either minor or major effects in the biochemistry of the organism. Most amino acids have more than one mRNA codon controlling their insertion into a protein. For example, glutamic acid is coded by both GAA and GAG in the mRNA. Therefore, a mutation of GAG to GAA or from GAA to GAG would have no effect on the protein being made. However, a mutation of GAG to GAU would result in *aspartic acid* being inserted instead of *glutamic acid*.

Mutations are happening all the time.

In Exercise #3, "Genes and Protein Synthesis," you learned that a single nucleotide change results in sickle-cell anemia. Single nucleotide mutations don't always create radical changes in the protein's basic chemical and physical properties. Sometimes the new protein is equal to the original protein in its performance for the organism. However, if a nucleotide is *deleted* during DNA replication, or if an extra nucleotide is *inserted* into the DNA, then the effects can be major. The following questions will clarify this point.

? QUESTION

1. Define *mutation.*

2. If a three-letter code is changed in the DNA, what happens to protein synthesis? (Be specific.)

3. Does a single nucleotide replacement always create a different protein?

4. Does a single amino acid change in a protein always radically change the properties of the protein?

5. Consider the consequences if an extra nucleotide is inserted into a gene. (The consequence is the same if a nucleotide is deleted.)

Original AGATCGGACATAGCCA

Mutation AGGATCGGACATAGCCA

Extra Nucleotide Inserted

What happens to all of the three-letter codes in the gene following that one nucleotide change? (Draw brackets around each codon in both the original and mutated gene to illustrate the consequences.)

6. How many amino acids in the protein will be changed by the type of mutation described in the previous question?

7. Researchers estimate that you have 30,000 genes. If the chance of any of your genes being mutated before it ends up in a gamete (egg or sperm) is 1 in 30,000, how many new mutations would be in each gamete you make?

8. Is there any way that the DNA can stay exactly the same from generation to generation?

Alleles in a Population

Gene Pool

You learned in the "Genetics" chapter that humans are diploid, which means that you have two copies of each gene. When using the word *gene,* we are usually referring to a function—for example, "the *gene* for eye color." The word ***allele*** refers to the particular form of a gene. There is a blue-eye allele and a brown-eye allele. We also know that there are variations of each allele. (Remember our examples of minor mutation events?) In fact, there are many minor variations of the blue-eye allele and the brown-eye allele.

Biologists haven't identified all of the 100,000 individual human genes or the millions of possible alleles of these genes, but eventually they will discover what each of them does. In order to continue our discussion of evolution, the term *gene pool* must be defined. All of the alleles in a species (or a population) alive at a particular time is called the ***gene pool.***

? QUESTION

1. What is the difference between a *gene* and an *allele*?

2. Assume that a particular human gene is 100 nucleotides long. How many different variations of this gene are possible? (circle your choice)
    - **a.** less than 100
    - **b.** two or three hundred
    - **c.** many thousands
3. Define *gene pool.*

Sexual Reproduction

Sexual reproduction stirs the mix of alleles that are passed to the next generation.

Certain events during sexual reproduction increase the genetic combinations of alleles in a species. ***Crossing-over*** recombines chromosomes into new mixes of alleles. ***Independent assortment*** creates variety in combinations of maternal and paternal chromosomes that are then segregated into separate gametes. ***Random fertilization*** further increases the mixing of alleles from one generation to the next. Review Exercises #6 and #7 in "Sameness and Variety" if you have difficulty remembering the sequence of events that take place during sexual reproduction.

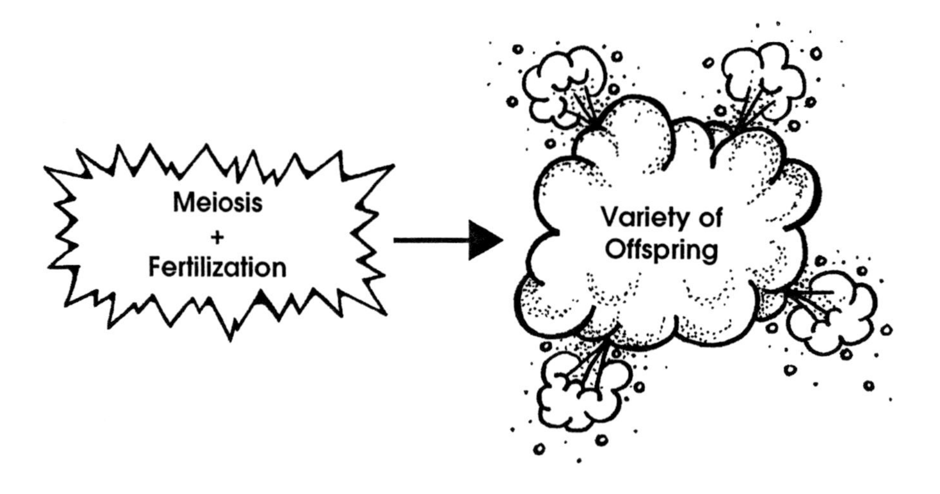

? QUESTION

1. What process makes it impossible for all genes to remain unchanged from generation to generation in a species?

2. What process makes it impossible for any one chromosome to be passed on to the next generation with exactly the same set of alleles?

3. You received 23 chromosomes from your mother and 23 chromosomes from your father. What process during meiosis makes it nearly impossible for you to pass on all of those chromosomes to your children?

4. If you don't reproduce, do your genes get passed on to the next generation?

5. Is it possible for one generation of humans to pass on all of its alleles in exactly the same form and frequency (percent) to the next generation?

Micro-Evolution

Evolution is the unstoppable process.

Evolution in its smallest scale is called ***micro-evolution***. It is a change in the *frequency* of alleles in the gene pool over time. You should have concluded from the previous questions that it is impossible for the gene pool *not* to change over time. Therefore, genetic change (the most specific definition of evolution) does happen—there is no question about it. Two aspects of evolution that should be considered are: "How fast can it happen?" and "How much change can it create?"

? QUESTION

1. Define *micro-evolution*.

2. From a scientific point of view, what is the one question about evolution that is no longer considered appropriate in an argument?

3. From a scientific point of view, what are two questions about evolution that are appropriate to ask?

EXERCISE #2

"How Fast Can Evolution Happen?"

You learned in Exercise #1 that genetic change over time is inevitable. In this Exercise you will examine several easy-to-understand events that increase the rate of genetic change in a population. (Your textbook and lecture will spend time discussing more complicated factors related to the rate of micro-evolution.)

Migration

Evolution can be very fast.

Migration is a simple example of how micro-evolution can happen rapidly in a population. We can represent this situation with two alleles (A and B) of a gene that exists in a species. Assume that the population in one region has a higher frequency of the "A" allele, and in another region there is a higher frequency of the "B" allele.

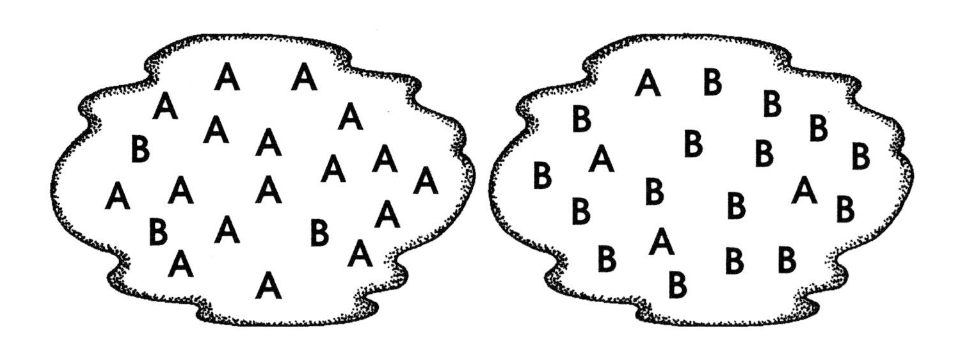

Population of Region 1 (85% have "A" allele).

Population of Region 2 (20% have "A" allele).

Now assume that there is a large migration of organisms from Region 1 into Region 2.

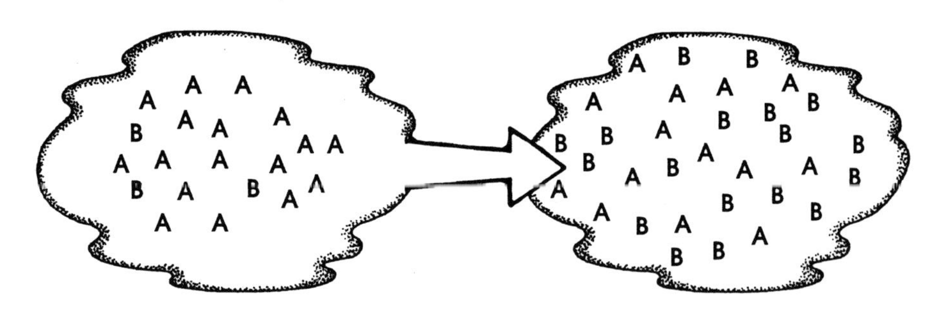

The frequency of alleles changes in Region 2. This change in the gene pool of Region 2 is micro-evolution, and it is a simple example of how genetic change happens on our planet.

? QUESTION

1. List some modern examples of micro-evolution created by human migration.

2. Describe a situation in which organisms in one region would suddenly be able to migrate into a new region.

Luck

In this example, assume there are three alleles for a particular gene in a species.

- ***Allele "A"*** is a real "clunker," and organisms with it survive and reproduce, but don't compete well against allele "B." Only a few of them live along the edge of the distribution of this species.
- ***Allele "B"*** produces a trait that works well for the species, and has been the most common allele for thousands of years.
- ***Allele "C"*** is a newly evolved mutation, and has been doing fantastically well for the last 100 years. This allele has produced the "best" animal of this species to ever have been on the planet.

If it weren't for bad luck, some genes would have no luck at all.

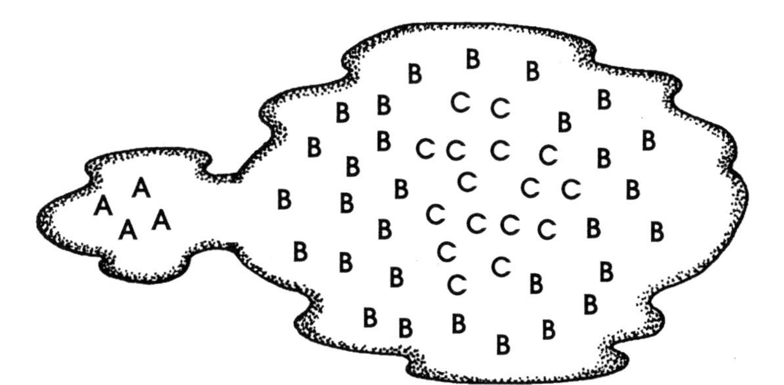

Scenario: An asteroid is heading for the earth and scientists have calculated it will hit at the black spot. All life in the area will be vaporized for 500 miles in every direction from the impact point.

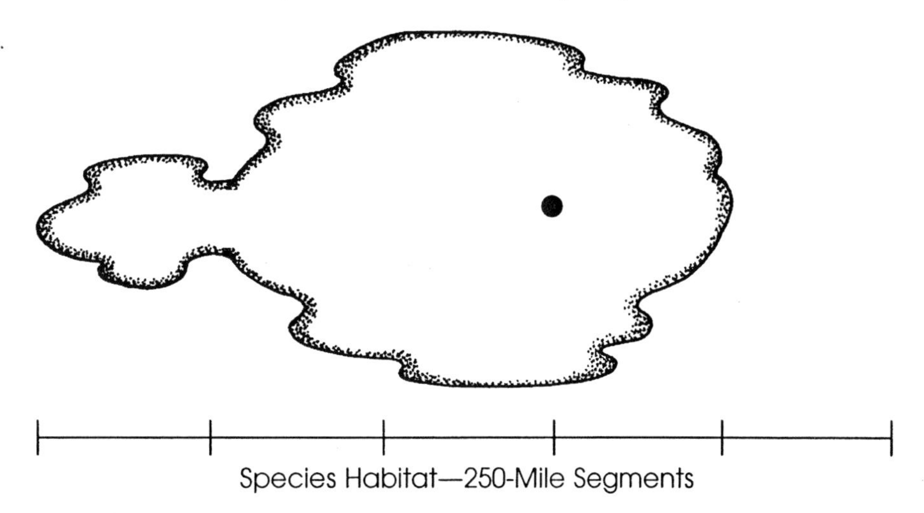

? QUESTION

1. Draw the asteroid zone of destruction on the species habitat map.

2. Which allele will be passed on to future generations?

3. Did the "best" allele survive?

4. Did the most common allele survive?

5. Which allele had the "good luck"?

6. Does this event represent micro-evolution?

7. Does the idea of "survival of the fittest" explain this example of micro-evolution?

8. How would disasters like fire and flood speed micro-evolution?

Small Group Phenomenon

Change can be fast and the direction of change is by chance when small groups are involved. Small groups do not necessarily represent the same percent of alleles found in the total species. *Think about it!* If 30 people got in a boat tomorrow and sailed away to a distant island and started a new population, there is no way that all of the existing human alleles would be in that boat. Furthermore, compared to the total species, it is very likely that there will be a significant difference in the frequencies of alleles that get in the boat (perhaps several red-haired families). The future generations of people on that faraway island won't be like the original group. Micro-evolution occurred because of the ***small group phenomenon***.

Fast evolution happens in small groups.

Whenever small groups survive a catastrophe, or a small group migrates to a new region and starts another population, there will be genetic change. Evidence suggests that this type of micro-evolution may be much more common than biologists of a century ago thought.

? QUESTION

1. How does the small group phenomenon lead to fast changes in the gene pool?

2. How could natural disasters like fires and floods influence evolution?

3. What would you expect organisms to look like that live on islands compared to those of the same species living on the mainland?

Natural Selection

Natural selection occurs when some particular aspect of the environment causes organisms with certain alleles to survive better and reproduce more than organisms with other alleles. It is the situation in which "survival of the fittest" applies. Review the last section of Exercise #3 in "Genes and Protein Synthesis." This section covers the natural selection of grasshoppers in different soil habitats.

There are two very important aspects about natural selection that you should remember. First, it does not create new alleles. Mutation does that! Natural selection only changes the frequencies of alleles already existing in the gene pool. Second, there is no "one direction" for natural selection. In the grasshopper scenario, brown alleles do better in areas of brown soil. In areas of red soil, red alleles do better.

Genetic change in the gene pool of a species can be visualized as a young growing bush. Mutation creates many new branches of variety. Some of those branches are pruned back by natural selection, migration, small group phenomenon, or just bad luck. The remaining branches of variety continue to reproduce, and more variety is added to them by new mutations. This process repeats over and over through the eons of time.

? QUESTION

1. Define *natural selection*.

2. Does natural selection create new alleles in the gene pool?

3. How is natural selection different from the small group phenomenon?

4. Is there one direction to natural selection? Explain.

EXERCISE #3

"Evolution of One Species Into Two Species"

Genetic change is created by random mutations and factors that increase or decrease the frequencies of alleles in future generations. The key to understanding the evolution of one species into many species is the event of ***separation***. If two groups become separated, each will change over time, but they will not change in exactly the same way.

Genetic Consequences of Separation

If two groups become separated, they will evolve into different species.

When the organisms of a species live together, they mate, exchanging and blending their genes over many generations. What happens if one group splits into two separate groups and they migrate far away from each other, never having a chance to interbreed again? As you would expect, the two groups no longer mix their genes or their mutations, and over time they will begin to look somewhat different from each other.

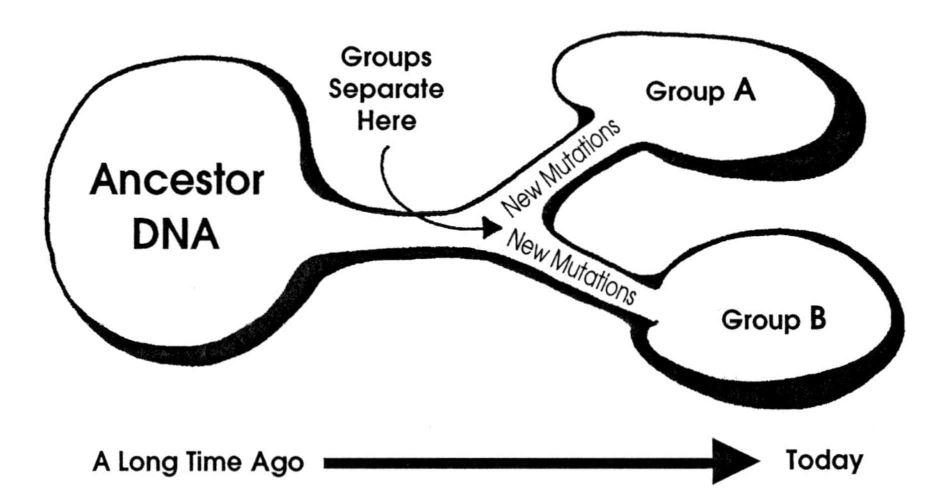

By comparing DNA patterns, chemists can detect the new mutations that have been added to the DNA. If an ancestral group splits into two groups, then both groups will begin to accumulate different mutations from each other. Each group is genetically separated from the other group when they no longer interbreed.

The easiest way to show that two groups have separated in the past is to count the number of new mutations found in the DNA of one group and not found in the DNA of the other group. The greater the number of new mutations, the longer the time span that the two groups have been separated. If however, the DNA of the two groups is quite similar, then the groups have not been separated for very long.

Diagramed below is a comparison of the DNA of three geographically separate human racial groups. Each bar represents a mutation.

African DNA

Northern Asia DNA

Native American DNA

? QUESTION

1. Which two groups are the most similar?

2. Which of the three groups is the most different from the other two groups?

DNA continues to change. No one is original stock!

3. Which group has been separated from the others for the longest time?

4. Which of the two groups haven't been separated from each other for very long?

Events that Cause Separation

Your textbook and lecture will present several different situations that lead to genetic separation of two populations. One of the situations is created by changes in the geography. ***Geographical separation*** leads to genetically isolated populations and eventually to different species.

Mountains, Rivers, and Highways

Graduate students at the local college studied the mouse populations around campus and in the neighboring communities on the opposite side of a major freeway. White-faced mice, purchased for physiological and behavioral studies, routinely escaped from the lab and became part of the campus population of wild mice. For years, the students sampled the neighboring communities across the freeway, but no white-face varieties were collected. The freeway acted as a geographical barrier between the two mouse populations.

The college mouse story is a fun example of small scale isolation, but many studies have been done on populations separated by rivers and mountain ranges. These investigations demonstrate that geographical separation leads to new species. For example, the squirrels on the north and south sides of the Grand Canyon are separate species now. Separation leads to ***speciation***.

Drifting Continents—Plate Tectonics

Sometimes geological processes create major opportunities for new speciation. North America and Europe were once connected, but following their separation, the species on each continent changed in different directions. Today, many species on both sides of the Atlantic are related, but each has evolved new traits that make them separate species.

Even a few decades ago, the idea of plate tectonics was controversial.

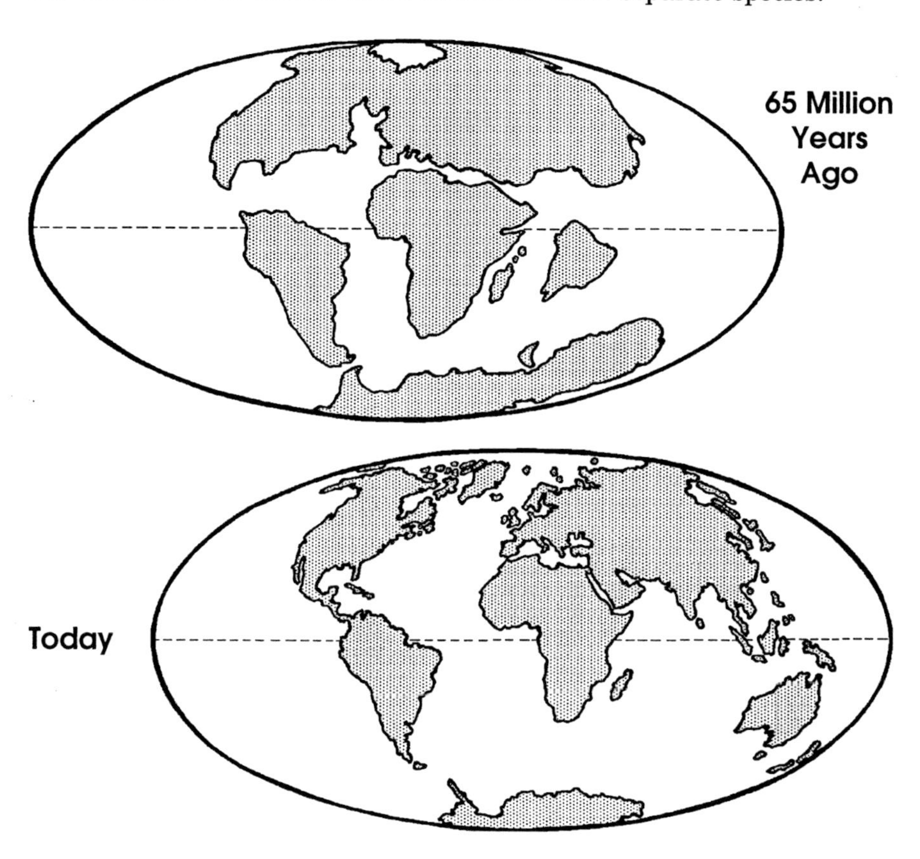

The African East-Side/West-Side Story

We now have an explanation for the split of humans from a common ancestor.

A fascinating geological separation event occurred in Africa starting about 8 million years ago. As a result of plate tectonics, a large rift valley and a high mountain range formed, separating East Africa from West Africa. By about 4–5 million years ago, East African fossils of elephants, antelope, rhinoceros, and other animals showed clear adaptations to a drier environment. East Africa was becoming drier, while West Africa remained wet as it is today.

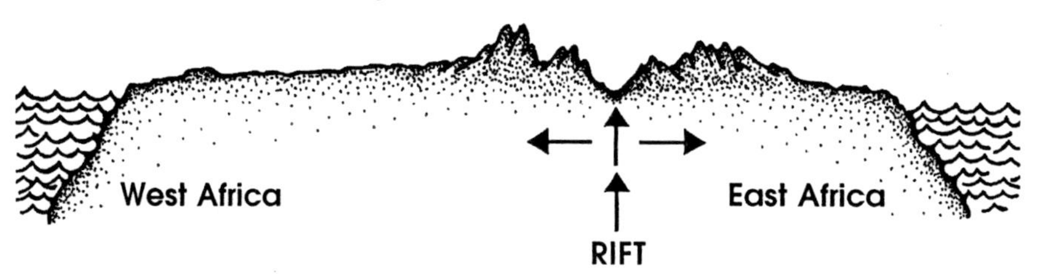

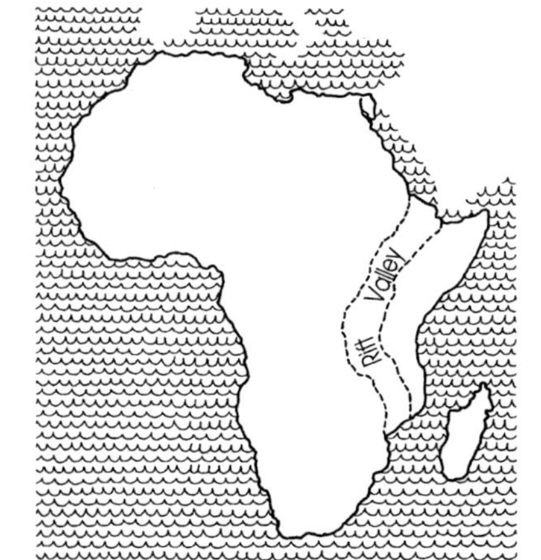

The earliest hominids—the *Australopithecines*—appear only in the fossil record of East Africa. None of these fossils have been found in West Africa. It is as if the two worlds were separated by the rift valley. Genus *Pan* evolved in West Africa, and led to today's chimpanzee. Genus *Homo* evolved in East Africa. (Apparently we are the kids from the east side of the tracks.)

? QUESTION

1. List five geographical changes that could lead to genetic isolation of populations.

2. How long would it take for micro-evolution to begin changing geographically separated groups of a species?

3. How much change is required before there would be two distinct species?

4. Explain how the geographical separation of primates in Africa could have led to two separate lines—chimp and human.

EXERCISE #4

"The Role of Mass Extinctions"

Uta's Story

Some rocks were once alive.

Ngoma poked at the curious rock that had been split open by the heat of the fire. Quickly Uta's hands jabbed at a piece until it was positioned near the edge of the cooking pit. Then, she scooped it up and tossed it into the clay water crock where it sizzled and cooled. After a few minutes she removed the rock and held it up to the fire's light. She had seen this rock before in a distant place near the edge of a quiet bay with her fisherman father. Tonight, the story was hers.

Uta pointed at the large mushroom shape contained in the rock. The edge of her nail traced each of the inner layers of the mushroom shape. With the flames reflected in her eyes she began her tale.

"This rock is long dead! There are rocks like these today that are alive at a place where my father and I go to fish." She paused. "Each layer is a different age, like the rings of a tree. The parts towards the center are the oldest, and each new layer grew over the top of the inner layers."

Masango and Ngoma looked at each other enthusiastically. This would be a great night! Uta was telling another one of her astounding stories.

The Scientific Account

Uta's story, as far as we have told it, agrees with modern scientific research. These same layered rocks have been found by scientists, and are part of the fabric of a new story emerging from the fields of geology, archeology, and biology. The scientific account tells of creation itself, fantastic life-forms, mass extinction, opportunities gained, and opportunities lost. A textbook would be required to summarize the many known details of the history of life, but we can consider the general facts.

Science has discovered that life has been shaped by many processes. Four of these processes include: 1) plate tectonics, 2) impacts by objects from outer space, 3) new species evolving and destroying previously existing species, and 4) the destruction of habitats by modern humans.

Plate Tectonics

Crust

Molten Rock

The interior of the earth is extremely hot. This heat is generated by various chemical and physical processes that are not completely understood today. Heat maintains a cycle of molten rock moving slowly beneath the earth's crust. (If you have ever watched fudge slowly boiling, then you have the general idea.)

Currents of molten rock affect the crust in two important ways. First, molten rock rises to the surface through cracks in the earth's crust. This continually adds new material to the surface of the planet. Second, the cracks break the crust into pieces, called ***plates***, that are moved around by the currents of molten rock beneath them. The movements of the earth's plates is called ***plate tectonics***. It has been a very important factor in the history of life. Plate tectonics makes, destroys, and moves continents around the surface of the planet.

The location of a continent determines its climate. If a continent moves, then the environment will change for the organisms living there. When continents are scattered as smaller chunks of land, or are clumped in a large land mass (such as Pangaea of 240 million years ago) the weather conditions are profoundly different for life on land. If a continent drifts over a pole, then glaciation will occur and the sea level will drop, which transforms the environment for life at the edge of the seas. More details about the effects of plate tectonics are discussed later in this Exercise.

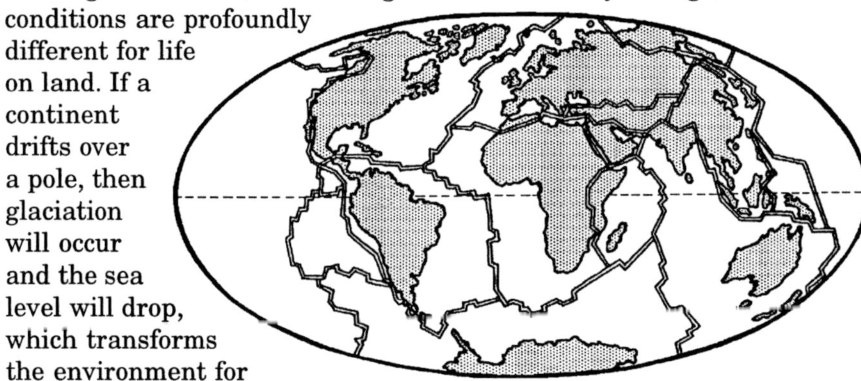

Impacts from Outer Space

The evidence for asteroid impacts as a force for mass extinction comes from the discovery of the element ***iridium*** in layers of sediment found in rocks. Iridium is a heavy metal that occurs naturally deep in our planet's interior, where it sank from the surface when the earth was forming. In other words, we would not expect to find any of this element near the surface of the planet. However, iridium is one of the metals found in asteroids moving around in outer space. When an asteroid impacts the earth, its mass and iridium is vaporized and scattered as dust in a thin layer around the impact zone.

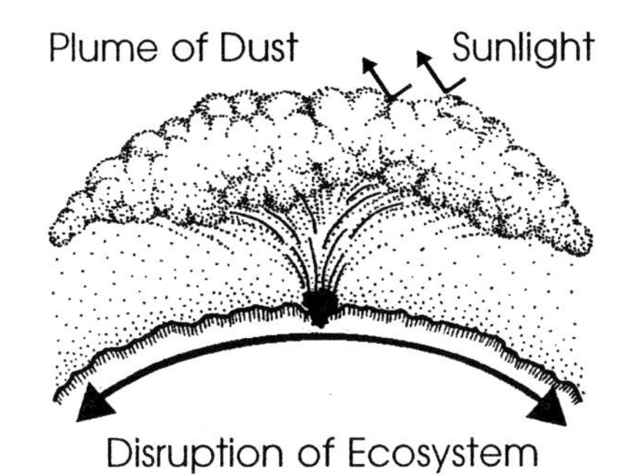

Impacts from asteroids or comets explode huge amounts of earth into the atmosphere, and the resulting atmospheric plume blocks much of the sunlight. In addition, there is evidence that large impacts can trigger ongoing volcanic activity for a period of time. The resulting darkness drastically shuts down the world ecosystem, and can create a mass extinction. This type of extinction closes the door of opportunity for some species, but opens it for others.

? QUESTION

1. What force moves the crust plates across the surface of the earth?

2. How does plate tectonics influence the history of life?

3. What specific evidence suggests that an asteroid impact has occurred in the past?

4. How is an asteroid impact catastrophic to life?

In the Beginning

The dream world of humans has created thousands of stories about the origin of life. These are fascinating tales, but little of what they say is testable from a scientific perspective. Although science cannot provide all of the details of how life began, a general understanding has emerged that is supported by a growing record of scientific evidence.

The first life on earth evolved in an environment without oxygen. There was no oxygen in the air because it had combined with other elements like hydrogen, nitrogen, and carbon to form molecules. This meant that early lifeforms had to gain their energy from a type of metabolism that did not require oxygen.

Science has a unique story to tell.

The oldest fossils are found in rocks nearly 4 billion years old. These fossils were originally living mats of bacteria, and look somewhat similar to the stromatolites discussed later. The fossilized bacteria also look similar to the anaerobic bacteria we find in today's world. Two contemporary types of anaerobic bacteria are called iron bacteria and sulfur bacteria. These are examples of organisms that can derive energy by metabolizing minerals. They do not require oxygen. Simple ***anaerobic bacteria*** are proof that life really can exist in an environment similar to that of the early earth.

Some questions arise. If anaerobic bacteria were the first lifeforms, then why aren't they the only type of organism on the planet now? And where did the oxygen in our atmosphere come from?

The answer to these questions is one of the first big lessons of life: *Somebody else can put you out of business!* In biology, we call this extinction. Let's finish Uta's story of the mushroom rocks.

Stromatolites

Uta's Rocks

Anaerobic bacteria had this planet to themselves for a billion or more years. Then, about 3 billion years ago new kinds of organisms evolved in the shallow seas. These colonial micro-organisms produced rocklike deposits. They included two new types of bacteria called ***cyanobacteria*** and ***aerobic bacteria***. The cyanobacteria produced oxygen gas during their biochemistry, and aerobic bacteria used oxygen during theirs. These bacteria grew in layers. Each new generation covered the previous one, eventually forming a large mushroom-shaped deposit that we call a ***stromatolite***.

Once the stromatolites appeared, the anaerobic bacteria began to disappear from the fossil record. These "newcomer" stromatolites changed the world. The cyanobacteria transformed the atmosphere by using a kind of energy-trapping process called ***photosynthesis***. Photosynthesis produces oxygen, a substance that is *poisonous* to anaerobic life. This new gas killed all the early anaerobic life-forms except those living in areas of the planet where oxygen gas could not reach. The anaerobes were restricted to these limited habitats. Today we can still find these very old types of anaerobic bacteria (called archaebacteria) in the black mud of swamps, in salt marshes, and in volcanic hot springs. These smelly, hot places might seem disgusting to us, but they are a paradise for anaerobes. To the anaerobes, the whole planet was once a wonderful smelly place—until stromatolites evolved.

As stromatolites oxygenated the world, they changed it. They covered the floor of shallow seas until about one billion years ago. Then another new group, the ***invertebrates***, evolved. These multicellular animals became very diverse and included some that ate stromatolites. What the stromatolites did to the anaerobes was done to them by the invertebrates, and today there are only a few places where stromatolites can exist. We find them in isolated salty bays where they are protected from the appetites of invertebrate predators. This is the scientific story of Uta's rocks—alive for billions of years, and still living reminders of a previous chapter in life's long history.

? QUESTION

1. What were the first organisms and where do they live today?

2. There was an almost complete extinction of anaerobes about 3 billion years ago. What gas in the atmosphere was responsible for their extinction?

   What biological process produced that gas?

   Which organisms evolved that process?

3. How do stromatolites form, and what two groups of bacteria are part of their formation?

4. There was an almost complete extinction of stromatolites between 500 and 700 million years ago. Which group of organisms was responsible for eating stromatolites and out-competing them?

   Where can living stromatolites be found today?

Glaciation and Sea Level

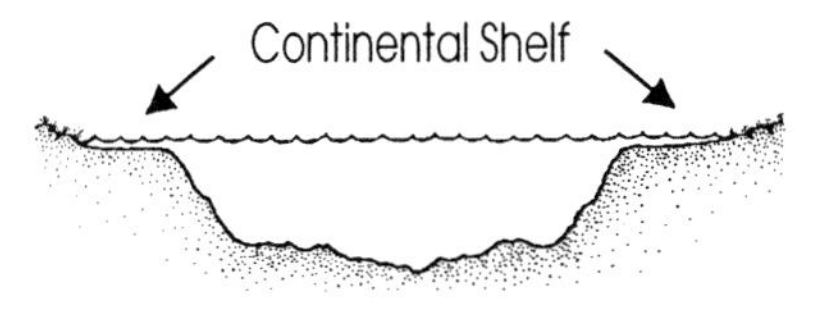

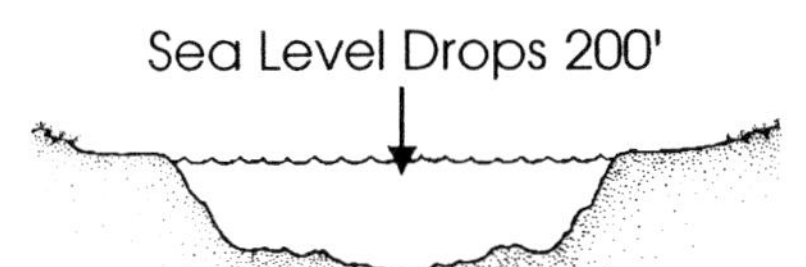

When thinking of life in the ancient seas, we usually visualize sharks and fish. But those early seas were filled with organisms far beyond our imaginations. Again, the history of life is best told by examining the fossil record. Invertebrate animals first evolved as very simple creatures somewhat like today's sponges and jellyfish. By about one-half billion years ago they blossomed into many varieties and bizarre forms. They shared the seas with a strange chordate group, the ***jawless fish***. One hundred million years later, a few ***cartilage fish*** and ***bony fish*** evolved. They were now the newcomers. Millions of species lived in the rich continental shelf zone. Today there are only a few jawless fish, and many of the early invertebrates are gone. What happened to them?

A big evolutionary event occurred ***370 million years ago*** that created a mass extinction of many invertebrate species and most of the jawless fish. Following this event we find new creatures in the fossil record—early types of amphibians. What could cause trouble for invertebrate sea life and jawless fish, yet open up opportunities for amphibians? Here we get some help from geologists.

The oceans of the world are shaped like deep bowls with thin shallow rims. These rims are actually 200 feet below water and are called the ***continental shelves***. Ocean life is concentrated in the continental shelf waters. The shelf zone is the richest nutrient area in the oceans because ocean currents and runoff water from land constantly stir the bottom sediments. These two processes increase the availability of inorganic minerals that are necessary for photosynthesis. High nutrient levels are resources for plants—the basis of the food chain.

Knowing this, what would happen if the sea level dropped 200 feet? All of the continental shelf zone would be exposed, and the species that depended on food from those waters would vanish. This did happen. As the continental shelf habitat was eliminated, many marine species died out, and a new environment was left behind for the remaining groups.

But what caused the sea level to drop? The answer begins with the ***water cycle***. First, sunlight evaporates water from the oceans forming large clouds. Next, those clouds rain somewhere on the planet. This rain creates rivers that eventually return to the sea, and the cycle repeats.

If water doesn't return to the ocean, then the sea level will drop by an amount equal to the mass of the new glacier.

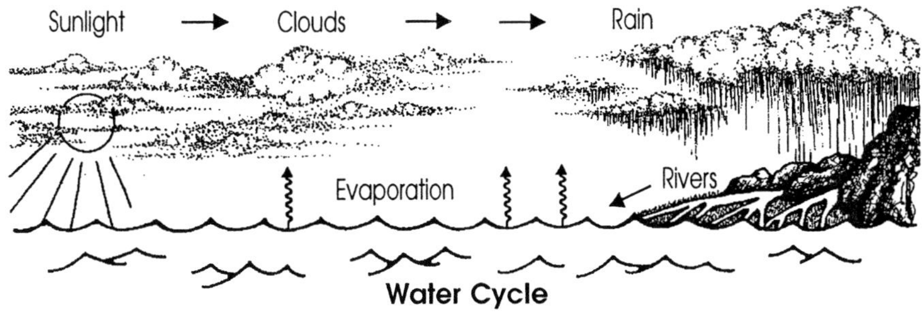

Water Cycle

Now, consider how the position of continents might affect sea level. Modern geology has learned that geological forces constantly form and move continents. In this planet's history, continents were sometimes small and scattered, and at other times continents were crunched together into bigger pieces. About 370 million years ago a large continental mass drifted into position over the South Pole. Before then, the winter snow fell into the ocean and quickly melted. With a continent over the South Pole, the snow fell onto land and did not melt. The snow piled up forming huge ***glaciers***. This interrupted the water cycle. (If water doesn't return to the ocean, then the sea level will drop by an amount equal to the mass of the glacier.) This is the probable explanation for the lowering of the sea level that caused the extinction.

However, what was bad for invertebrates and jawless fish was a great opportunity for whomever was left behind. The surviving groups included some of the ***cartilage*** and ***bony fish***, from which many new species evolved. One group that adapted to the slowly changing conditions was an air-breathing fish with simple limbs. We call them ***early amphibians***. Amphibians were to have the stranded pockets of isolated lakes and shallow seas to themselves until the next big calamity on the planet.

Pangaea—The Biggest Desert That Ever Was

The amphibians of ***240 million years ago*** were challenged by a continent bigger and drier than anything in our modern experience. Geologists call this mega-continent ***Pangaea***. Pangaea formed as land masses drifted and merged together from the forces of plate tectonics. Compared to the comforts of the swamps, this land was hot!

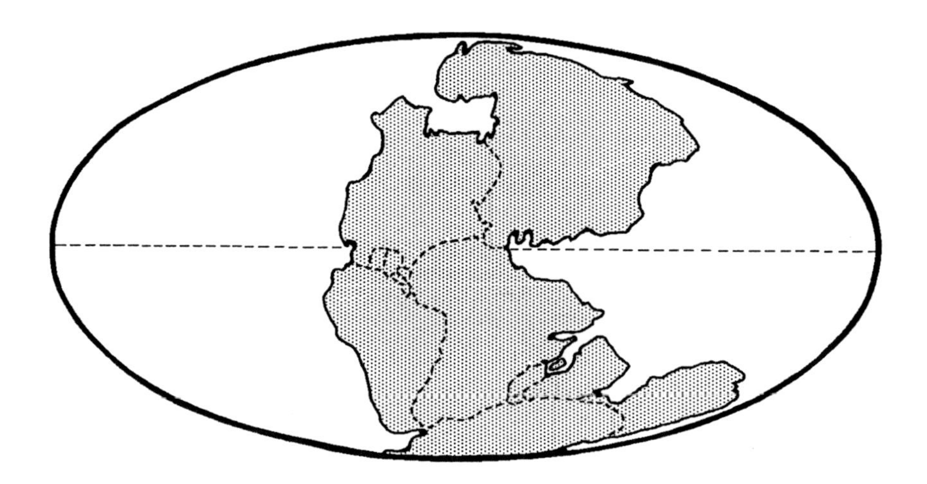

The amphibians were incredibly successful, but their days were numbered.

When continents are small and surrounded by water, the rain clouds can reach most of the land and the environment is moderate. The smaller, scattered continents before 240 million years ago were suitable for early land vertebrates like amphibians. But, Pangaea was a huge land mass. Most of it was far from the ocean and the rain clouds. This supercontinent was the largest dry land environment that we know of in the last half-billion years of the Earth's geologic history.

The change from small swampy continents into one huge mass of dry land produced the first big land extinction during which ninety-five percent of the land animals were eliminated. All of the larger amphibians died. But, a newer group survived—one that had dry skin and a different method of reproduction. These animals could mate and raise their young on dry land. The new traits evolved from the old traits of the amphibians, and produced what we call a ***reptile***. The innovation was so profound for land habitats, and the opportunities for reptiles on dry land were so great, that every new species flourished.

With this new game, it seemed as if nothing could go wrong—well almost nothing—unless you are unfortunate enough to be hit by a 10–mile-diameter rock traveling at 150,000 miles an hour.

? QUESTION

1. About 370 million years ago an extinction of jawless fish and many invertebrate species occurred. Explain how this extinction happened. (Include in your discussion continental movements, glaciation, the water cycle, and sea level.

2. Which group evolved and flourished in the new environment provided by the receding seas?

3. Does the sea level have to drop thousands of feet before sea life is affected? Explain.

4. Another extinction event (95% of known species) occurred about 240 million years ago. This is considered to be the first big land extinction. Explain how this event was created by continental movements, weather patterns, and reduced shorelines.

5. What name is given to the supercontinent of 240 million years ago?

6. Which animal group was almost exterminated by the reduction of swampy environments?

7. Which new animal group greatly benefited from the large amount of available dry land on the planet and the drastic reduction of competitors?

Bad Times for Reptiles

We know that many asteroids have struck our planet during its history, including one at about ***210 million years ago***. The crater made by this impact is located in eastern Quebec, Canada. It measures nearly 50 miles across. The crater has been dated near the time that an extinction event occurred for many of the early reptiles.

So, approximately 30 million years after the opportunity first developed for reptiles, about 60% of them became extinct. The reptiles that remained alive after this extinction would have the next big opportunity. Several groups, including crocodile, turtle, and snake-lizard types, remain until today. But, the first groups to flourish were the predecessors of the ***dinosaurs***, and dinosaurs did very well indeed. There was every type and every size of dinosaur that you could imagine during their 150-million-year dynasty. But, their days of being the dominant land vertebrate were numbered.

A rock 10 miles in diameter traveling at 150,000 miles per hour can ruin your day.

Scientists now have evidence that about ***65 million years ago***, a huge object from outer space hit the earth. Preliminary evaluation of the evidence suggests that the dinosaur extinction was partly caused by the impact. Whether the object was a comet or an asteroid, there is a 100–mile-diameter crater at the tip of the Yucatan Peninsula dated near the time of this extinction. An impact that large would create total destruction within at least 1,000 miles from ground zero. In addition, several thousand cubic miles of vaporized rock would be carried into the atmosphere, blocking sunlight and photosynthesis for months or longer. During this dark time the world's ecosystem was radically interrupted, destroying many species. Dinosaurs and about 75% of all marine species were eliminated near the time of this asteroid impact. Many of these species were instantly extincted, while others died out during the years that followed.

As with previous extinctions, whatever species survived had a special opportunity for new evolutionary experiments. Almost anything will work when there are few competitors. The next two successful land groups that flourished were ***mammals*** and ***birds***. From 65 million years ago until today, these two groups and the ***flowering plants*** have taken advantage of the land environments.

Human Hunters

The previous examples are large-scale events that radically affected the evolution of life. Fossil evidence also suggests that thousands of ***small-scale extinctions*** have happened on this planet. One such small-scale extinction of large mammals about 20,000 years ago followed the migration of humans from the Eastern Hemisphere into the Western Hemisphere. Hunting artifacts are dated at the same time that a dozen or so large mammal species disappeared. This type of extinction occurs whenever a new species moves into a region where there is easy prey. This is also an example of how much impact one successful species can have on another species, and serves as a warning to us concerning our destructive actions of today.

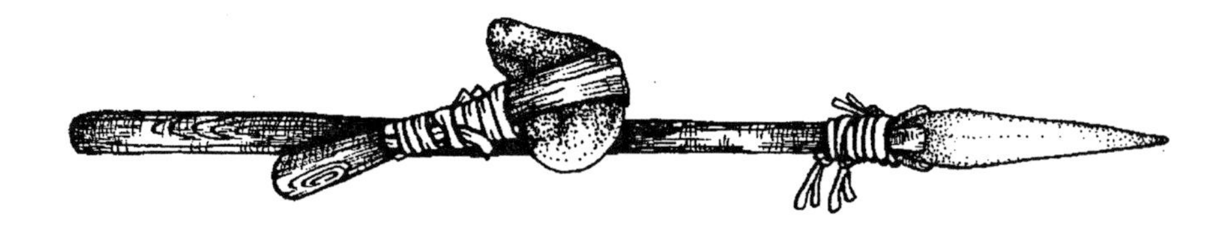

Habitat Destruction

When habitats are destroyed, the species that depend on those habitats are doomed for extinction.

The most recent extinctions have been caused by humans eliminating the natural habitats of other species. When habitats are destroyed, the species that depend on those habitats are destroyed. This process is occurring at an alarming rate wherever there are dense populations of humans.

Consider for a moment all of the forests that humans have destroyed or substantially changed in our country during the last 200 years. Many forest species are gone forever because of that habitat destruction. Today there is an even faster rate of extinction occurring in the tropical rain forests of the world. These are the richest ecosystems that exist on land. As many as 100 or more species become extinct daily in these special forests. Rain forests are being cleared and burned for their wood, to grow crops, and to graze cattle. These ventures usually destroy the fragile top soil within a few years, and all benefit is lost for indigenous people as well as for the world ecosystem.

? QUESTION

1. Explain how the presence of the element iridium in sediment is related to asteroid impacts.

2. About 210 million years ago, an asteroid or meteorite hit the earth (perhaps creating the crater in Quebec), and about 60% of the existing species became extinct. Many early reptiles were eliminated, which allowed one of them the opportunity to become very successful. What group was that?

3. Explain how asteroid impacts can cause extinctions.

4. Another asteroid hit the planet (perhaps creating the crater in Yucatan) about 66 million years ago. Which reptile group was eliminated?

   Which two vertebrate groups benefited from the reduction of reptilian competitors?

5. What caused the extinction of most of the large mammal species in the Western Hemisphere?

6. What specific change accounts for today's rapid rate of species extinction?

Summary

Science has discovered some of the truths about the history of life. This short story covers only a partial description of all we know and suspect. The example events we have discussed include:

- an organism changing the environment for every other species on the planet,
- a new group of organisms that eats and eliminates other species,
- glaciation and the lowering of sea levels,
- what happens when swamps dry up,
- impacts by asteroids and comets,
- the effects of early human hunters, and
- the destruction of ecosystems and habitats by humans today.

The challenge to humanity is to find a balance— before Nature does.

What once began as exciting stories around the fire pit has expanded into a more complete understanding of Earth's history. Now, the challenge for humanity is to find a balance—to limit reproduction rates and develop lifestyles that are not as damaging to ourselves and other species. If we don't, the planet will continue to experience major changes in evolution because of us. And the new direction that evolution takes could cause serious problems for humans—even our own decline, possibly our extinction.

Human Evolution

People are very curious about where things come from, and what they mean. This apllies especially to human history. The many written stories about humans are all we had to tell us about our past, until some scientists got together to discover what could be learned about humans before writing was invented.

Now, we are able to present the scientific story of man's evolution on this planet. All that you will need to join in this discussion is practice with archeological and biological information. During this lab you will learn to develop a scientific history of humans, using several techniques of thinking, map-making, and group discussion.

Exercise #1 "Trails of the Cross-Country Hikers" 205
Exercise #2 "The Mitochondrial DNA Clock" 207
Exercise #3 "Time Trails of Modern Humans" 210
Exercise #4 "The Story of the Really Old Human-Like People" 212

EXERCISE #1

"Trails of the Cross-Country Hikers"

Who were they with last?

This first problem will develop your thinking method for understanding a scientific history of modern humans. The information provided may seem to be brief. But, as you think through the problem, you will be able to draw a map of the different paths taken by the five students, indicating when they separated from each other, and the trails they took.

Your time map will look something like this, using branches to represent when the different hikers separated from each other.

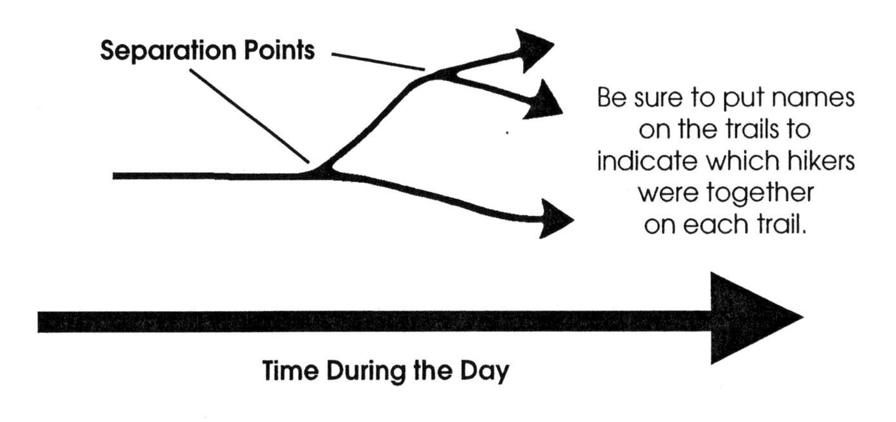

Procedure

[1] Divide into groups of 3 or 4 people.

[2] Your group may need some quiet space to do this Exercise, so don't hesitate to work outside or in another room if necessary.

Materials

- Refer to the worksheet at the end of this lab titled, "Trails of the Cross-Country Hikers."
- A five-card set of names. You can use these name cards as an aid to physically keep track of the hikers during your analysis of the problem.

"Cross Country Hikers"

The Problem

Five members of the Cross-Country Running Team decided to have a rugged day of fun last Saturday. They drove out a long dirt road, and parked the car at a grove of trees that was 12 hours of hard hiking away from the freeway. All they had to do was to walk west and they were guaranteed to reach the Highway 8 Freeway, where they could hitch a ride back to their parked car.

Your task is to figure out the general paths that were taken by the individual people using the limited information provided below, and to draw those paths on a map.

Information

- ▶ The five hikers started out together, but divided into smaller groups, taking different paths as they went along.
- ▶ Below you will find clues about the different people. The clues tell you when particular hikers separated from each other and took different paths. You will have to use those clues to figure out and draw their approximate paths to the highway.
- ▶ The hikers names are: Bill, Hector, Julie, Tom, and Maria.
- ▶ The hikers all began to walk at about 6:00 a.m. All of the hikers reached Highway 8 at around 6:00 p.m., but they did not necessarily arrive together.

Clues

- ▶ Tom and Maria arrived at Highway 8 together.
- ▶ The last time Hector was with Julie was 5 hours before he reached the highway. (This clue does not tell you whether or not Hector or Julie were with anyone else when they separated from each other. You will have to figure that out—read on!)
- ▶ The last time Julie was with Tom was 8 hours before she arrived at the highway. (Again, this clue does not tell you whether or not Julie or Tom were with anyone else when they separated from each other.)
- ▶ The last time Bill was with Hector, they were 10 hours away from the highway. (Was Hector or Bill with anyone else at the time they last saw each other? Again, this clue does not tell you. Work it out!)

Procedure

[1] Draw a map of the trails taken by these five hikers. Show when they split up during the day, and how they were grouped when they arrived at the highway at 6:00 p.m.

[2] Make a practice map to work out the problem, then draw your final map on the "TRAILS" worksheet at the end of this lab.

[3] You may find it helpful to use the name cards to keep track of the hikers. Lay them on your paper, on the ground, or on the table, and move them along the trails as you work through the clues.

In Conclusion

1. Check with your instructor to make sure you got the correct answer to this problem before going on.
2. The next problem is going to take much longer, and will use the same kind of thinking you used for the Cross-Country Hiker problem.
3. **Good advice:** Have the best reader in your group read aloud as the rest of your group reads Exercise #2, "The Mitochondrial DNA Clock."

   Work together.

   Work slowly.

   When you have questions, check with your instructor.

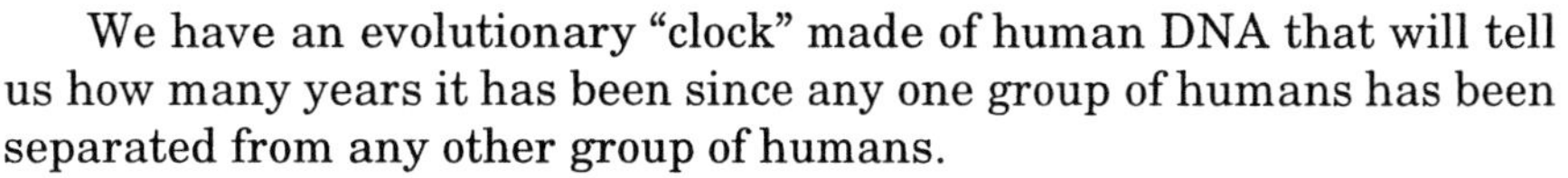

EXERCISE #2

"The Mitochondrial DNA Clock"

We have an evolutionary "clock" made of human DNA that will tell us how many years it has been since any one group of humans has been separated from any other group of humans.

Think about it this way: When humans live together, they mate, exchanging and blending their genes over many generations of time. This is how sexual reproduction creates common traits in human DNA. Now, what happens if one group splits into two separate groups that migrate far away from each other, and they never get a chance to interbreed with each other again?

As you would expect, the two groups are no longer mixing genes (and DNA), and over time they will begin to look somewhat different from each other.

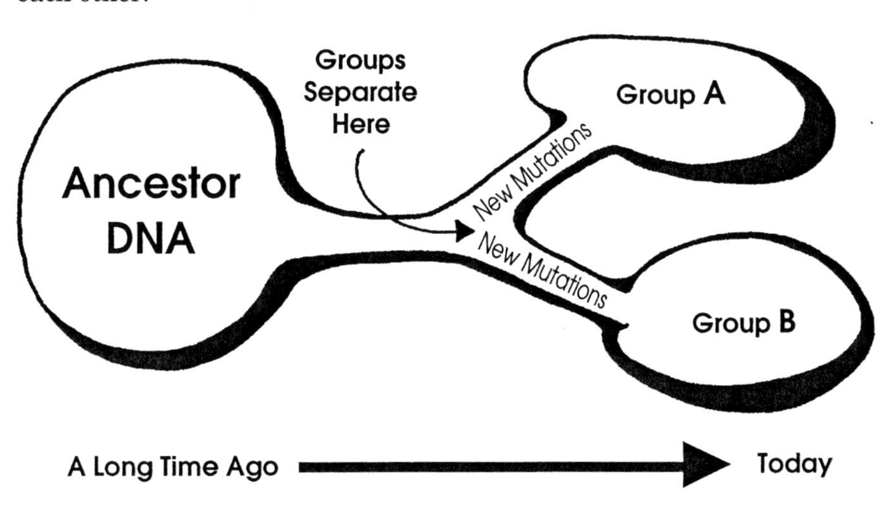

Mitochondrial DNA

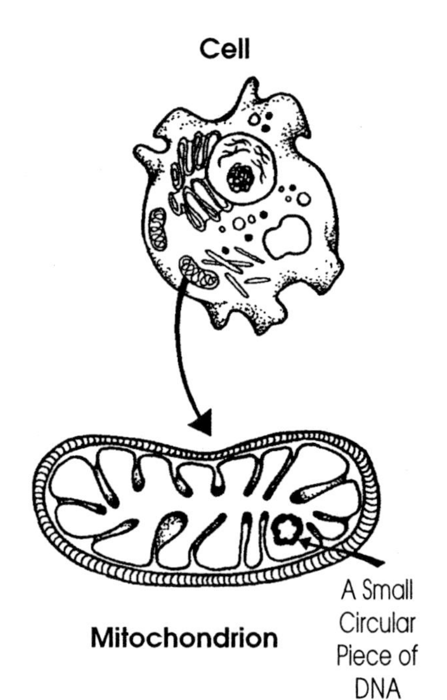

Humans have DNA in the mitochondria. This mitochondrial DNA does not mix with the DNA in the cell nucleus during sexual reproduction. So, the only way it gets passed onto the next generation is through the mitochondria in the *mother's egg*. The father's sperm does not give mitochondria to the egg during fertilization.

This strange situation means that mitochondrial DNA (passed on by the mother) stays the same, and cannot change except by "new" mutations. These mutations occur randomly as a natural process of living on this planet.

By comparing DNA patterns, chemists can detect any new mutations that have been added to the small piece of original mitochondrial DNA that was passed on from mother to child through thousands of human generations.

If a group of humans splits off from a common ancestral group, then both groups will begin to accumulate different mutations from each other. Each group is genetically separated from the other group. This starts at the point where they no longer interbreed.

The easiest way to show that two groups have separated in the past is to count the number of *new* mutations found in the DNA of one group and *not* found in the DNA of the other group.

The *greater* the number of *different* mutations, the longer is the amount of time that the two groups have been separated from each other. (**Remember:** Mutations take a lot of time to happen.) If however, the DNA of the two groups is quite similar, then they have not been separated for very long.

Comparative DNA Analysis

Diagrammed below is a comparison of the mitochondrial DNA of three geographically separate racial groups. Each *bar* represents a mutation.

African DNA

Northern Asia DNA

Native American DNA

? QUESTION

1. Which two groups are the most similar?

2. Which one of the three groups is the most different from the other two groups?

3. So, which group has been separated from the others for the longest time?

4. And, which two groups haven't been separated from each other for very long?

The "Clock"

Biologists have studied many different species, including humans, and have estimated that it takes 500,000 years for 1% of the mitochondrial DNA to be changed by mutation. (This estimated mutation rate is currently being debated, but we will base our calculations on it. Perhaps more research will clarify our understanding of evolutionary timelines.)

This mutation rate is a kind of "clock." Researchers have gone around the world collecting mitochondrial DNA samples. They have counted the total number of small mutations that have occurred in the humans of today. Each one of those mutations represents an amount of "time" on the mitochondrial clock.

? QUESTION

1. A representative sample of different human geographical/racial DNA has been collected and analyzed. When all of the different mutations were counted, scientists found that only 0.4% of the DNA of modern humans has been mutated. How many years have *modern* humans been on this planet? **Remember:** It takes 500,000 years for 1% of the DNA to be mutated.

____

2. Can you estimate how long geographical/racial groups have been separated from each other?

HOW TO DO IT

If you want to estimate how long it has been since two racial or geographical groups have been separated, you must:

1. Use the figure for the % of genetic difference between the two groups (as determined by genetic researchers).

2. Multiply the % of genetic difference by 500,000 years. (It takes 500,000 years for 1% of the mitichondrial DNA to be mutated.)

3. *Practice Example:* If Group A and Group B have 0.1% genetic difference, then it has been about __________ years since they separated, and have not had the opportunity to interbreed.

500,000 years x 0.1% = 50,000 years

Answer: 50,000 years

Procedure

Calculate the answer to the following problem:

[1] Group A, the "Altitudinals," have lived in the high mountain ranges of Nepal for longer than anyone can remember. Group B, the "Basinals," have lived on the flat river delta plains of Southern India for hundreds of centuries according to their written history.

[2] A genetic researcher has measured a 0.05% genetic difference in the mitochondrial DNA of Group A and Group B.

[3] How long would you say it has been since these two groups lived together and had the opportunity to interbreed? In other words, how long ago was it that the Altitudinals and the Basinals split company, one moving to the highlands and the other moving to the lowlands?

__________ years
Show your work.

In Conclusion

It's time to move on to Exercise #3, the "Time Trails of Modern Humans," and test your understanding of the "Mitochondrial Clock." If you are having trouble understanding the mitochondrial clock, then *now* is the time to get help from your instructor!

EXERCISE #3

"Time Trails of the Modern Humans"

The migration of humans can be retraced just like the trails of the cross-country hikers.

Archaeologists have found skeletal evidence of a modern-type human (looks like people today) who lived on this planet about 100,000 years ago. There is a question as to exactly how long modern humans have been here, and where they might have originated.

You will investigate a partial answer to these questions. Modern human types are known as *Homo sapiens*. They are somewhat lighter in build, with a bit larger brain size than the previous human fossils. Evidence also suggests that they lived much like the hunter-gatherer peoples of today.

If you were to bring some of these skeletal remains back to life, put some modern clothes on them, and put them on a bus, you could not tell them apart from anyone else on that bus.

Where did modern humans originate on this planet?

What is the "trail" that modern humans took when spreading out to the different continents on this planet?

Figuring out the answers to these questions is somewhat like doing the "Trails of the Cross-Country Hikers" problem. Here, you will follow five geographical/racial groups instead of five hikers.

You must calculate when these groups separated from each other, and relate those times of separation to a world map. From that information, you will be able to tell a story of modern human migration and origin.

? QUESTION

Answer the following questions using the information you've learned about the Mitochondrial DNA Clock.

1. When genetic researchers compare the DNA of *Northern Asian* populations versus *Native American* populations, they find a 0.07% difference. The "Mitochondrial Clock" formula tells us that there has been about __________ years since the separation of these two groups.
2. When comparing the DNA of *Northern Asian* and *European* populations, they discover a 0.10% difference. The "Mitochondrial Clock" formula tells us that there has been about __________ years since the separation of these groups.
3. When comparing the DNA of *Indonesian* peoples to either the *European* group or the *Northern Asians*, they find a 0.12% difference. The "Mitochondrial Clock" formula tells us that there has been about __________ years since separation.
4. When researchers compare the DNA of *African* populations to any other group, they find a 0.20% difference. The "Mitochondrial Clock" formula tells us that there has been about __________ years since the separation of Africans from other groups.

Materials

- Refer to the "Time Trails of Modern Humans" worksheet at the end of this lab.
- Refer to the "World Map" worksheet at the end of this lab.

Procedure

[1] Use the "years since separation" information you just calculated on the previous page for the different geographical/racial groups, and . . .

[2] Fill out the map of the "Time Trails of Modern Humans" on your worksheet. Show when each group split off from the original group (just like you did for the Cross-Country Hikers).

[3] Using the "World Map" worksheet and the "Time Trails" map you just completed, draw your best interpretation of the trail taken by humans as they separated and spread out on this planet. Title your map "Migrations of Modern Humans." Show where the original population started and show the sequences of where they separated and where they went after that. There may be more than one possible trail based on the genetic information presented in this Exercise.

In Conclusion

Below is a table of information about the general anthropological evidence of the earliest settlements of modern humans that have been excavated so far.

Region	Time of the Earliest Settlements
Americas	less than 35,000 years ago
Indonesia	50,000 years ago
Europe	35,000 years ago
Asia	60,000 years ago
North Africa	100,000 years ago

? QUESTION

How does your story, illustrated by the Migration Map, compare to the anthropological evidence?

If you think that all of this is hard to figure out, remember that many researchers have spent their careers just to give you the small amount of information you are working with in this lab. There is a great deal more to know that will make the picture more accurate. Anthropology is one of the sciences of biology. Archeology is another. If you are enjoying what you are learning in this lab, consider studying in these other fields of biology as well.

EXERCISE #4

"The Story of the Really Old Human-Like People"

Dating back to a time before *Homo sapiens* (modern humans) appeared on this earth, there are a number of very old fossils that give clues about early humanoid evolution and migration.

1. **Australopithecus**
   The Australopithecines are the oldest humanoid fossils to be found. They were smaller than the *Homo sapiens* species, standing approximately four feet high, walked slightly bent over compared to modern man, and they had a smaller brain. Other skeletal differences suggest that they may have lived a somewhat varied kind of life from each other and from *Homo sapiens.*

2. **Homo habilis**
   The *Homo habilis* group was a tool maker. They had a somewhat larger brain than *Australopithecus*, stood more erect, and were slightly bigger in overall skeletal size than *Australopithecus*. Compared to *Homo erectus*, who also made tools, the *Homo habilis* group had a smaller brain capacity.

3. **Homo erectus**
   These fossil groups, designated as Peking, Java, and Heidelberg, stood fully erect (for which they are named), and had the same general height and brain size as modern man. These fossils have been found with more artifacts, telling us something about how they lived. Apparently, they were very much like *Homo sapiens*, but perhaps had less ability to speak because there were some throat structure differences.

The story of human evolution is very old indeed.

Materials

- Refer to the second World Map worksheet at the end of this lab.

Procedure

[1] Using the information provided below and the World Map worksheet, put a dot on the map for each of the groups. See if the dots suggest a story of how these human-like species might be related to each other, and where they developed as they separated and migrated.

[2] We have no ability with current technology to get DNA samples from very old fossils, so there is no Mitochondrial Clock to give us data; we can only tell the age of the fossils themselves. Use the fossil dates as approximate dates of separation, and put them on your map story. Title this map, "Trails of Really Old Human-Like People."

Information

Fossils	Age of Fossils	Where Found
Ardipithecus ramidus	4.5 million years	Africa
Australopithecus anamensis	4.2 million years	Africa
Australopithecus afarensis	3.5 million years	Africa
Australopithecus africanus	3.5 million years	Africa
Homo habilis	2.5 million years	Africa
Homo erectus	1.8 million years	Africa
Homo erectus (Peking)	500,000 years	China
Homo erectus (Java)	1.5 million years	Southeast Asia
Homo erectus (Heidelberg)	900,000 years	Germany

? QUESTION

1. Based on available fossil evidence, where did the various old human-like people originate?

2. Which group first migrated out of the continent of origin?

   How long ago? ___________

   How far did that group spread over the world?

3. Was there another group that later came out of the continent of origin? (See Exercise #3.)

   Which species was that group? ___________

   When did the migration of the new species happen? ___________

   Indicate that event on your map.

In Conclusion

Show your completed map to your instructor.

"TRAILS of the CROSS-COUNTRY HIKERS"

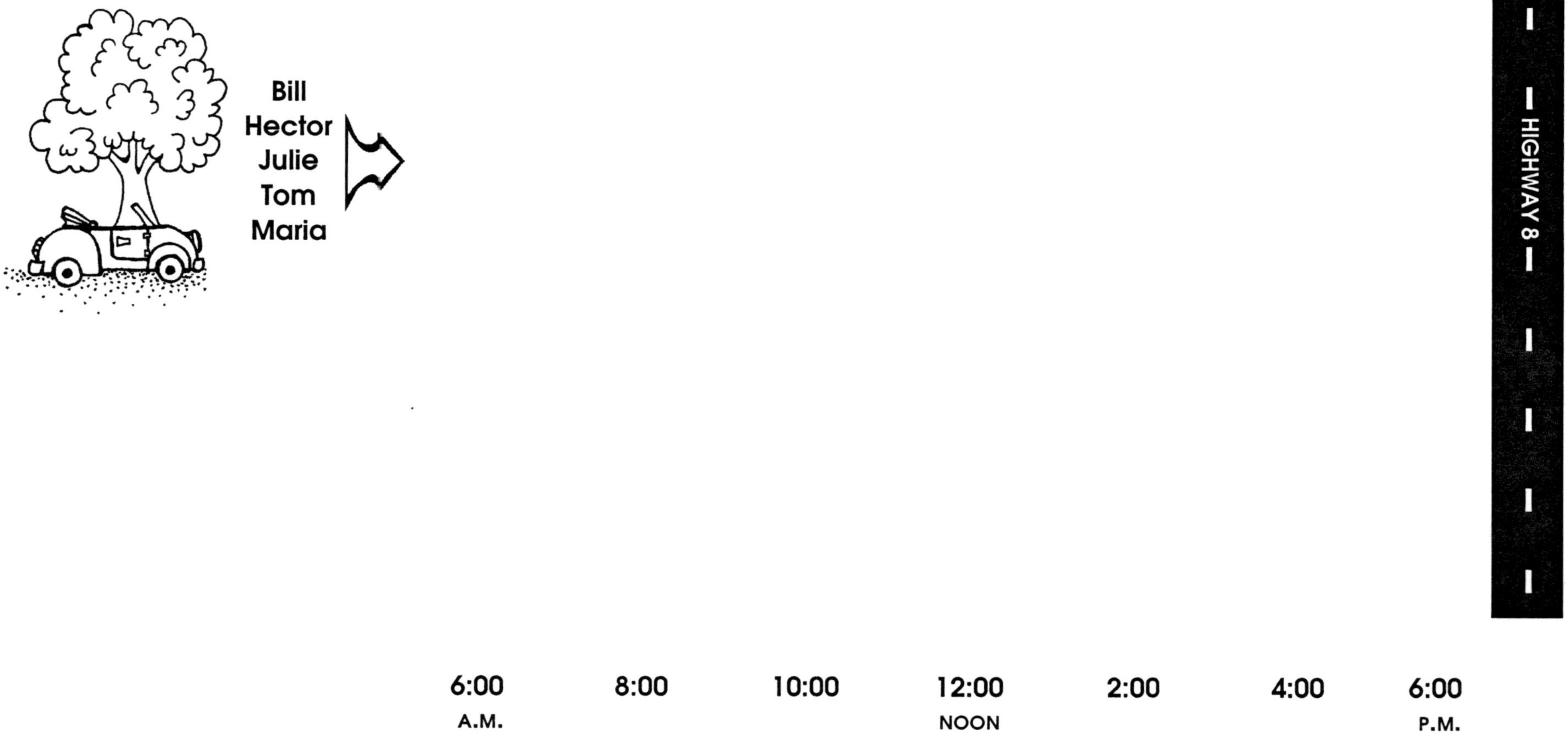

Bill
Hector
Julie
Tom
Maria

HIGHWAY 8

6:00 A.M. | 8:00 | 10:00 | 12:00 NOON | 2:00 | 4:00 | 6:00 P.M.

“TIME TRAILS of MODERN HUMANS”

Early
Homo
sapiens

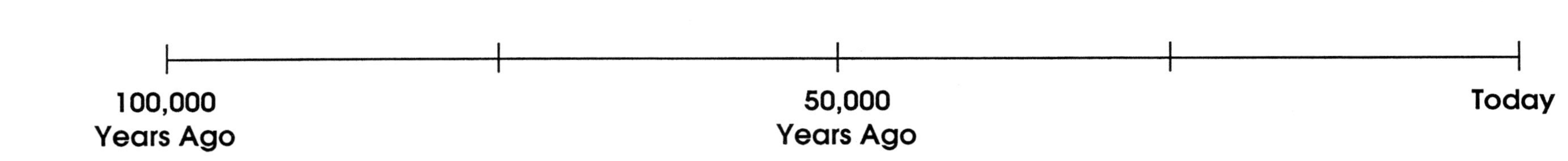

WORLD MAP

"   "

WORLD MAP

“ ______________________________________ ”

Surrounded by Microbes

It has been about 80 years since this country created a free and open public education beyond the 8th grade. The early high school curriculum offered the first opportunity for all citizens to learn about science. And, as part of that curriculum, biology students were taught about the amazing diversity of known lifeforms. Today's college biology teachers continue with that tradition, and this lab begins our general survey of organisms on this planet.

The Five Kingdoms

The similarities and differences between organisms are used by biologists to establish probable evolutionary relationships. Using this method, they have placed each lifeform into one of the ***five biological kingdoms***. A general evolutionary timeline for these groups is shown in the illustration below.

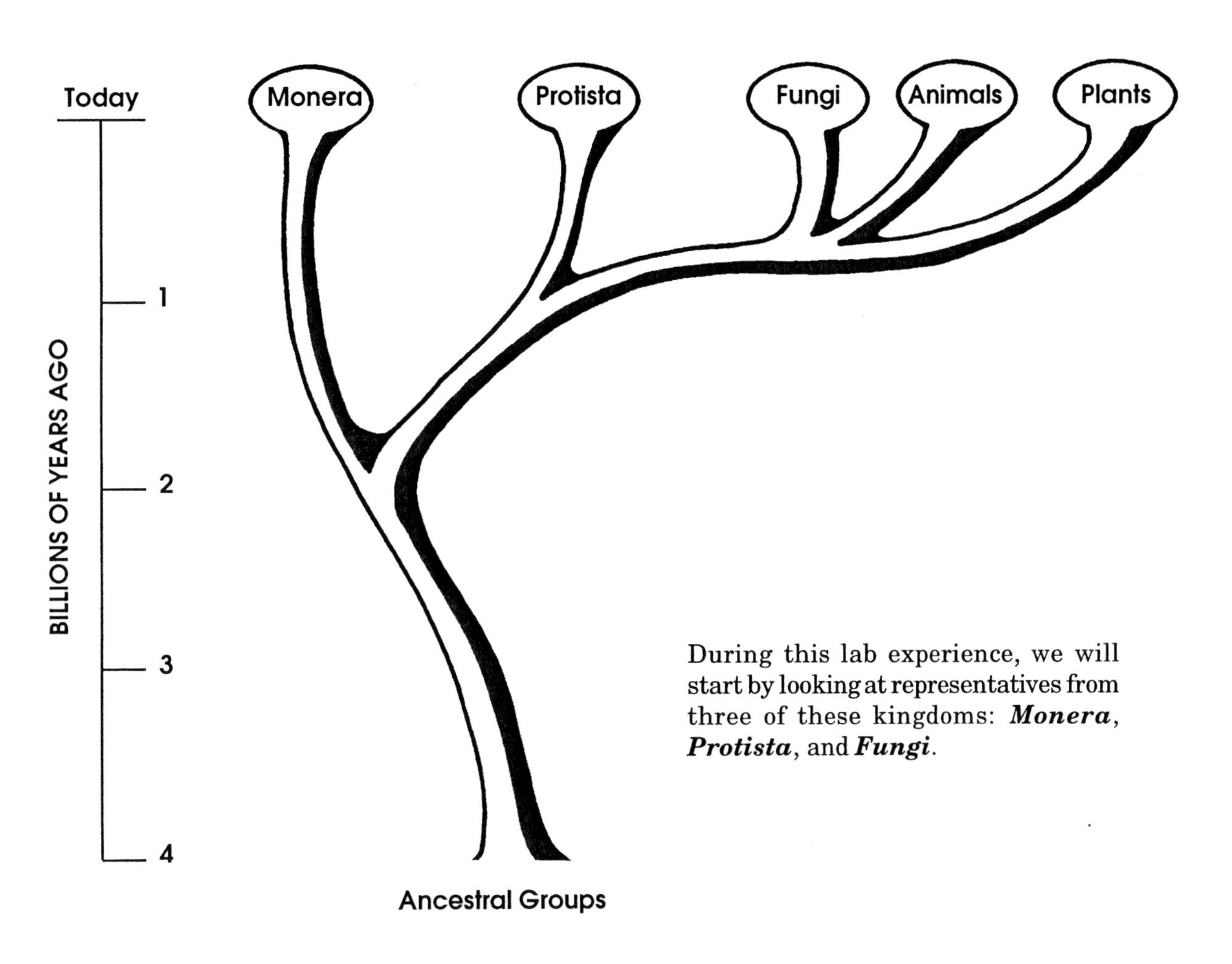

During this lab experience, we will start by looking at representatives from three of these kingdoms: ***Monera***, ***Protista***, and ***Fungi***.

Exercise #1 "Bacteria in Our Environment" 220
Exercise #2 "How Do You Tell Which Kingdom an Organism is In?" 222
Exercise #3 "Bacteria" 223
Exercise #4 "Cyanobacteria and the True Algae" 224
Exercise #5 "Protozoans" 226
Exercise #6 "Bread Mold and Mushrooms" 226
Exercise #7 "Basic Characteristics of Monera, Protista, and Fungi" 228
Exercise #8 "Lichens" 230

EXERCISE #1

"Bacteria in Our Environment"

This Exercise must be completed during the week prior to the "Surrounded by Microbes" Lab so that there will be enough time for the bacterial colonies to grow and be seen.

Where Are the Bacteria in Our Environment and How Many Are There?

Today your class will pose one or two questions related to bacteria that live with you on this campus. Then you will design a sampling procedure that answers your questions.

Procedure

[1] Divide the class into three large groups.

[2] Each group has 5 minutes to discuss interesting bacteria questions that you think are possible to answer by sampling with agar plates.

[3] Decide which question is the most interesting to your group.

[4] Each group will present its favorite question to the entire lab class.

[5] As a class you will decide the final experimental questions.

[6] You will have sterile agar plates for growing bacteria.

[7] As a class, decide how you will use these plates in your experimental design.

Agar Plates

The special glassware used for agar sampling are called Petri dishes. The top of the Petri dish loosely covers the bottom. This permits oxygen gas in the air to enter the plate without allowing contamination by spores and other microbes in the environment.

Agar has been poured into the bottom of the Petri dish. Agar is a derivative of a red algae called *agar-agar*. It is a solidifying agent. The true nourishment is added to the agar, and consists of partially digested protein, carbohydrates, minerals, and other nutrients.

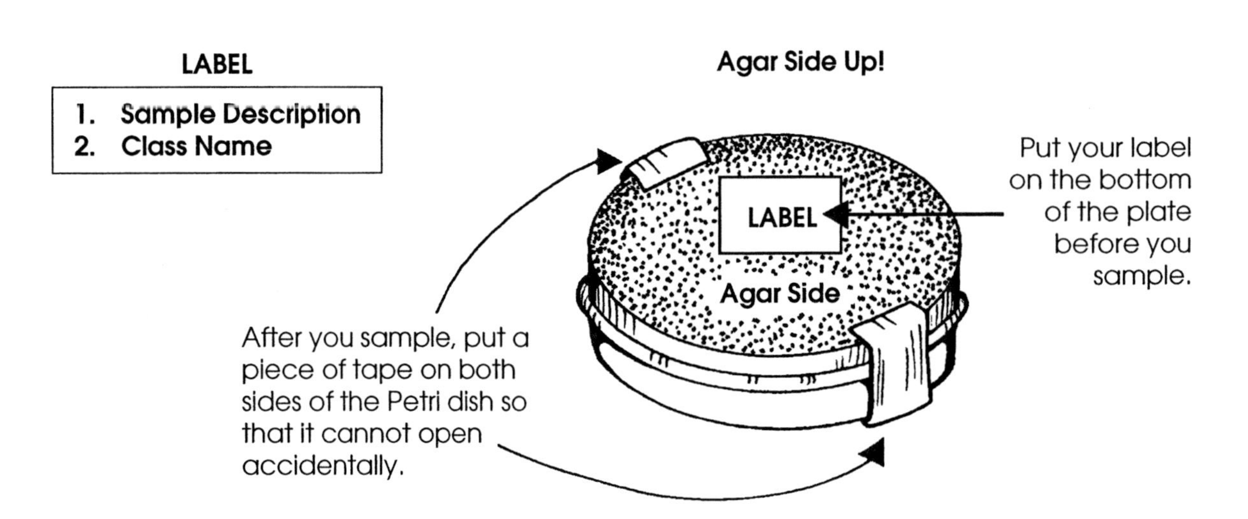

Important!

It is important that the label be placed on the bottom half of the Petri dish (the one containing the nutrient agar). The Petri dishes will be incubated in an inverted (bottoms up) position. When the label is placed on the bottom of the dish, it will be easily identified and will be on the same side as the bacteria.

Another reason that the Petri dish is inverted during incubation is to prevent drops of moisture from collecting on the agar and permitting the bacteria to swim out into your environment.

Sampling for Bacteria

Do not obtain sampling specimens from your mouth, throat, groin, nose, or any area of your body except for healthy skin.

Be very careful when carrying out this experiment. The environment contains a large variety of different microbes (bacteria and fungi). Some of these microbes may be ***pathogenic*** (disease causing).

Procedure

[1] You have labeled your dishes before sampling. Check your label for complete information. Print the Sample Description, and your Class Name.

[2] A sterile moistened swab will be used to sample the environment and then to inoculate the sterile agar in the Petri dish.

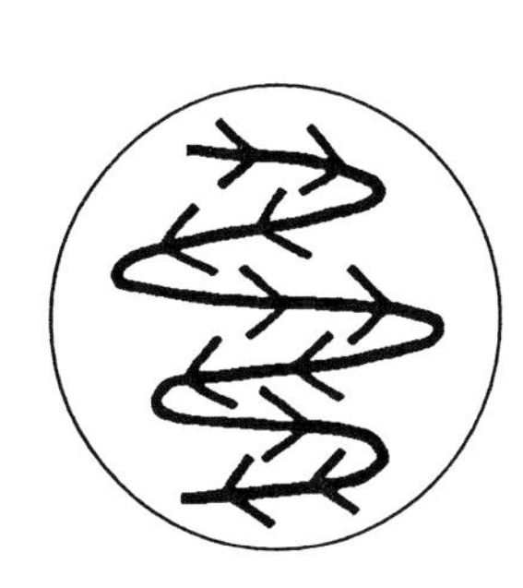

[3] *Swab the plate in this fashion:*

Gently roll the swab across the surface of the agar. *Do not dig!*

Skimming the surface as you roll the swab back and forth will do the trick.

[4] Tape the two sides of the Petri dish together.

[5] Put a piece of tape around the entire class's agar plates so they don't get mixed up with those from other lab classes.

[6] Place the *inverted plates* in the tub labeled "To Be Incubated."

Be sure that your plates are all inverted in the same direction (agar side up). The Petri dishes will be incubated during the next week.

Important!

After you obtain your sample and inoculate your agar plate, discard the used swabs in the test tube or envelope provided. Then place the test tube or envelope in the laboratory area marked:

Danger—Contaminated Materials to be Autoclaved

A special piece of equipment called an *autoclave* will be used to decontaminate these materials by using a process that involves steam sterilization under pressure.

EXERCISE #2

"How Do You Tell Which Kingdom an Organism is In?"

By the end of this lab you will know three very basic characteristics for the majority of organisms in each of the three kingdoms of ***Monera***, ***Protista***, and ***Fungi***.

First Characteristic

All cells can be divided into one of two types.

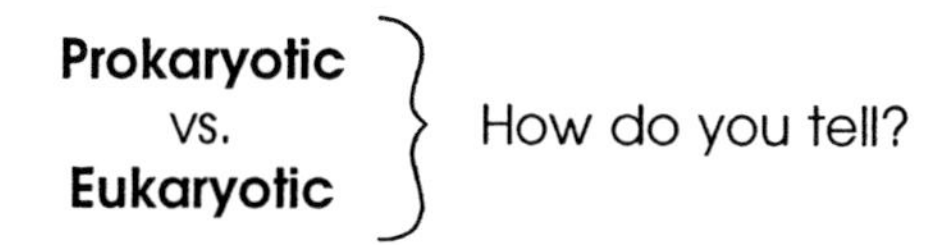

Information

- A prokaryotic cell is very small compared to a eukaryotic cell.
- A prokaryotic cell does not have a nucleus.
- A prokaryotic cell does not have any chloroplasts. If the cell does photosynthesis, then the light-catching pigments are spread throughout the cell water and not in plastids.
- A eukaryotic cell has many organelles (including a nucleus, chloroplasts, mitochondria, and others).
- A eukaryotic cell is big compared to a prokaryotic cell.
- By the way, the word ***prokaryotic*** means "before nucleus," and the word ***eukaryotic*** means "true nucleus."

? QUESTION

1. Remember the onion cells that you saw in an earlier lab? What kind of cells were they—prokaryotic or eukaryotic? ____________________
2. How do you know?

Second Characteristic

All living organisms are composed of one or more cells.

- ***Unicellular*** means that the organism is a single cell, or it is a small chain or cluster of identical cells.
- ***Multicellular*** means that the organism is made of more than one kind of cell type.

? QUESTION

1. The plant leaves that you saw in an earlier lab are part of what type of living organism—unicellular or multicellular? ____________________
2. How do you know?

Third Characteristic

All living organisms get their energy for their life processes from a specific source.

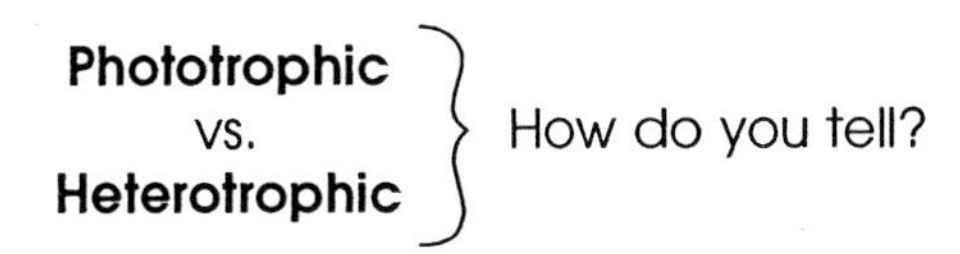

- The word ***phototrophic*** means "light-eater," and the word ***heterotrophic*** means "other-eater."
- A phototrophic organism will either have chloroplasts or photosynthetic pigments spread throughout the cytoplasm of the cell. These organisms require light for their survival.
- A heterotrophic organism does not have photosynthetic pigments or chloroplasts. It cannot do photosynthesis. This type of organism must eat some other organism or the by-products of that organism to get its energy. Sometimes, this category of organism is called a *chemotroph* because it gets its energy from chemicals.

If you go on to advanced biology classes, then you will study a more detailed distinction in energy categories. For this class, we will combine all non-phototrophs under the one word *heterotroph.*

EXERCISE #3

"Bacteria"

Bacteria are the most numerous organisms on this planet. They exhibit more variety in the way they get their energy than any other group on Earth. They are an essential organism for the recycling of dead plants and animals.

Last week, various parts of the environment were sampled for different kinds of bacteria, and Petri dishes were inoculated.

Materials

- The agar plates from last week's sampling experiment.

DANGER! Do not open the Petri dishes under any circumstances!

The bacteria and molds inside could be *very* dangerous to you in these high concentrations. These agar plates were kept in the dark ever since last week. This should help you decide whether bacteria are heterotrophic or phototrophic.

Procedure

1. Determine the answer to your experimental question.
2. You should be able to tell the difference between bacterial colonies and fungus colonies. (Fungus looks "fuzzy" and bacteria usually don't.)
3. Look at your bacterial colonies and describe their growth (color and shape).
4. When you are finished, *put all the agar plates in the container labeled* "To Be Sterilized."
5. *Wash your hands!*

Materials

- A compound microscope.
- A prepared microscope slide of bacterial shapes.

Procedure

[1] Examine the three different samples on this slide and determine the three basic shapes of bacteria. (The bacteria may be grouped together, so look at the individual cells.)

[2] The 3 shapes of bacteria are: ______________________

____

____

[3] *Return the slide when you are finished.*

Basic Characteristics

You should have enough information to determine the three basic characteristics of bacteria. (circle your choices)

Prokaryotic	Unicellular	Phototrophic
vs.	vs.	vs.
Eukaryotic	Multicellular	Heterotrophic

EXERCISE #4

"Cyanobacteria and the True Algae"

The ***cyanobacteria*** and the ***true algae*** are very different kinds of organisms, but *both* of them do photosynthesis.

The objective of this Exercise is to determine the difference between the algae and the cyanobacteria. You will need to work with another microscope group in order to make comparisons of the two kinds of organisms.

Materials

- One microscope group is to get a prepared microscope slide of *Spirogyra*.
- The other group is to get a prepared microscope slide of *Nostoc*.

Nostoc	Spirogyra

Procedure

1. Work as a team with another group at your table. Each microscope group is to find their organism, and put it under the *same* high magnification.
2. These two organisms grow in chains of identical cells.
3. Look back and forth between the microscope views until the most basic differences become obvious. (Ignore the color of the stain.)
4. ***Nostoc*** is a cyanobacteria and ***Spirogyra*** is a true algae.
5. **Remember:** You are going to have to answer the three basic characteristics questions about both of these organisms. Make some quick notes and a sketch of both organisms.

   Ask your instructor about the ribbon-shaped structure inside the *Spirogyra* cells.
6. *Return the slides to where you found them.*

Basic Characteristics

Talk it over with your group and decide which features represent each group. (circle your choices)

Cyanobacteria	Prokaryotic vs. Eukaryotic	Unicellular vs. Multicellular	Phototrophic vs. Heterotrophic
True Algae	Prokaryotic vs. Eukaryotic	Unicellular vs. Multicellular	Phototrophic vs. Heterotrophic

Test Your Skills!

Materials

- Use your compound microscope.
- Make a slide using a sample from the “Live Mixed Algae” jar. (One drop of the green stuff will do it.)

Procedure

1. Find two different cyanobacteria in the sample. *Show them to your instructor.* What characteristics tell you that they are cyanobacteria?

2. Find two different true algae in the sample. *Show them to your instructor.* What characteristics tell you that they are true algae?

EXERCISE #5

"Protozoans"

We've seen the Protozoans before in previous labs. Now, it's time for you to determine their basic characteristics.

Materials

- Make a slide using a sample from the "Live Mixed Protozoa" jar. (One drop from the bottom of the jar.)

Procedure

1. Find at least two different kinds of Protozoans.
2. *Show them to your instructor.*
3. You may need to use Protoslo® to slow the swimming organisms enough to follow them with your microscope. Ask your instructor how to use Protoslo®.
4. You may also find some attached Protozoans in the sample.
5. We have a booklet that will give you the names of some Protozoans you are seeing. Try to identify your Protozoans.

Basic Characteristics

Talk it over with your group and decide which features represent Protozoans. (circle your choices)

Prokaryotic	Unicellular	Phototrophic
vs.	vs.	vs.
Eukaryotic	Multicellular	Heterotrophic

EXERCISE #6

"Bread Mold and Mushrooms"

Bread mold and mushrooms are strange beasts indeed! During this Exercise you will continue to determine the basic characteristics, and you will discover some interesting structural features of this group, along with one of its strange methods of reproduction.

Spore Reproduction

Some organisms have a special type of reproduction to get through very tough times in their environment. These organisms produce ***spores***, which are small cells with a very thick protective coating. (They look like little balls.) When the environment changes, these spores will grow into the next stage in the life cycle of these organisms.

Bread Mold

Procedure

1. Look at the bread mold samples growing on the bread.
2. Make observations about how you think it is "making a living" and talk this over with your group. What are your conclusions?

Materials

- Go to the bread mold cultures growing on agar plates, and cut out a small part ($\frac{1}{4}$" x $\frac{1}{4}$") of the organism, and put it on a slide (no coverslip).

Procedure

1. Examine this organism with the *dissecting microscope* and make observations about its basic characteristics.
2. Find those "little balls" called *spores*.
3. Draw a simple sketch of the structure of the bread mold.
4. Make a wet mount slide of the balls and a few of the filaments. Use a compound microscope, and examine under higher magnification.
5. Make a simple sketch of what you see. (The filaments are used for feeding, and the very little balls are spores.)

Bread Mold	Filaments and Spores

Mushrooms

Materials

- Remove a small piece of the brown-colored tissue from under the mushroom cap.

Procedure

1. Make a slide of the mushroom tissue and determine what that brown-colored tissue is.
2. Make a simple sketch of what you see.

Mushroom

? QUESTION

1. How do you think that this mushroom organism gets its energy for life? Talk it over with your group. What are your conclusions?

2. Do you think that this organism does photosynthesis? Why or why not?

Basic Characteristics

Talk with your group and decide what features represent the *bread mold* and *mushroom* organisms. (circle your choices)

Prokaryotic	Unicellular	Phototrophic
vs.	vs.	vs.
Eukaryotic	Multicellular	Heterotrophic

EXERCISE #7

"Basic Characteristics of Monera, Protista, and Fungi"

We designed this lab to include organisms that would define the basic characteristics of each of the *three kingdoms (Monera, Protista, and Fungi).* You have answered the three characteristics questions about each of the groups presented.

? QUESTION

1. Which ***cell type*** do you think is the most complex? (circle your choice)

   Prokaryotic or Eukaryotic

2. Which ***cell arrangement*** do you think is the most complex? (circle your choice)

   Unicellular or Multicellular

Procedure

[1] Using your answers to the above questions, list the four possible combinations of cell type and cell arrangement, placing them in order from *simple* to *more complex.* (Fill in the Kingdoms after reading Procedure [3]).

Simple	Cell Type		Cell Arrangement	Kingdom
	____________	&	________________	______________
	____________	&	________________	______________
	____________	&	________________	______________
More Complex	____________	&	________________	______________

[2] Refer to the descriptions of basic characteristics that you chose in earlier Exercises for each of the subgroups examined in this lab. You should see that only one of the four possible combinations above is not represented. Cross out that combination of traits—it doesn't exist!

[3] You are left with three combinations. The order of those combinations from simple to more complex represents the evolution of:

Monera ⟶ *Protista* ⟶ *Fungi.*

Put those kingdom names on the correct lines in Procedure [1].

[4] Place the following five subgroups under the correct kingdom in the table below, and list whether that subgroup is a *heterotroph* or a *phototroph.*

Bacteria
Cyanobacteria
True Algae
Protozoans
Bread Mold and Mushrooms

Kingdom	Subgroup	Method of Energy
Monera		
Protista		
Fungi		

In Conclusion

You should be able to name the subgroups in each kingdom that you studied, and list the three basic characteristics of each of the three kingdoms.

Your textbook or lecture class may define the basic characteristics of the kingdoms in more detail than this lab did. Also, your textbook or lecture class may point out some of the exceptions to the generalizations presented in this lab.

EXERCISE #8

"Lichens"

A little more than 100 years ago, botanists discovered that ***lichens***, a strange plant-like organism, are actually a cooperative relationship between *two* very different organisms. One of the organisms provides the basic structural framework of the lichen, and the other organism does the energy collecting (photosynthesis). Lichens are often found growing on rocks, tree trunks, or fallen branches on the forest floor.

The metabolic lives of these two very distantly related species are intimately intertwined in many different ways. Lichens exhibit a high resistance to unfavorable environments including the cold polar regions, barren mountain rocks far above the timberline, and fully exposed rocks in the hottest of desert areas.

There are several different forms of lichens, and we have some in the lab today: a crust-like form growing very close to the rock's surface, a leaf-like form, a shrub-like or hair-like form, and a flat-sheet form with tiny cups coming out of the surface.

Materials

- A probe.
- A small sample of lichen from the station marked "Lichen Sample."

Procedure

[1] Crush a small amount of lichen with a probe in a drop of water on the slide. Cover and examine it with your compound microscope.

[2] Search the slide for evidence of two very different kinds of organisms. *Show your instructor* the two organisms, and answer the next six questions based on your sample.

? QUESTION

1. Which organism does the photosynthesis?
2. Draw a quick sketch of the organism.
3. What is your evidence that this organism does the photosynthesis?

Photosynthetic Part

4. Which organism makes up most of the body structure of this lichen?

Structural Part

5. Draw a quick sketch of this organism.
6. What is your evidence that this organism does *not* do photosynthesis?

In Conclusion

We hope you've enjoyed discovering some of the many organisms that inhabit the kingdoms of Monera, Protista, and Fungi. The study of these organisms is what *microbiology* is all about.

Mosses and Ferns

This week's lab is a continuation of the evolutionary story of the plants on this planet. Last week we looked at algae, and saw that they started as one-celled organisms and developed into more complex algae. The advanced types of algae were the most probable ancestors of land plants.

Exercise #1 "Plant Life Cycles" 231
Exercise #2 "Which Works Better on Land—Haploid or Diploid?" 233
Exercise #3 "The Moss Plant" 237
Exercise #4 "The Fern Plant" 241

EXERCISE # 1

"Plant Life Cycles"

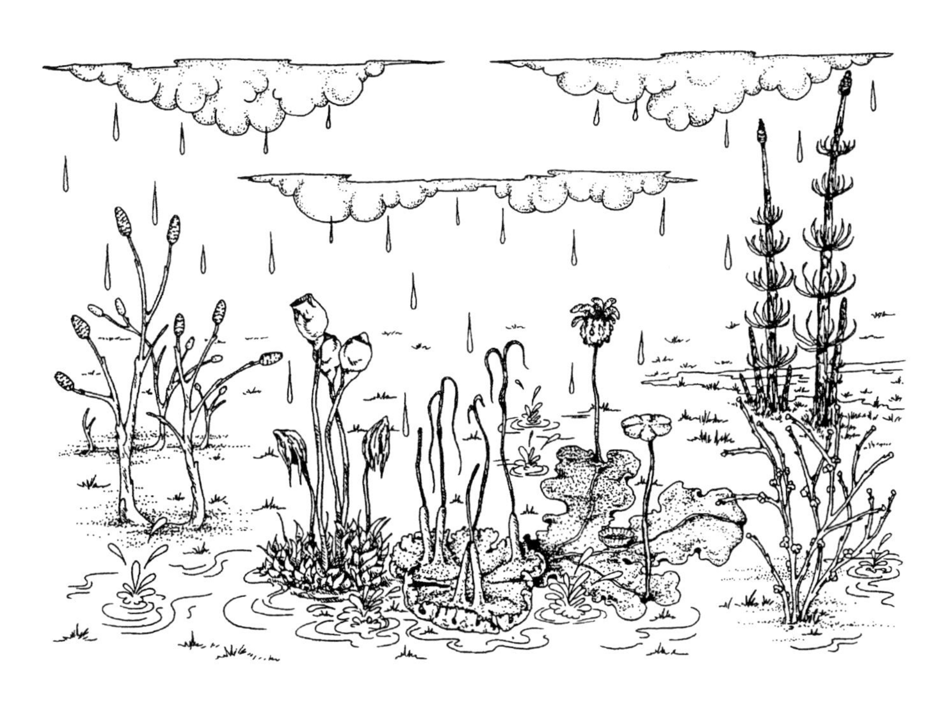

It was the time of Rhyniophytes and whisk ferns, club mosses, and horsetails. And the world was a swampy place. Ah, those were the days!

There was a lot of swampy land about 400 million years ago, and natural selection picked traits that allowed aquatic plants to move into that environment. Land would have been an excellent new opportunity for any plant. It provided "unlimited access" to sunlight for photosynthesis.

The two basic questions to understanding this evolutionary event are:

- *What are the traits that would be needed by land plants?* and
- *What traits did the algae already have before the land invasion?*

We will consider the traits needed on land as we look at ***mosses*** and ***ferns*** later in the lab. But, now we will begin with a basic understanding of the algae life cycle on which all land plant reproduction is based.

Algae Life Cycle

Before plants were capable of invading land, the algae had already evolved a unique life cycle called ***alternation of generations***. During this cycle, the plant alternated between a stage called ***gametophyte*** (which means "gamete plant") and a stage called ***sporophyte*** (which means "spore plant"). These two stages repeated one after the other, generation after generation.

Several advantages may be derived from this type of life cycle, but two aspects are of particular interest to the understanding of land plants.

- The *alternation of generations* allowed plants to reproduce by using two different methods—***spores*** and ***gametes***.
- One of the generations can live in a part of the environment (perhaps in shallow water) and the other generation can live somewhere else (perhaps in deeper water). Sometimes these two generations lived at different times of the year, due to different seasonal conditions. **Remember:** We are talking about algae that lived in the water.

For Practice

1. Put the names of the correct stage inside the *Life Cycle* boxes below.
2. Then, complete the *Life Cycle* diagram by writing the word *"spores"* above the arrows that indicate the stage that produces them. Also, put the word *"gametes"* above the other appropriate arrows, indicating the stage that produces eggs and sperm.
3. Don't worry about the lines underneath the boxes; they'll come later.

Alternation of Generations

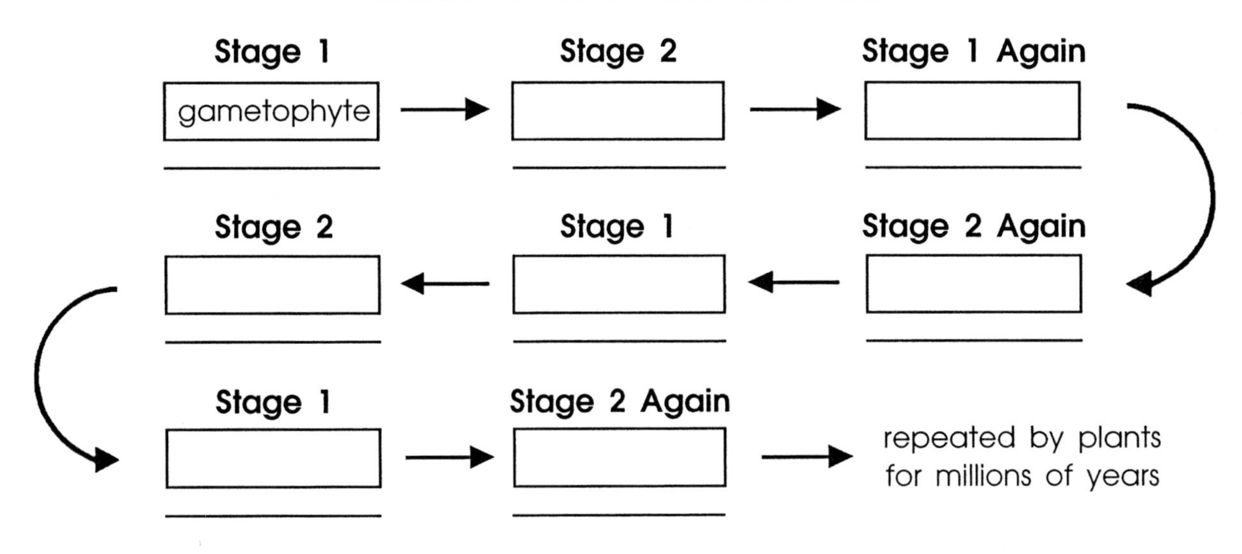

Information

- The gametophyte stage is *haploid.* This means that the cells have only *one* set of chromosomes.
- The sporophyte stage is *diploid*. This means that the cells have *two* sets of chromosomes.

? QUESTION

1. Which stage in the life cycle of algae produces eggs and sperm?

2. How many sets of chromosomes does an egg have? ______________
3. How many sets of chromosomes does a sperm have? ______________
4. When the egg and sperm unite, we are back to . . . (circle your answer)
   - **a.** one set of chromosomes
   - **b.** two sets of chromosomes

5. The fertilized egg will grow into one . . . (circle your answer)
   - **a.** gametophyte plant
   - **b.** sporophyte plant

Procedure

Go back to your *Life Cycle* diagram, and write the words *"diploid"* and *"haploid"* on the line under the appropriate stages (boxes).

Summary

We have seen that the algae have two stages in their life cycle—the gametophyte stage and the sporophyte stage. One of these stages is haploid and the other stage is diploid. The algae used these two stages for different environmental and reproductive purposes.

Now, the question is: ***Which of these two stages would prove to be best able to adapt to a harsher land environment?*** Any thoughts? Well, let's proceed to Exercise #2.

EXERCISE #2

"Which Works Better on Land—Haploid or Diploid?"

Genes and Mutations

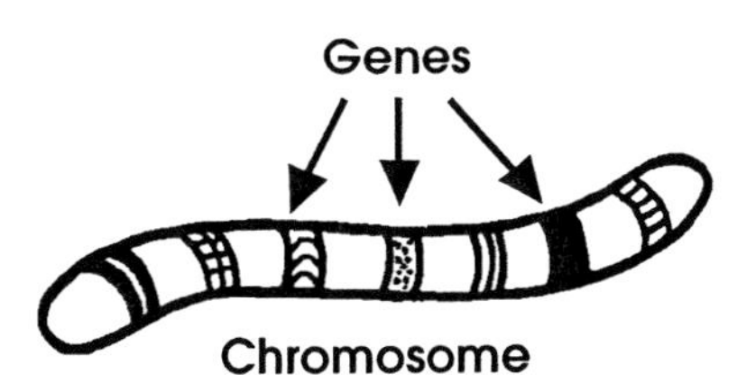

Plants have many genes, and these genes are carried on the chromosomes. Each gene is involved with the making of a specific protein. Each protein is used in one of two ways:

- A protein might be part of a plant structure.
- A protein might be an enzyme.

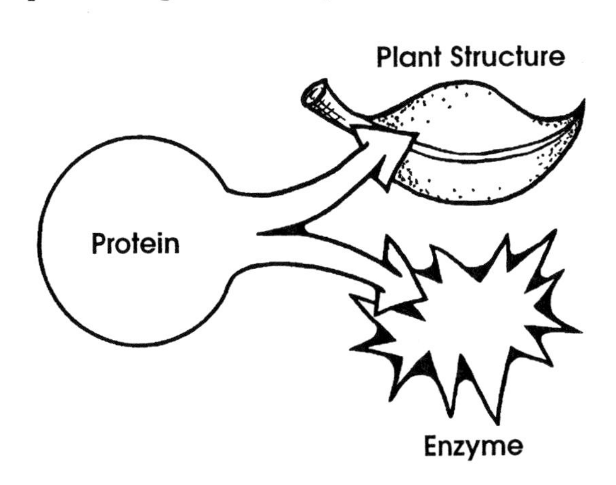

? QUESTION What will be lost if a gene is destroyed? ____________ or ____________

Information

A plant has a certain number of chromosomes (the number depends on the species) and each chromosome has many genes. Each gene produces a plant trait. The more genes that a plant has, the more traits it has. Presumably, more traits give the plant a better chance to adapt to dry land conditions.

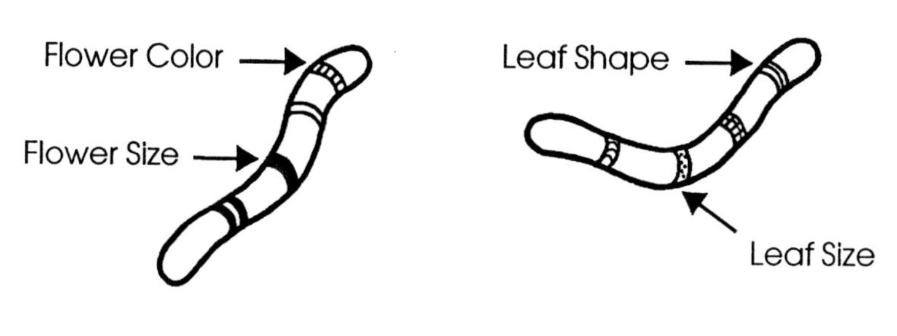

A ***mutation*** often results in a destroyed gene. A gene can be destroyed by radiation, chemicals in the environment, or other spontaneous events that are surprisingly common on this planet.

? QUESTION Are mutations almost always a good thing, or a bad thing? Explain your answer.

Procedure You are ready for the next two puzzles comparing the limitations and advantages of haploid and diploid plants.

CASE 1

"Limitations of the Haploid"

Information

- Imagine that you are a plant.
- You have a single set of chromosomes. (That is, you are *haploid*. **Remember:** Haploid means that you have only *one copy* of each different chromosome in your species.)
- Assume that mutation events occur at a rate of one mutation for every 1000 genes. The probability of a mutation is, therefore, $\frac{1}{1000}$. (The actual probability of mutations is dependent on many factors, and the value of $\frac{1}{1000}$ has been chosen so that you can make comparison calculations during this lab.)
- Assume that every gene (every trait) that you have is necessary for your survival as a plant.
- Assume it takes *more than* 10,000 genes to be a *complex* plant with lots of structures and enzymes.

? QUESTION

1. Assuming the previous information, what would be the safest number of genes for you if you lived as a haploid plant? (circle your choice)

a. 1000

b. more than 1000

c. less than 1000

Check your answer to question #1 with your instructor before going on. This "clue" must be properly understood or you will spend the rest of this case chasing a "red herring"!

2. Describe your basic structure as a haploid plant using the conclusion in question #1. (circle your choice)

a. You are a complex plant (lots of structures).

b. You are a simple plant (fewer structures).

3. If you were an algae plant that was starting to move onto land through the process of evolution, then what would be the most likely requirement to ensure your survival in a land environment? (circle your choice)

a. Don't change any genes.

b. Add more genes (new traits).

4. You should have decided that more genes enable you to develop more traits to help you survive on land. But, you are still *haploid!* What risk would you be taking if you were to add many more genes?

5. Some of the algae plants that evolved onto land developed an emphasis on the haploid stage. These plants could not add very many more genes without risking extinction. Therefore, the algae plants that used this strategy were restricted to being . . . (circle your choice)

a. simple plants (few structures).

b. complex plants (many structures).

In Conclusion

Some plants did move onto land with an emphasis on their *gametophyte* stage. From what you have discovered, you should be able to deduce what their structure would be like.

Now you know why we called this first puzzle, "Limitations of the Haploid."

CASE 2

"Advantages of the Diploid"

All of the assumptions of Case 1 are also true for Case 2 *except* that now you are a plant with a double set of chromosomes. (That is, you are *diploid.*) Diploid means you have *two copies* of each different chromosome in your species. You have two of every gene.

? QUESTION

1. Consider what would happen to a diploid plant when genes are destroyed by mutation. How many genes have to be destroyed in a diploid plant for one trait to be lost? ________________

2. Case 1 assumed that the probability of a gene mutation was one chance for every 1000 genes. A rule in statistics states that:

   > *"If you want to know the probability of two events happening together, then* ***multiply*** *their separate probabilities."*

   Based on your answer to the previous question, as a diploid plant, what is the probability of your losing both genes for the same trait because of destruction through mutation? ________________

3. Consider your answer to question #2, and that mutation does destroy genes. How many genes can a diploid plant "safely" have? (circle your choice)

   a. about the same number as a haploid plant

   b. fewer genes than a haploid plant

   c. many more genes than a haploid plant

4. Based on your conclusions in the previous questions, describe yourself as a diploid plant. (circle your choice)

   a. You are a complex plant (lots of structures).

   b. You are a simple plant (fewer structures).

5. How does your description above relate to your chances for doing well in a land environment? Explain.

In Conclusion

Some plants did move onto land with an emphasis on their *sporophyte* stage. You should be able to deduce what their structure would be like.

Now you know why we call this puzzle, "Advantages of the Diploid." The time has come to test your detective skill in Exercises #3 and #4 with the moss plant and the fern plant.

EXERCISE # 3

"The Moss Plant"

During this Exercise you will figure out the life cycle of the moss plant, and you will identify some main features, adaptations, and limitations of the plant.

When you see green moss growing on a river bank or on a fallen tree, you are actually seeing only one of the two alternating generations in its life cycle. You must look closely at this generation to determine which it is—***gametophyte*** or ***sporophyte***.

Materials

- A dissecting microscope.
- A moss plant.

Observations

1. Look at the green parts of the moss plant under the dissecting microscope. Moss leaves don't have veins for transporting nutrients or stomata for regulating air flow like typical plants. Does this suggest that moss leaves are as complex as tree leaves? ________

   How many "leaf genes" would the moss plant have compared to a tree (many or few)? ________

2. Does the moss plant have much root? ________

   Assuming that large roots require more genes than small roots, how many "root genes" would the moss have (many or few)? ________

3. In general, being bigger requires more genes than staying small. How many "size related genes" would the moss plant have (many or few)?

   ____

4. By looking at the general features of the moss plant you should be able to conclude that it has (many or few) ________ genes for land traits.

? QUESTION

In Exercise #2 you concluded that being haploid restricts the plant to being simple, whereas being diploid allows for much more complex structures. Now you are ready for the important questions about *this generation* in the life cycle of the moss plant. (circle your choices)

1. This stage of your moss plant is . . .
   - **a.** a complex plant
   - **b.** a simple plant
2. Therefore, this stage of the moss plant is probably . . .
   - **a.** haploid
   - **b.** diploid
3. This means that you are looking at the . . .
   - **a.** *sporophyte* generation
   - **b.** *gametophyte* generation
4. Which means that this stage produces . . .
   - **a.** *gametes* (eggs and sperm)
   - **b.** *spores*
5. Draw a quick sketch of your moss plant.
6. *Save your moss plant.* You'll be needing it later.

Moss Plant

Moss Sex Organs

It is not possible to see the details of the moss sex organs in a living plant, so you'll need . . .

Materials

- A compound microscope.
- A microscope slide marked "Moss Archegonia."
- A microscope slide marked "Moss Antheridia."

Procedure

1. Don't worry about the fancy names of these sex organs. You will discover which one is male and which one is female as you continue.
2. Female organs usually have one large ***egg*** cell in the middle of the structure.
3. Male organs usually have a dozen or more small round cells inside called ***sperm***. After a rain the moss sperm must swim to the egg for fertilization. This means that moss reproduction depends on a *wet* environment.
4. Work with a partner so that one of you has the "Antheridia" slide, and the other has the "Archegonia" slide. Finding the female organ is a bit difficult. Looking back and forth between slides may help.
5. Search both slides of the moss *stem tips* and find a female organ. Show your instructor! Draw a quick sketch of what you see.
6. Search the other slide and find a male organ. Show your instructor! Draw a quick sketch of what you see.
7. These sex organs are considered to be very simple in structure, which shouldn't be a surprise to you. The sex organs are part of the gametophyte, which is haploid. And, by now, you realize that haploid forces a simple design (fewer genes). You can't make a fancy sex organ unless you are using the diploid stage to do so.

Swimming Sperm

Female

Male

Funny-Looking Stalk

Materials

- A fine-pointed probe and small tweezers.

Procedure

[1] Using your dissecting microscope, search for a funny-looking stalk coming out of the top of the moss plant you saved.

[2] Look at the top of the stalk. You should see a small capsule with very small balls attached to it. Draw a simple sketch of what you see.

[3] If there is a cap on the capsule, pry it off and look at what is inside.

[4] Make a wet mount of the small balls, and look at them with your compound microscope.

[5] Draw a simple sketch of the balls.

Dissecting Scope View	Compound Scope View

? QUESTION

1. What are the little round balls? ______________________

Hint: If you have trouble figuring out the answer, then look back at the *Life Cycle* in Exercise #1.

2. Now that you know what the little balls are, you also know that the "funny looking stalk" is actually the . . . (circle your choice)

a. *gametophyte* stage.

b. *sporophyte* stage.

Moss Ancestor

Materials

- A microscope slide marked "Moss Protonema."

Procedure

[1] Using your compound microscope, search the slide for some branching filaments. The very first product of a growing *spore* is a branching chain of cells called the ***protonema***.

Normally, spores are produced by mosses when the environment gets tough. When the environment gets better, these spores grow into the protonema, which eventually grows into the next generation gametophytes.

[2] Draw a quick sketch of the protonema.

[3] Compare the protonema structure to that of the filament algae (like Spirogyra) you saw in last week's lab. Explain how this is evidence that mosses may have evolved from the algae.

Protonema

[4] Diagram the *Life Cycle* of the moss plant below. Include: spores, gametes, sporophyte, and gametophyte. Draw a simple sketch of each generation as a visual reminder.

Life Cycle—Moss Plant

EXERCISE # 4

"The Fern Plant"

During this Exercise you will figure out the life cycle of the fern plant, and you will identify some main features, adaptations, and limitations of the plant.

When you look at a fern plant growing in a forest or in your yard, you are seeing only one of the generations in the fern life cycle. You must look closely at this generation to determine which it is—***gametophyte*** or ***sporophyte***.

Materials

- A fern plant.

Procedure

[1] Remember the logic steps we used in Exercise #3. Look at the plant very carefully.

[2] Decide whether it is simple or complex. ____________________
Therefore, whether it is haploid or diploid. ____________________
And conclude which generation it is (gametophyte or sporophyte). ____________________

[3] Finally, what will it produce in this stage—spores or gametes? ____________________

Materials

- A microscope slide of a cross-section of a "Fern Stem."

Procedure

 or 

[1] Find evidence of vascular tissue in the fern stem using your compound microscope. Vascular tissue is like the circulatory system in animals. It moves food and water up and down the plant through specialized tubes. The cells of these tubes will have *thicker cell walls* than other stem cells. The vascular tubes may be grouped in clusters that show a cylinder or a horseshoe-shaped arrangement just outside the center of the stem.

[2] Look for the vascular tissue and see if you can see the difference in the cell wall thickness between it and the other stem cells. Show your instructor! *Return the slide to where you got it!*

? QUESTION

1. The fern plant that you normally see is . . . (circle your choices)

Simple	Diploid	Sporophyte
or	or	or
Complex	Haploid	Gametophyte

2. Go back to the demonstration table and examine the fern plant. Based on your conclusions, what does this stage in the life cycle of a fern plant produce?

3. *Find them!* And, show them to your instructor.

4. Draw the part of the fern plant that produces these structures.

5. What do these structures grow into?

Fern Sex Organs

We will have to depend on slides and preserved specimens for the next stage in the life cycle of ferns.

Materials

- A prepared microscope slide of "Fern Prothallium" (Antheridia and Archegonia).

Procedure

Prothallium ______________

1. Use your compound microscope to examine the general structure of this stage of the fern plant life cycle. This very simple fern generation called the ***prothallium*** is normally found growing on the forest floor, and may be only 1 or 2 centimeters in size. You should be able to conclude which stage you are looking at. Is it the sporophyte or the gametophyte? ______________ As a reminder, write your answer next to the word "prothallium."
2. Find the male and female organs. Show your instructor!
3. Draw the prothallium showing the male and female organs.
4. The egg and sperm unite to form the beginning of the next stage in the life cycle. (You already learned that stage as the fern plant you normally see.)

   Important! The ferns have swimming sperm just like the mosses. This means that fern reproduction is also dependent on a wet environment.
5. Diagram the *Life Cycle* of the fern plant below. Include spores, gametes, sporophyte, and gametophyte. Draw a simple sketch of each generation as a visual reminder.

Life Cycle—Fern Plant

Dry Land Plants

We finish the story of plants with the two groups that are most successful living in dry environments—the ***cone plants*** and the ***flowering plants***. Both of these groups developed different structural and reproductive mechanisms to master the challenges of living on land.

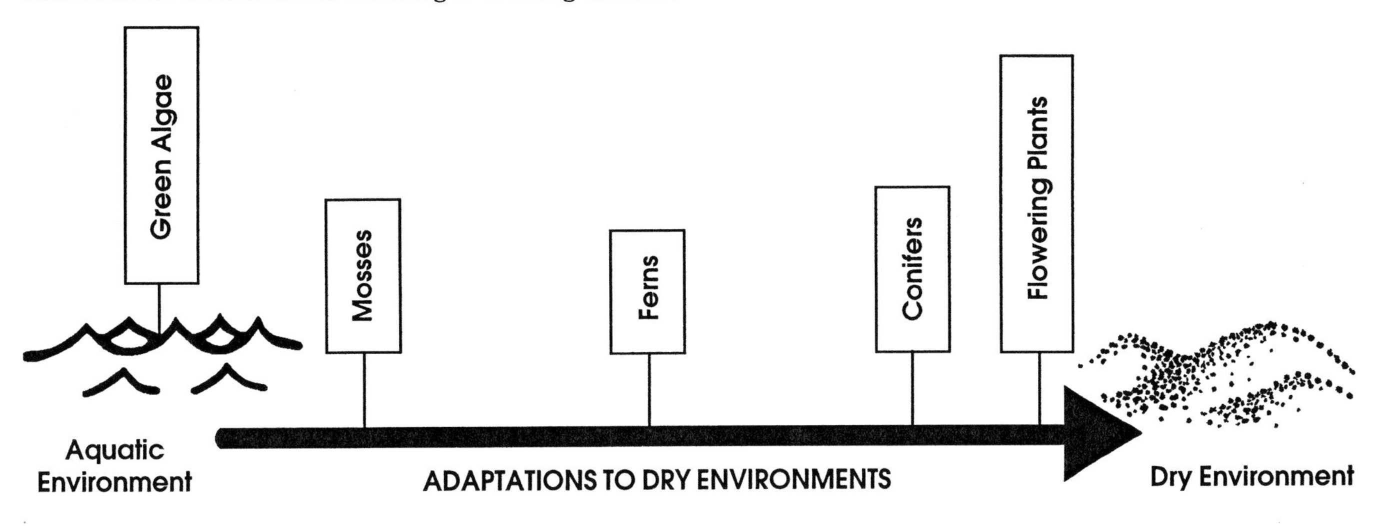

Exercise #1 "Review of Mosses and Ferns" 243
Exercise #2 "Cones" 245
Exercise #3 "Flowers and Fruit" 248
Exercise #4 "Vascular Tissue in Stems and Roots" 250
Exercise #5 "Wood" 251
Exercise #6 "Leaves" 252

? QUESTION

Can you remember two features of moss and fern plants that would limit their success in very dry environments?

__________________ and __________________

EXERCISE #1

Review of Mosses

? QUESTION

1. The moss plant that we see growing on a log is haploid. How many *sets* of chromosomes does a haploid plant have? ____________

2. A haploid plant is limited to a comparatively simple structure. How does the mutation of genes particularly hurt a haploid plant?

3. What is one example of a simple *structure* in moss plants?

4. What is *simple* about moss reproduction?

Review of Ferns

? QUESTION

1. The fern plants that we see growing in the forest are diploid. How many *sets* of chromosomes does a diploid plant have? ______________

2. A diploid plant can possess more complex structures. How does having a diploid set of genes contribute to a plant possessing more complex structures?

3. What is one example of a complex structure in fern plants that helps them to be better adapted to dry environments?

4. Remember that the ferns use a simple approach to reproduction just like the mosses. How does this limit fern plant success in dry environments?

Cone Plants and Flowering Plants

Both the cone plants and the flowering plants owe their reproductive success to an emphasis on the *diploid* stage of the life cycle. As you remember, an emphasis on the diploid stage allows the organism to have many more genes without taking a serious risk to the destruction of traits by mutation. (Refer to the Moss and Fern Lab for a more thorough review of mutation risks.)

Mosses and ferns drop their spores to the ground, and these spores grow into the gametophyte stage. *The gametophyte is the stage that is vulnerable in a dry land environment.* It is this stage that requires water for fertilization in the mosses and ferns.

Conifers and flowering plants solve some of the reproductive challenges of dry land by *not dropping spores to the ground, but by keeping the spores inside of cones or flower parts,* which allows the spores to grow into gametophytes in a *safe* environment. Actually, the gametophyte is being *protected by the sporophyte* of the cone plants and flowering plants.

The diploid sporophyte is the dominant stage in the life cycle of conifers and flowering plants.

Conifers and flowering plants also evolved a new reproductive method called ***pollination***. They don't have swimming sperm. The male sex cell (called ***pollen***) is carried to the female sex cell (called the ***egg***) by wind or insects or birds or mammals. (Refer to the diagram on the next page.) How is this new method a giant step forward for land plants?

The conifers and flowering plants have ***seeds***. A seed is an embryo sporophyte plant with a protective and nourishing covering. They are able to live a very long time in the environment until it rains. Then the embryo plant will grow into the adult plant. (Mosses and ferns do not have seeds. They only have a single-cell spore to weather the dry season.)

Study the diagram on the following page to be sure you understand the differences in the general life cycles between mosses and ferns as compared to cone plants and flowering plants. There are some structural and reproductive differences between the conifers and the flowering plants. You will discover some of those differences during the Exercises in this lab.

Comparison of Life Cycles

THIS PART OF THE MOSS AND FERN LIFE CYCLE
IS VULNERABLE TO DRY LAND CONDITIONS.

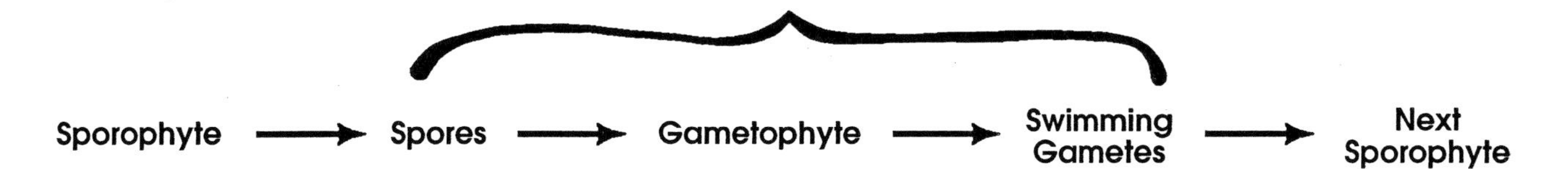

THIS PART OF THE CONIFER AND FLOWER LIFE CYCLE
IS PROTECTED FROM DRY LAND CONDITIONS.

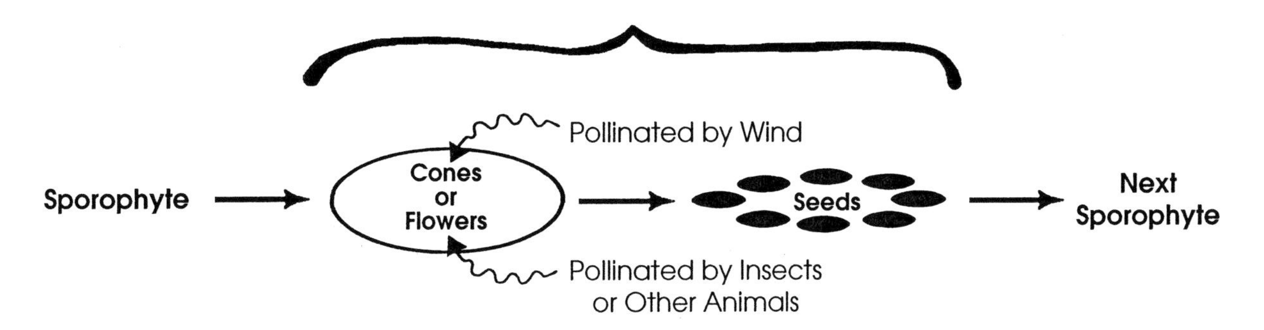

EXERCISE #2

"Cones"

The word ***conifer***, which means "cone-bearing," refers to any plant that uses ***cones*** as its characteristic reproductive structures. Basically, the cone is where the *gametophyte* generation is produced. Remember, in mosses and ferns the gametophyte has to live independently on the forest floor where it is in danger of drying out and dying.

In conifers, the vulnerable gametophyte stage develops inside the protective cones.

There are both *male* and *female* cones (although the female cone is what people call a "pine cone"). The male cone produces *pollen,* and the female cone produces *eggs*. Pollen grains contain the male sex cell, and are carried by the wind to the female cones. This is called ***pollination***. After pollination, the pollen grains produce tubes that grow towards the eggs. Fertilization happens when a pollen nucleus (haploid) is transported through the pollen tube and fuses with an egg nucleus (haploid). A diploid seed (pine nut) grows from that union.

Male Cones

Male cones are small (about 1 to 2 cm in length), and can be found in clusters at the ends of pine branches. They are designed for wind pollination.

Materials

- A compound microscope.
- A small sample of pollen from the male cones on display. Look at the design of the male cones when you are at the demonstration table.

Procedure

1. Draw a simple sketch of a cluster of male cones.
2. Make a wet mount of the pollen and put it under high magnification.
3. Draw a simple sketch of the pollen.

Male Cones

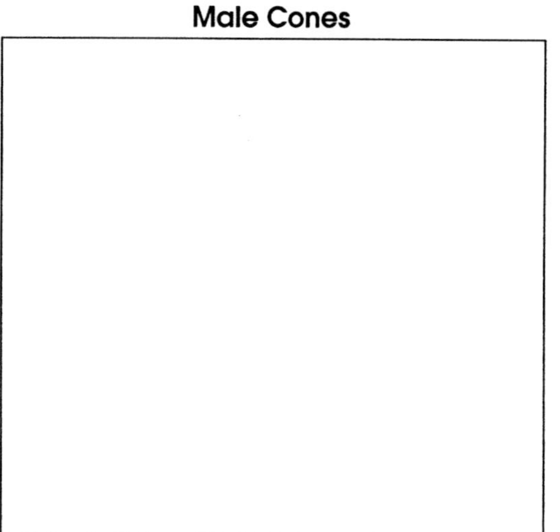

Pollen

? QUESTION

1. Why is so much pollen produced by each male cone?

2. How does the pollen get to the female cones?

3. What is the value of the two "ears" on the conifer pollen?

4. Which generation is the pollen? (sporophyte or gametophyte)

Female Cones

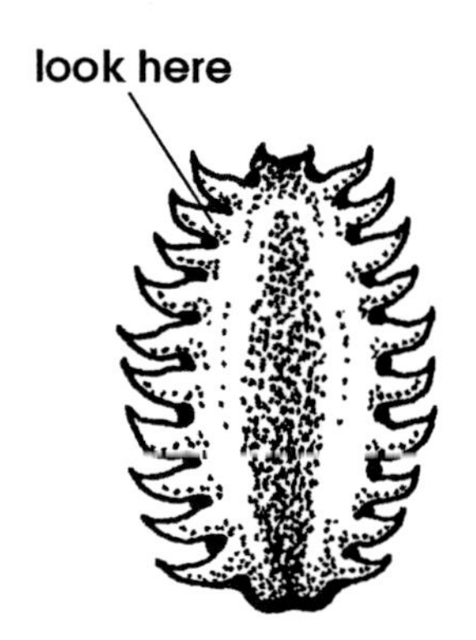

Female cones are found near the tips and along the branches of the pine tree. They start out very small, but there may be more than 100 eggs in each of these cones.

The pollen from the male cone blows onto these female cones when they are very small and the eggs are fairly close to the outside of the cone. Fertilization happens with the union of pollen and egg.

The fertilized egg becomes part of the seed and continues to be protected by the female cone as it grows larger. Eventually the female cone is quite large (what you usually see on a pine tree), and the seeds are now ready to be released into the environment.

Materials

- A prepared microscope slide marked "Pine Megasporangia" (very young female cones).

Procedure

1. Put your compound microscope on low power and position the slide.
2. See if you can find some eggs in this cross-section of a female cone. The eggs should be found around the outside of the cone. **Hint:** Look for a *big* cell! Look at the base of the projections from the female cone. (Refer to the picture of a sectioned cone on the left.)

[3] Draw a simple sketch of the egg and where it is found.

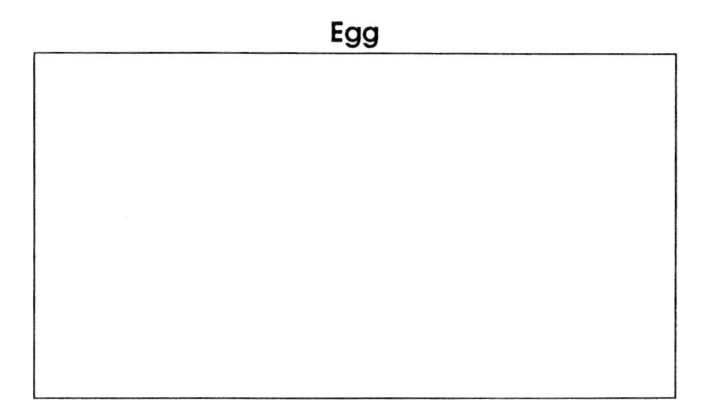

[4] *Return the slide to where you got it.*

? QUESTION

Which generation (sporophyte or gametophyte) of the conifer life cycle actually produces the egg, and is being protected inside the female cone?

Hint: This is the same generation of the moss and fern life cycle that has to live unprotected and on its own.

Seeds

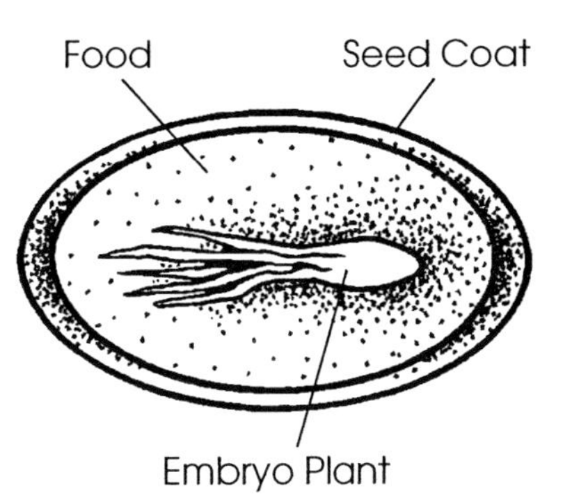

Seeds are a "land habitat" feature of both conifers and flowering plants. (There are no seeds in mosses and ferns.) The seeds of conifers can be found inside the female cone about one year after pollination.

The mature seeds are released into the environment and must survive on their own. Because conifers do not have nutritive fruit surrounding the seeds, they are called ***gymnosperms*** (meaning "naked seeds").

The seeds of flowering plants are found inside the fruit, which gives these plants their name—***angiosperms*** (meaning "covered seeds"). The fruit is discussed in more detail in the next Exercise.

Materials

- A dissecting microscope.
- A pine seed (soaking in water).
- A dissecting probe.

Procedure

[1] Carefully crack the seed coat with pliers, then under the microscope, use your probe and carefully remove the seed tissue from the outer part of the pine seed until the ***plant embryo*** can be seen.

[2] Show your instructor.

? QUESTION

1. What function does the seed coat provide?

2. Why is this a better adaptation to a dry land environment?

3. Which generation of the conifer life cycle (gametophyte or sporophyte) is the seed?

EXERCISE #3

"Flowers and Fruit"

In flowering plants, the gametophyte stage develops inside the flower.

Flowers are like "neon signs" to attract pollinators (usually insects, birds, and mammals). Color, fragrance, nectar, and shape are some of the more important features that determine which pollinators will be attracted to a particular flower species.

Examine each flower in this lab to see if you can determine what kind of pollinator might be involved with the reproduction of the plant. If a plant can involve another species to help it with reproduction, then that plant will be more successful than a plant that must rely on the wind for pollination.

Materials

- A tree-tobacco flower.
- Dissecting materials—a probe, a pair of tweezers, and a razor blade.

Procedure

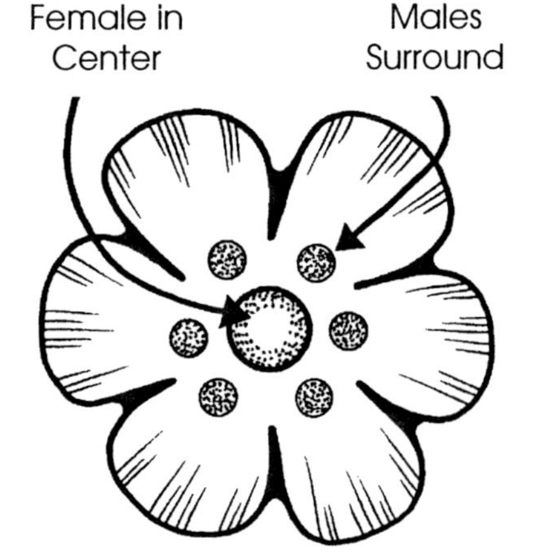

Hypothetical Simple Flower

[1] The basic theme of most flowers is: The male organs surround the female organ. The male organs are called the ***stamens***, and they contain the pollen. The female organ is called the ***pistil***, and it contains the eggs.

[2] Use the dissecting microscope to look at the open end of the tree-tobacco flower. Does any reproductive organ stick up higher than the others?

What color is the tip of that organ?

[3] Carefully tear open the top half of the flower so that you can better see the reproductive organs. If you see several identical filaments, then you have found the males (stamens). **Remember:** There will only be one female (pistil). Can you now determine which organ sticks up highest in the flower?

What would be the value of the female organ being higher than the stamens?

[4] Continue to tear the yellow petal until it is removed from the green base of the flower. Draw a simple sketch of the arrangement of pistil and stamens in the box provided.

Tree-Tobacco Flower

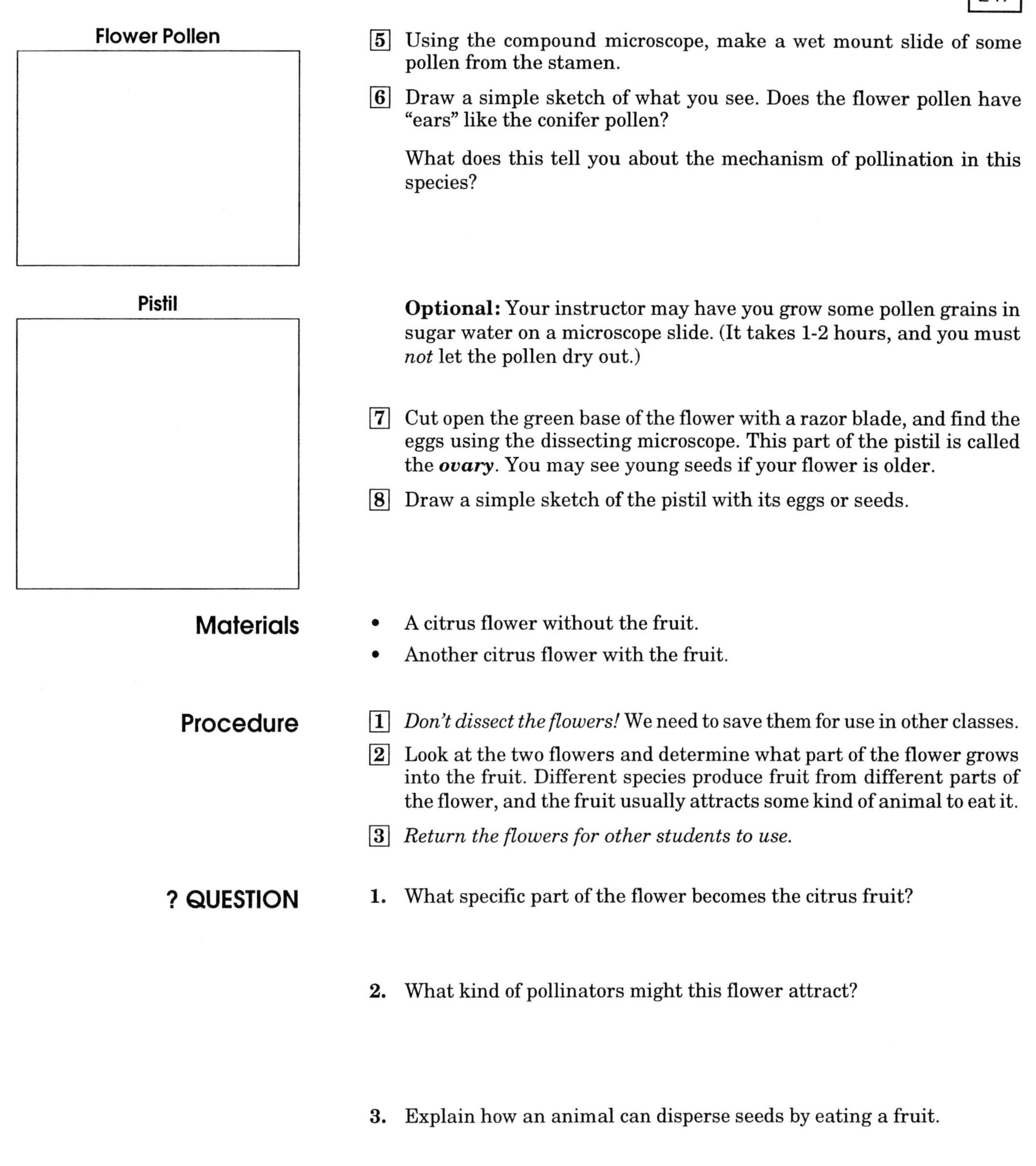

[5] Using the compound microscope, make a wet mount slide of some pollen from the stamen.

[6] Draw a simple sketch of what you see. Does the flower pollen have "ears" like the conifer pollen?

What does this tell you about the mechanism of pollination in this species?

Optional: Your instructor may have you grow some pollen grains in sugar water on a microscope slide. (It takes 1-2 hours, and you must *not* let the pollen dry out.)

[7] Cut open the green base of the flower with a razor blade, and find the eggs using the dissecting microscope. This part of the pistil is called the ***ovary***. You may see young seeds if your flower is older.

[8] Draw a simple sketch of the pistil with its eggs or seeds.

Materials

- A citrus flower without the fruit.
- Another citrus flower with the fruit.

Procedure

[1] *Don't dissect the flowers!* We need to save them for use in other classes.

[2] Look at the two flowers and determine what part of the flower grows into the fruit. Different species produce fruit from different parts of the flower, and the fruit usually attracts some kind of animal to eat it.

[3] *Return the flowers for other students to use.*

? QUESTION

1. What specific part of the flower becomes the citrus fruit?

2. What kind of pollinators might this flower attract?

3. Explain how an animal can disperse seeds by eating a fruit.

4. Flowers typically grow in small clumps scattered in their habitat, whereas conifers typically grow in large groves of trees. How does this difference in distribution relate to your answer in question #3?

EXERCISE #4

"Vascular Tissue in Stems and Roots"

Cross-Section Cut

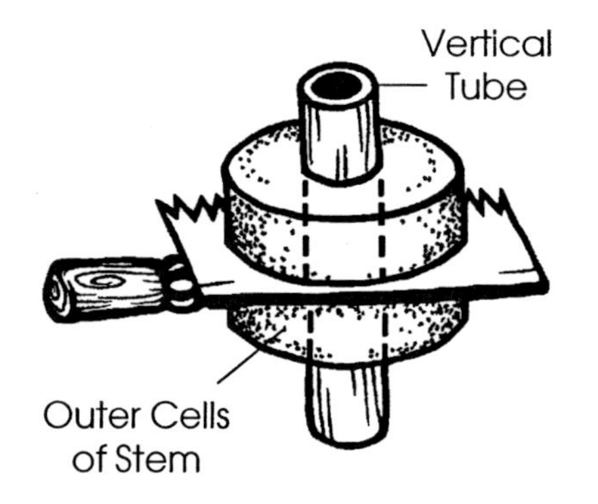

Vascular tissue is an evolutionary invention occurring near the time of the first fern plants. In these early ferns, the conducting tissue was only moderately developed, which is one reason that fern plants can't get water from deep in the ground.

The cone plants and flowering plants developed a much better system of vascular tissue, and their resulting success in dry land environments is obvious.

The vascular system in plants is analogous to the circulatory system in animals. It is a network of vertical tubes that extend from the leaves to the roots. As a general rule, the cell walls forming the tubes are very thick when compared to other cells in the leaves, stems, and roots.

You will be looking at prepared *cross-sections* of the plant parts. You will have to imagine the three-dimensional vertical tube structure that makes up the vascular tissue.

Materials

- A prepared microscope slide of a "Root Cross-Section."
- A prepared microscope slide of a "Herbaceous Stem Cross-Section."

Procedure

Root

Stem

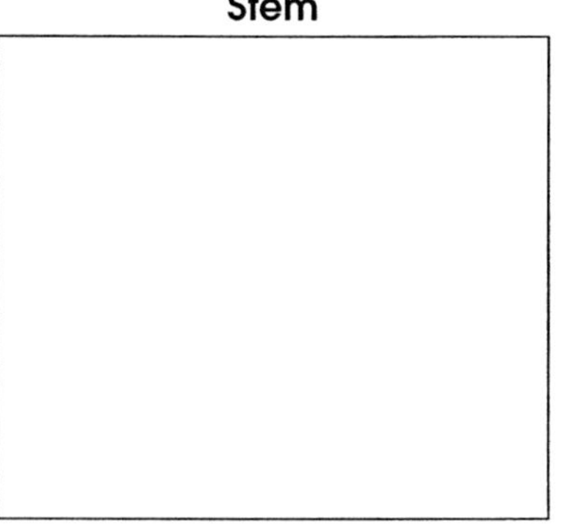

[1] Look at the root cross-section under low magnification, and determine where most of the vascular tissue is located. (The vascular tubes are *thick-walled* cells.)

[2] Draw a simple sketch of the root showing the location of the vascular tissue.

[3] The nutrients and water in the soil must diffuse into the vascular tubes. Now that you know where the vascular tubes actually are, what does this tell you about the size of the roots that do the absorption for a plant? (Is it the big roots or the small roots?)

[4] Look at the herbaceous stem cross-section under low magnification, and determine where the vascular tissue is located. (circle your choice) **Hint:** The arrangement in the stem is called "vascular bundles."

Center of the stem	or	Towards the outside of the stem

[5] Draw a simple sketch of the distribution of vascular tissue in the stem. This arrangement of vascular bundles is common in plants.

? QUESTION

An unthinking person made a cut with a pocket knife all the way around the trunk of a tree. The tree died soon after. Explain why the tree died.

EXERCISE #5

"Wood"

The wood that we use to build our houses comes from the stiff cell walls that are a part of the plant's vascular system. This part is called the ***xylem***, and it conducts water from the roots in the ground up to the leaves at the tips of every branch. **Remember:** Water is one of the necessary ingredients for photosynthesis, which takes place in the leaves.

The xylem continues to be produced on the inside of an ever-expanding growth ring called the ***cambium***. Cambium is just inside the bark of a living trunk or branch. It is shown as a "dashed" circle in the diagram.

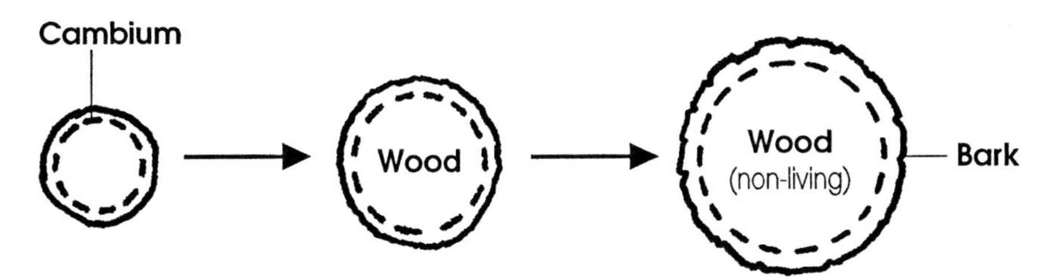

The xylem continues to conduct water even after it dies. This tissue is somewhat like thousands of tiny straws that transport water and provide support for the tree at the same time. The water moves up the xylem by a process called ***capillary action***. Your textbook explains the details of capillarity and the osmotic "pushing" forces in the roots. In larger trees, the middle section fills with resin and no longer transports water, but functions only as a supporting structure.

A thin zone of living vascular tubes called the***phloem*** is produced on the outside of the cambium. These tubes carry sugars from the leaves to the roots for storage. Sometimes the outermost part of the bark is the remains of dead phloem that has been pushed out by the ever expanding cambium.

? QUESTION

1. In a forest fire, sometimes a tree can be badly burned, forming a hollow in the inside, yet the tree doesn't die. How do you explain that?

2. What specific part of the vascular tissue is the wood of the tree?

3. What specific cellular feature of vascular tissue makes it ideal as a building material for our houses?

4. What specific part of the vascular tissue is produced on the outside of the cambium?

5. What specific part of the vascular tissue is produced on the inside of the cambium?

Materials

- A prepared microscope slide marked "Woody Stem."

Procedure

[1] Look at the slide under the lowest power of the dissecting microscope and find the annual rings of tissue. These are called ***tree rings***.

[2] Find these same rings with your compound microscope and look carefully at the individual cells of the rings.

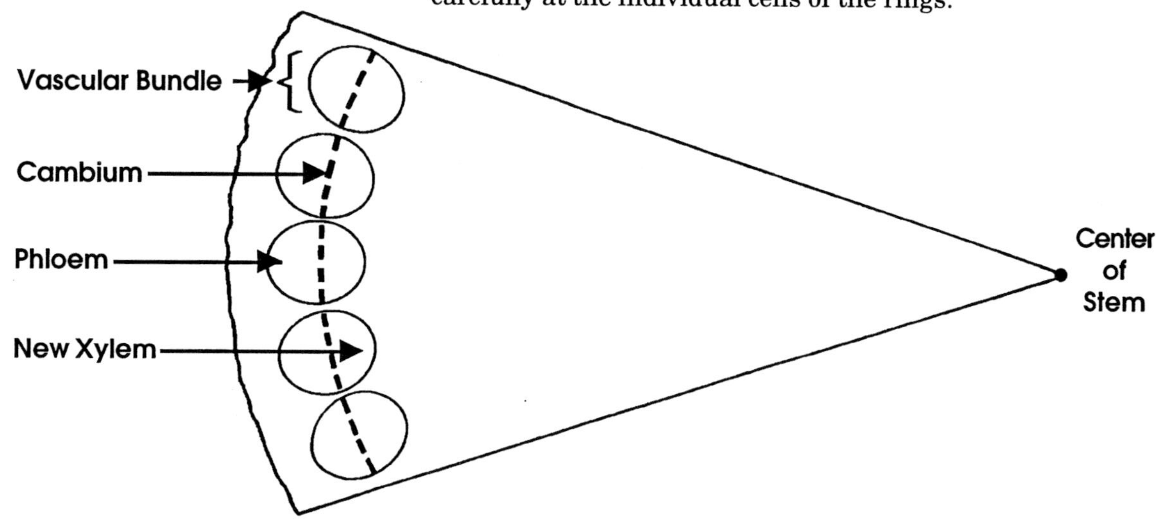

[3] Draw the tree rings in the above diagram, and show how the cell *size* changes in these rings. By the way, you should be able to see the vascular bundles that are towards the outside of the stem.

? QUESTION

1. What is the direction of time as you move outward from the center of the tree stem?

2. Where are the youngest tree rings?

3. What would cause these annual rings to show the cell size change that you see?

   a. When would you see big cells? ____________________

   b. When would you see small cells? ____________________

EXERCISE #6

"Leaves"

The most basic purpose of a leaf is to expose chlorophyll to light so that the plant can perform photosynthesis. There are a number of problems that a leaf has to overcome to be successful. Some of those challenges include:

- A leaf has to get water from the roots as a raw material for photosynthesis, and it has to send sugars made during photosynthesis to the rest of the plant.
- A leaf has to get CO_2 as a raw material for photosynthesis, and this carbon dioxide has to be delivered to all of the cells in the leaf.
- A leaf has to catch light energy for photosynthesis, but not overheat from absorbing too much sunlight. (A leaf must not lose too much water from evaporation, or the plant will dry up.)

Leaf Veins

Leaf veins are vascular bundles leading to various parts of the leaf. The pattern of veins depends on the type of plant and the shape of the leaf. However, in all cases, these veins are used for carrying water to the leaf cells and for carrying photosynthetic sugars away from the leaf cells.

Materials

- A broad leaf from the *mallow tree*. (We will have it in the lab, or your instructor will tell you where you can find it.)

Procedure

1. Look at the leaf under the dissecting microscope, and determine the pattern of leaf veins. You already have some information about these structures from Exercise #3.
2. Draw a simple sketch of the pattern of veins in this mallow leaf.

Leaf Veins

? QUESTION

What is carried in these mallow leaf veins? (Be specific about the content and direction and purpose.)

Leaf Stomata

Leaf stomata are openings on the underside of the leaves. We looked at stomata in the *Zebrina* plant during the Cell Lab earlier in the semester. If you didn't get a chance to see these structures, then make a leaf peel now and search for them.

Materials

- A prepared microscope slide of a cross-section of a leaf.

Procedure

1. First, locate the vein so that you know where you are.
2. Look for the stomata on the underside of the leaf. The guard cells are round in cross-section instead of being like "lips" in the leaf peel. Some of the guard cells are sectioned through the middle so that you can see the stomata opening.
3. Draw a simple sketch of the structural relationship between the stomata and the clear spaces inside the leaf. Also, show the vein. Label each part.

Cross-Section

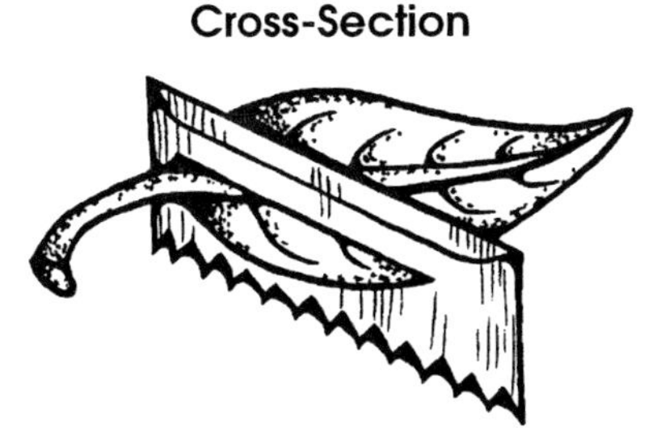

Leaf Cross-Section

? QUESTION

Describe how the CO_2 gets into the leaf and to the interior cells of the leaf.

Comparisons of Leaf Types

Procedure

We have selected four different leaf types to demonstrate several leaf adaptations. Carefully look at each leaf on display, and make a simple sketch of each.

Mallow	Pine

Jojoba	Buckwheat

? QUESTION

1. Which leaf design is best for catching the most light? ______________
2. Which leaf design would absorb the most heat? ______________
3. Which leaf design loses water the fastest, and must live in a wetter environment than the rest? ______________
4. The jojoba leaf does better than the pine leaf in very dry environments. What features of the jojoba help the leaf lose less water than the pine leaf?
5. Notice the orientation of the jojoba and buckwheat leaves. What time of the day do these two leaf types absorb more light? ______________

   What part of the day do they absorb less light and heat? ______________

   What advantage is there to this kind of leaf orientation to the sun?

Survey of Animals

During this lab you will explore a brief survey of the ***Kingdom Animalia***, which includes organisms that are *eukaryotic*, *muticellular*, and *heterotrophic*. You will investigate some of the defining characteristics and evolutionary relationships that divide the Kingdom Animalia into subgroups called ***phylum*** and ***class***.

The taxonomic story presented in most textbooks gives the impression that more complex forms of life were created sequentially during evolution. This portrayal is convenient for viewing life from a human perspective, but there is no evidence that evolution has always produced increasingly more successful forms of life. In fact, there is some evidence to the contrary. For example, the reptiles as a group have enjoyed a much longer and more diverse history than the mammals (the group to which you belong). And that poses an interesting question. Evolutionarily speaking, are you a more successful lifeform than a lizard?

But, before you spend a lot of time arguing about it, and before you place yourself at the top or the bottom of the evolutionary heap, you should know that there is a single group of animals that far out-numbers all of the other animal groups combined. That group is called the *arthropods*. They are by sheer *number* and *diversity* the dominant animal in both aquatic and land environments on this planet!

Although humans are out-numbered and out-diversified, and haven't been around on this earth for very long, it is interesting for us to understand and experience some of the other animals that we live with. Today's lab begins with the ***invertebrates*** (animals with no backbone), and continues through the ***vertebrates*** (animals with a backbone).

Exercise #1 "Basic Body Plan" 255
Exercise #2 "Type of Digestive System" 258
Exercise #3 "Segmentation" 259
Exercise #4 "Skeletons" 262
Exercise #5 "Land Adaptations" 264
Exercise #6 "The History of Life" 266

EXERCISE #1

"Basic Body Plan"

There are three basic body plans found in the animal groups—***no symmetry***, ***radial symmetry***, and ***bilateral symmetry***.

No Symmetry

(no directionality)

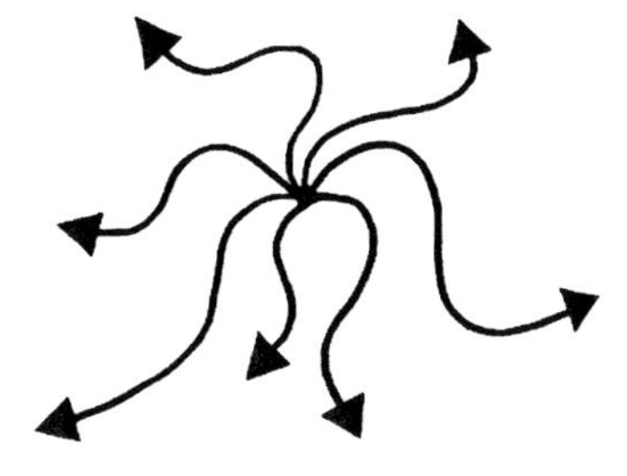

An animal with this design grows in all directions, usually on the surface of something else, and typically has no particular defining form. These organisms do not respond to stimuli in their environment because they don't have a nervous system. Each part of the organism can act independently of the other parts.

We will look at only one group with this design.

Radial Symmetry

(radial directionality)

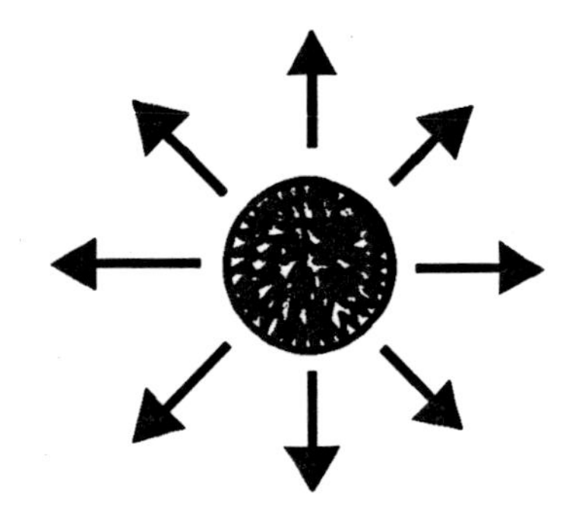

An animal with radial symmetry can be divided into two equal-looking halves by any cut through the longitudinal axis (like cutting a pie or a ball in half). These animals move or react to the environment equally in all directions.

This design does not have a head end and a tail end, and does not search for and locate its food like the next group does. To get food, this animal is usually stationary or is carried by water currents, capturing the food that passes by. There are some predators with radial symmetry, like the starfish, but their sense organs are dispersed over the surface of the animal and are not concentrated at one end of the animal.

Bilateral Symmetry

(head and tail ends)

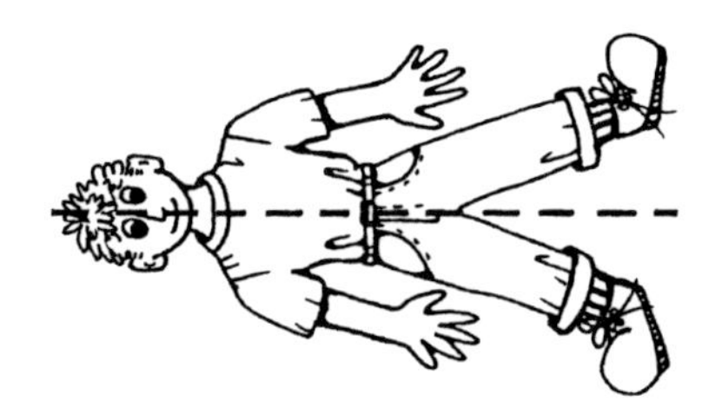

Bilateral means "two sides," and there is only one way to cut the animal into two equal-looking halves.

An animal with this design responds to its environment by using *directional movement*. Its head end is the location of the sense organs, and orients to different kinds of environmental stimuli such as light and food.

Materials

- A dissecting microscope and a compound microscope.
- Three small glass dishes.
- A small preserved *sponge*, a *flatworm*, and a *hydra*.

Procedure

1. Place each organism in a separate glass dish.
2. Put the sponge, the hydra, and the flatworm next to each other on the table.
3. Look at each organism under the dissecting microscope.
4. Draw a quick sketch of each organism:

Sponge	Hydra	Flatworm

? QUESTION

1. Which type of symmetry is displayed by the sponge?

2. The taxonomic name of the sponges is phylum ***Porifera,*** which means "hole bearing." Why do you think this term is appropriate?

3. Which type of symmetry is displayed by the hydra?

4. Jellyfish and hydras are in the phylum ***Cnidaria,*** which means "stinging cells." What does this name indicate about these organisms?

5. Which type of symmetry is displayed by the flatworm?

6. The flatworms are in the phylum ***Platyhelminthes,*** which means "broad, flat stem." What does this definition tell you about these organisms?

7. What are the two most obvious characteristics of these worms?

   ____________________ and ____________________

8. Put the three animal subgroups (hydra, flatworm, and sponge) in order of increasing complexity or symmetry.

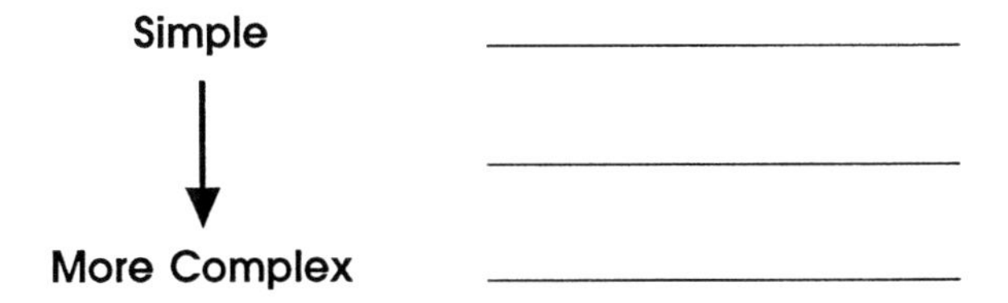

Please return all specimens to the specimen table so that the other labs may also use them. *Do not throw away any preserved animals.*

EXERCISE #2

"Type of Digestive System"

Animals with a complete gut are better at digestion.

There are two basic designs for the digestive system within the subgroups of Kingdom Animalia—***incomplete*** and ***complete gut***.

Incomplete Gut

Basically, this digestive system consists of a tube with only one opening to the outside. The food is taken into the tube through the mouth and it is digested. After digestion is completed, the waste products are ejected from the animal through the same opening.

Animals with an incomplete gut do not have excretory and circulatory systems. Therefore, they are dependent on diffusion to obtain nutrients from digestion and to get rid of waste products.

Complete Gut

This digestive system consists of a complete tube extending through the organism. There are two openings—one for taking in food, and another for eliminating the waste products of digestion.

These animals have excretory and circulatory systems to transport nutrients to all body cells and to remove wastes from those cells for transport to the outside of the organism.

Materials

- A preserved specimen of *Ascaris.*

Procedure

1. Look at the ***roundworm*** called *Ascaris*, which is in the phylum ***Nematoda,*** meaning "thread-like."
2. This animal has a complete gut.
3. Compare the general appearance of the roundworm to the form of the sponge, hydra, and flatworm.

   What main differences do you notice?

4. Draw a simple sketch of the roundworm.

Roundworm

? QUESTION

1. Sponges are full of holes, the hydra types are big open sacs, and the flatworms are flat. How are these features important in facilitating the process of diffusion?

2. Roundworms are round. Is a round shape a better shape for diffusion than a flat shape?

3. The shape of roundworms would favor . . . (circle your choice)

Swimming or Burrowing

4. Which digestive system would you call an "eating machine"—more efficient at getting nutrients? (circle your choice)

Incomplete gut or Complete gut

5. If an animal group is more efficient at getting nutrients, then it will probably be . . . (circle your choice)

More successful or Less successful

6. Draw a simple sketch of an incomplete digestive tract. Which animal groups have an incomplete gut?

Example	Phylum
____________________	____________________
____________________	____________________
____________________	____________________

7. Draw a simple sketch of a complete digestive tract. Which animal group has a compete gut?

Example	Phylum
____________________	____________________

8. Which digestive system type do you belong to?

Return the specimen of "Ascaris" to the Specimen Table.

EXERCISE #3

"Segmentation"

Segmentation is a kind of body design that results from a radical change in the normal development of an animal. This unique evolutionary feature allowed a "new" kind of animal to develop.

In segmentation, blocks of embryonic tissue are duplicated instead of remaining as one single block.

Segmentation was originally a profound error during embryonic development. But it led to a new line of animals.

Segmented Animal

These repetitive blocks result in an organism that is a combination of *segments*. This "new" kind of animal is like making a chain of the original animal, linked together, one after the other.

The segmented design eventually led to the most successful group of animals—the *Arthropods*.

Materials

- Go to the specimen table.

Procedure

[1] Put a roundworm, an earthworm, a millipede, and a crayfish next to each other in that order. ***You have just laid out an important evolutionary arrangement!***

[2] The earthworm is called the *segmented worm* (phylum ***Annelida***, meaning "ringed"). The millipede and crayfish are the phylum ***Arthropoda*** (meaning "jointed foot"), the most successful animal group today. Arthropoda includes body designs based on segmentation.

? QUESTION

1. A segmented animal would be __________________ than the early "pre-segemented" animal form. (circle your choice)

Smaller or Larger

2. What would be the "biggest" benefit to the animal that developed from the above change?

3. After simple segmentation evolved, what evolutionary changes happened to the individual segments?

4. Use your knowledge of *basic body plan*, *digestive systems*, and *segmentation* to put the animal subgroups in order of increasing complexity (hydra, sponge, arthropod, roundworm, flatworm, and segmented worm).

5. How do you think the evolutionary development of segmentation applies to human development?

Special Stuff

All organisms are either exceptions to, or examples of "the rule," depending on your point of view.

Evolution is not a "directional game," even though we humans find it convenient and easy to talk about the process that way. Evolution is like an experiment. Some experiments succeed; others fail. And some go on and on because there are no big obstacles to prevent their continuation.

So, many creatures resulting from evolution's experiments may be seen by us to be "exceptions" to the rule, or "special examples." But, in evolutionary reality, all organisms are either exceptions or special examples of "the rule," depending on your point of view.

The mollusks and echinoderms are two evolutionary groups that fit somewhere in our animal survey.

Mollusks

Phylum ***Mollusca***, meaning "the soft ones," includes animals that have a *shell* during some stage of their lives. They evolved at about the same time as segmentation was evolving, perhaps a bit before the segmented worms.

Echinoderms

Members of the phylum ***Echinodermata***, which means "spiny skin," are very complex animals that have returned to a radial symmetry. (Originally, they came from a bilateral marine group.) Echinoderms have an unusual arrangement of tube-feet that are moved by a hydraulic system within the body.

The larval stage of echinoderms is more like the larva of simple chordates than the early developmental stages of other invertebrates. This embryonic similarity means that these invertebrates are one of the closest relatives of the phylum ***Chordata*** animal group to which you belong.

Materials

- Go to the specimen table.

Procedure

1. Look at the examples of mollusks on the specimen table.
2. Look at the examples of echinoderms.
3. Draw a simple sketch of an example of each organism.

Mollusk	Echinoderm

EXERCISE #4

"Skeletons"

A ***skeleton*** is a supporting structure made of different pieces that can be individually moved by muscles. The evolutionary success of animals with skeletons is an expression of the capacity for refined movement. The better you are at moving on this planet, the better off you are in terms of survival.

There are two basic kinds of skeletons in the animal kingdom—the *exoskeleton* and the *endoskeleton*. Both types are successful. The main difference is in the size of the organism that can be supported by each.

Exoskeleton

Animals with an exoskeleton stay small; those with an endoskeleton can be big.

"Exo" means outside. An ***exoskeleton*** is on the outside surface of the animal. The animal group that has most successfully utilized the exoskeleton is the *arthropod*.

Name an example of an arthropod: ______________________________

Endoskeleton

"Endo" means inside. An ***endoskeleton*** is inside of an animal, and there are three variations of the endoskeleton.

- **Internal Flexible Rod (Notochord)**

  This doesn't exactly fit our definition of skeleton because it is a one-piece structure. But it does have many muscle fibers attached along its length for control of swimming movements.

- **Cartilage Skeleton**

  This type of skeleton is made of pieces of cartilage. Cartilage is a softer supporting material than bone. An example of cartilage is the supporting structure of your nose or ear.

- **Bony Skeleton**

  This type of skeleton is made of many hard bony pieces. Bone consists of molecules made from minerals such as calcium.

Materials

- Go to the specimen table.

Procedure

1. Put an insect, an *Amphioxus*, a jawless fish, a cartilage fish, and a bony fish in that order.
2. Put the cartilage skeleton and the bony skeleton plastic mounts next to the appropriate fish.
3. Study these animals carefully, observing their particular skeletal structures.

? QUESTION

1. Compare the insect to the other four animals. What is the most obvious difference?

2. What kind of skeleton does the insect have? ______________________

3. Which type of skeleton looks like it would present a problem to a growing animal? (circle your choice)

   Exoskeleton or Endoskeleton

4. Explain why you think it would be a problem.

5. Which skeleton is designed to grow along with the animal? (circle your choice)

   Exoskeleton or Endoskeleton

6. Can a fish have a *jaw* if it doesn't have a multi-piece skeleton?

7. What does that tell you about an animal we call a jawless fish?

8. Which two organisms are examples of animals with a one-piece skeleton (internal flexible rod)?

   ______________________ and ______________________

9. Which type of skeleton is capable of more refined swimming movements? (circle your choice)

   Cartilage skeleton or Bony skeleton

   Explain why you think so.

10. Do you think that *Amphioxus* can swim as well as an aquarium fish?

    Explain why or why not.

EXERCISE #5

"Land Adaptations"

About 350 million years ago the vertebrate fish had already evolved, and there were large areas of the planet with shallow seas and lowlands. These conditions favored the evolution of any group with features suited to land.

Several environmental challenges had to be overcome before any organism could survive in the land environment. Among those challenges were: *movement on land*, *breathing air*, *preventing water loss*, and *reproduction*.

Materials

- Go to the specimen table.

Procedure

Put a bony fish, a frog, a lizard, a bird, and a mammal in that order. This order represents a series of land adaptations that were made by the vertebrate groups.

? QUESTION

1. Compare the bony fish to the frog. What is the most obvious structural difference?

2. Those structures found on the frogs evolved from what parts of the fish? ____________

3. The frog is in the phylum ***Chordata*** (meaning "notochord"), is part of the subphylum ***Vertebrata*** (meaning "having a backbone"), and is a member of the class ***Amphibia***. The word *amphibian* means "living both lives." What process in the amphibian life cycle depends on water? (Recall the situation with moss and fern plants.)

4. The fact that a frog can live out of water and a fish cannot tells you that the frog has solved the problem of ____________.

5. What structure in the frog solves that problem? ____________

6. Compare the frog to the lizard. What is the most obvious difference?

7. The lizard belongs to the vertebrate subgroup class ***Reptilia***, which means "creeping." Modern reptiles have legs that are positioned out to the side of the animal, which makes them appear to creep as they move. One of the early reptiles had legs positioned under the body, and that group gave rise to what are now mammals.

   What two features allowed the reptiles to be so successful in dry environments?

   ____________________ and ____________________

8. What two special changes in reproduction were developed by the reptiles? (Ask your instructor.)

9. Compare the lizard to the bird and the mammal. What are the most obvious external differences?

   ____________________ and ____________________

10. Birds belong to the vertebrate subgroup class ***Aves***, which means "bird." Mammals belong to the vertebrate subgroup class ***Mammalia***, meaning "breast." Mammals are given their name because they are the only animal with mammary glands for feeding their young. The external features of birds and mammals are an indication of an important evolutionary development that separates them from the reptiles.

    How do these external differences allow birds and mammals to live in places where reptiles can't survive?

11. Using your knowledge of *skeletons* and *land adaptations*, put the subgroups of chordate animals in order of complexity (reptile, jawless fish, mammal, amphibian, bony fish, bird, and *Amphioxus*).

Simple ____________________

____

____

____

____

____

More Complex ____________________

In Conclusion

The classification system we have been using in this lab is a simplified version of the ***Kingdom Animalia***. The following listing is a review of what you have covered.

Kingdom Animalia

Phylum Porifera .... sponges
Phylum Cnidaria .... jellyfish and hydra
Phylum Platyhelminthes .... flatworms
Phylum Nematoda .... roundworms
Phylum Mollusca .... mollusks
Phylum Annelida .... segmented worms
Phylum Arthropoda .... arthropods
Phylum Echinodermata .... echinoderms
Phylum Chordata .... amphioxus

Subphylum Vertebrata

Class Agnatha .... jawless fishes
Class Chrondrichthyes .... cartilage fishes
Class Osteichthyes .... bony fishes
Class Amphibia .... frogs
Class Reptilia .... lizards
Class Aves .... birds
Class Mammalia .... mammals

EXERCISE #6

"The History of Life"

The information in this Exercise is presented to give you a general summary of life on this planet and a review of some of the features of the animal groups covered in the lab.

Procedure

1. Use the chart information on the next page to fill in the correct group names for each of the "arrows" on the History of Life chart on the following pages. The arrows on the History of Life chart indicate the time each group has existed, and the expanded box part of the arrow represents the time of greatest success for the group.
2. Fill in the invertebrate and vertebrate Family Trees and the matching features of the animal groups.

In Conclusion

This concludes your survey of the animal kingdom. You have covered several major evolutionary themes and have examined representatives of many animal subgroups.

If you feel a bit overwhelmed, then you are not alone. Biology majors usually spend one semester on the invertebrates, and another semester on the vertebrate animals. There are fascinating relationships to be discovered and revealed when more time is spent on comparative anatomy, physiology, and embryology.

Try it. You'll like it!

"HISTORY OF LIFE — CHART INFORMATION"

Times When Various Life Groups Occurred on this Planet

Group Name	Earliest Fossil Evidence	Time of Greatest Success for Group
Amphibians	350 million years ago	320 to 240 million years ago
Early Chordates	600 million years ago	————
Dinosaurs	210 million years ago	180 to 66 million years ago (extinct 66 million years ago)
Eukaryotic Cells (Protists)	1.5 billion years ago	1 billion years ago to today
Lung Fish	370 million years ago	————
Mammals	200 million years ago	66 million years ago to today
Jawless Fish	500 million years ago	480 to 370 million years ago
Anaerobic Life	4 billion years ago	3.9 to 3.3 billion years ago
Early Reptiles	300 million years ago	230 to 210 million years ago
Humans	1.5 million years ago	today
Cartilage Fish	400 million years ago	380 to 240 million years ago
Stromatolites (photosynthetic bacteria and aerobic bacteria)	3.5 billion years ago	3 billion to 700 million years ago
Birds	180 million years ago	66 million years ago to today
Lobe-Finned Fish	370 million years ago	————
Invertebrate Animals	700 million years ago	500 million years ago to today
Bony Fish	380 million years ago	66 million years ago to today

THE HISTORY OF LIFE

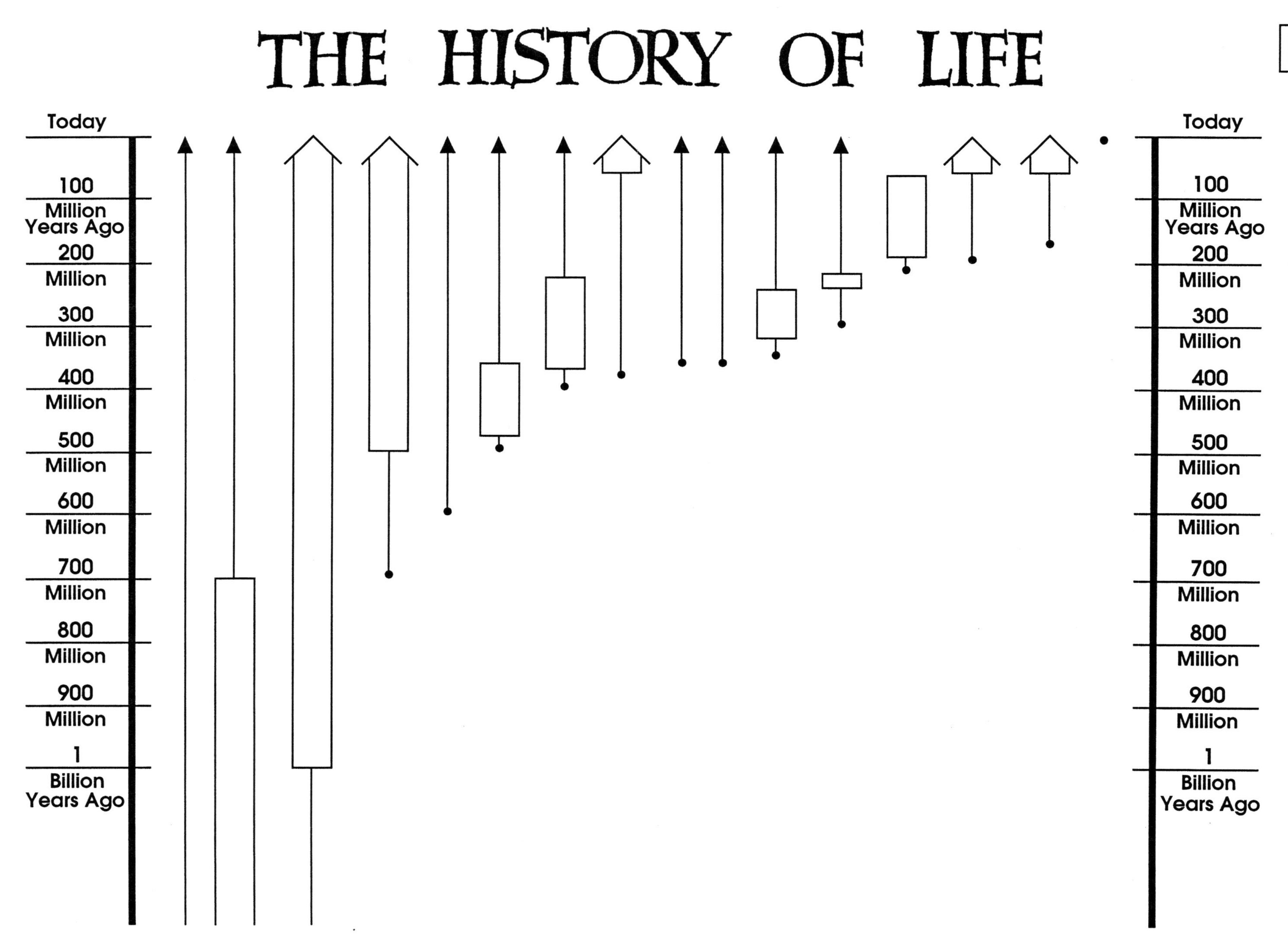

Today
100
Million
Years Ago
200
Million
300
Million
400
Million
500
Million
600
Million
700
Million
800
Million
900
Million
1
Billion
Years Ago

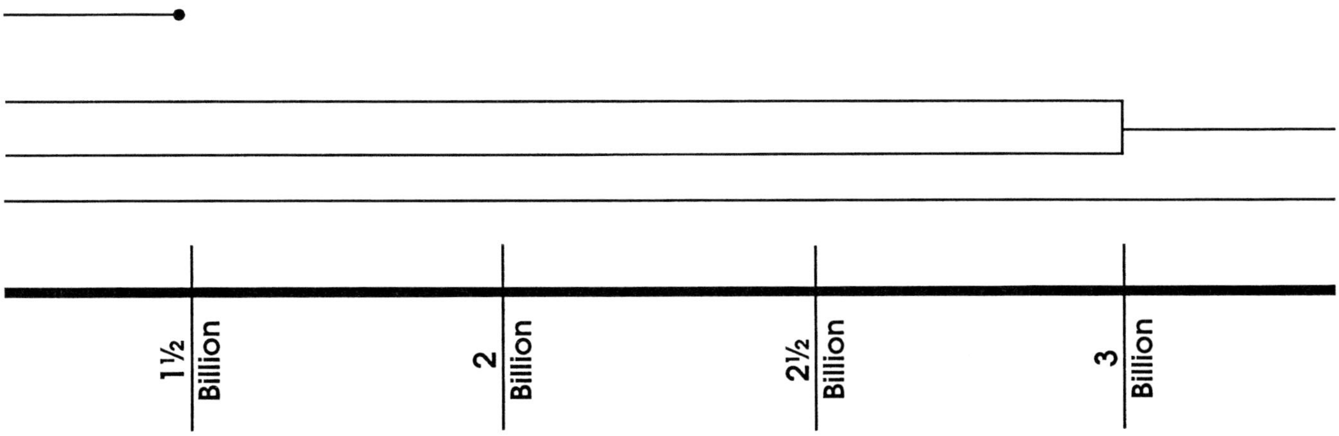

1½ Billion

2 Billion

2½ Billion

3 Billion

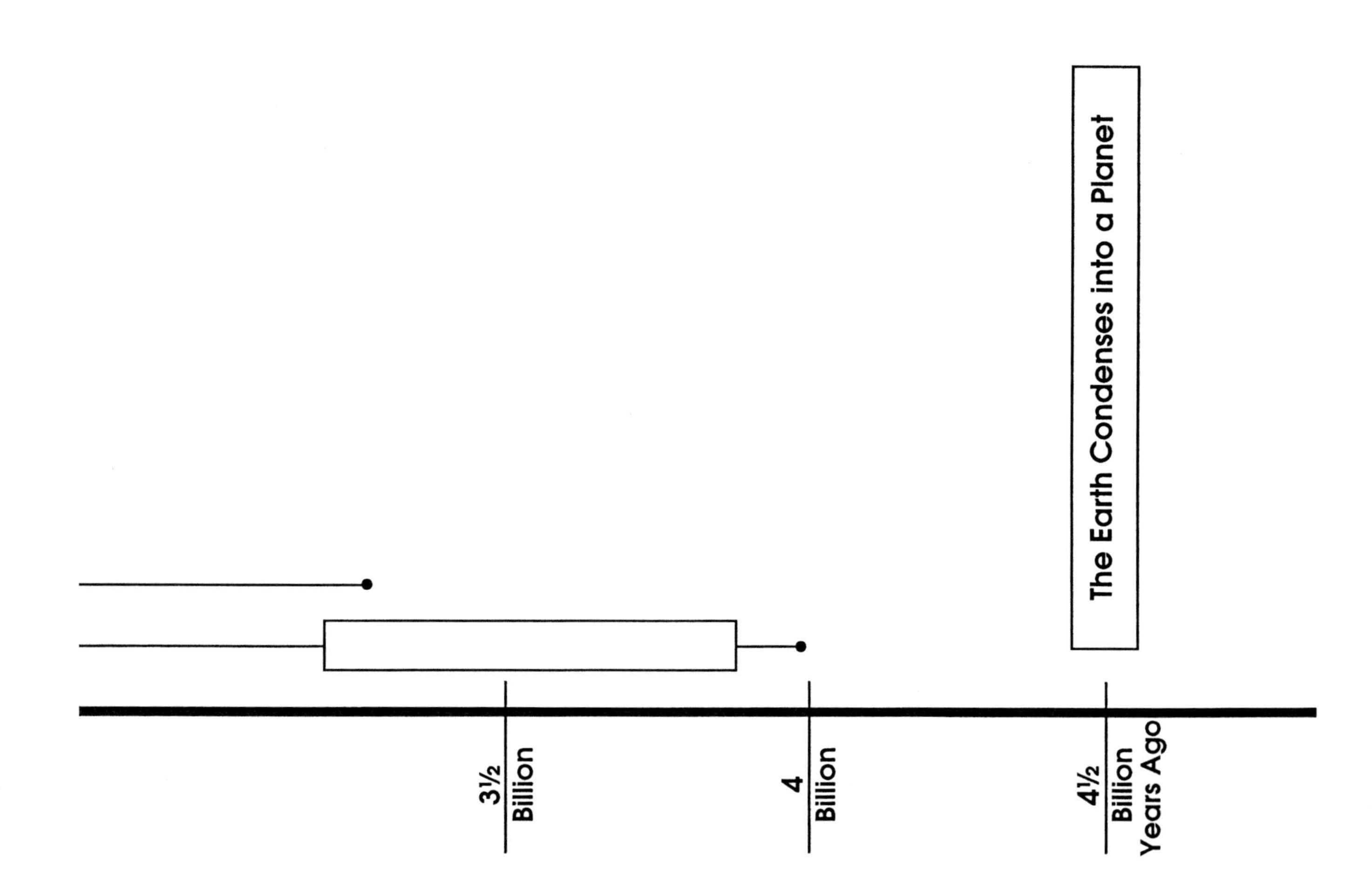
The Earth Condenses into a Planet
3½ Billion
4 Billion
4½ Billion Years Ago

Invertebrates

Fill in the boxes on this Invertebrate Family Tree.
(The bottom of the tree represents older groups, and the top represents newer groups.)

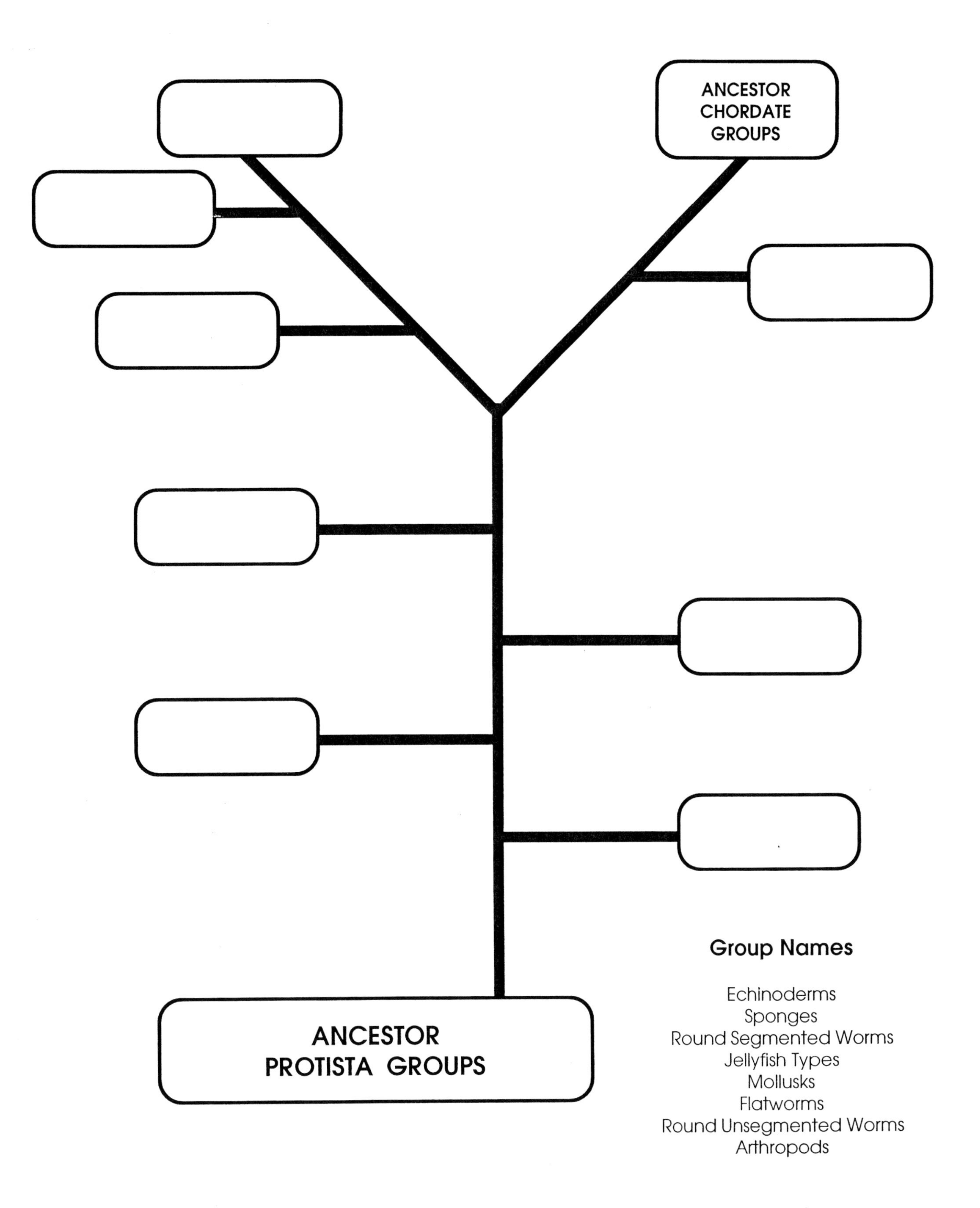

Matching Invertebrates

Make the correct match between the two columns.

Group Names		Defining Features
Sponge ____	A	This group has blocks of duplicated embryonic tissue that develop into different parts in the adult organism.
Jellyfish Types ____	B	This invertebrate group is more closely related to chordates than to other invertebrates. Their larval stage is bilateral, but the adult phase reverted to radial symmetry.
Flatworm ____	C	This group is by far the most successful animal group living on this planet today. And they have an exoskeleton.
Round ____ Unsegmented Worm	D	This group includes clams, squids, octopuses, slugs, and snails. There is a shell during some stage of the animal's development.
Mollusk ____	E	This animal has stinging cells, and has radial symmetry in the adult form.
Round ____ Segmented Worm	F	This animal is the first multicellular animal of the groups considered in this course.
Arthropod ____	G	This is the first group with a complete digestive tract—food in one end and waste out the other end.
Echinoderm ____	H	This is the first group with a head end and a tail end. They have bilateral symmetry.

Vertebrates

Fill in the boxes on this Chordate Family Tree.
(The bottom of the tree represents older groups, and the top represents newer groups.)

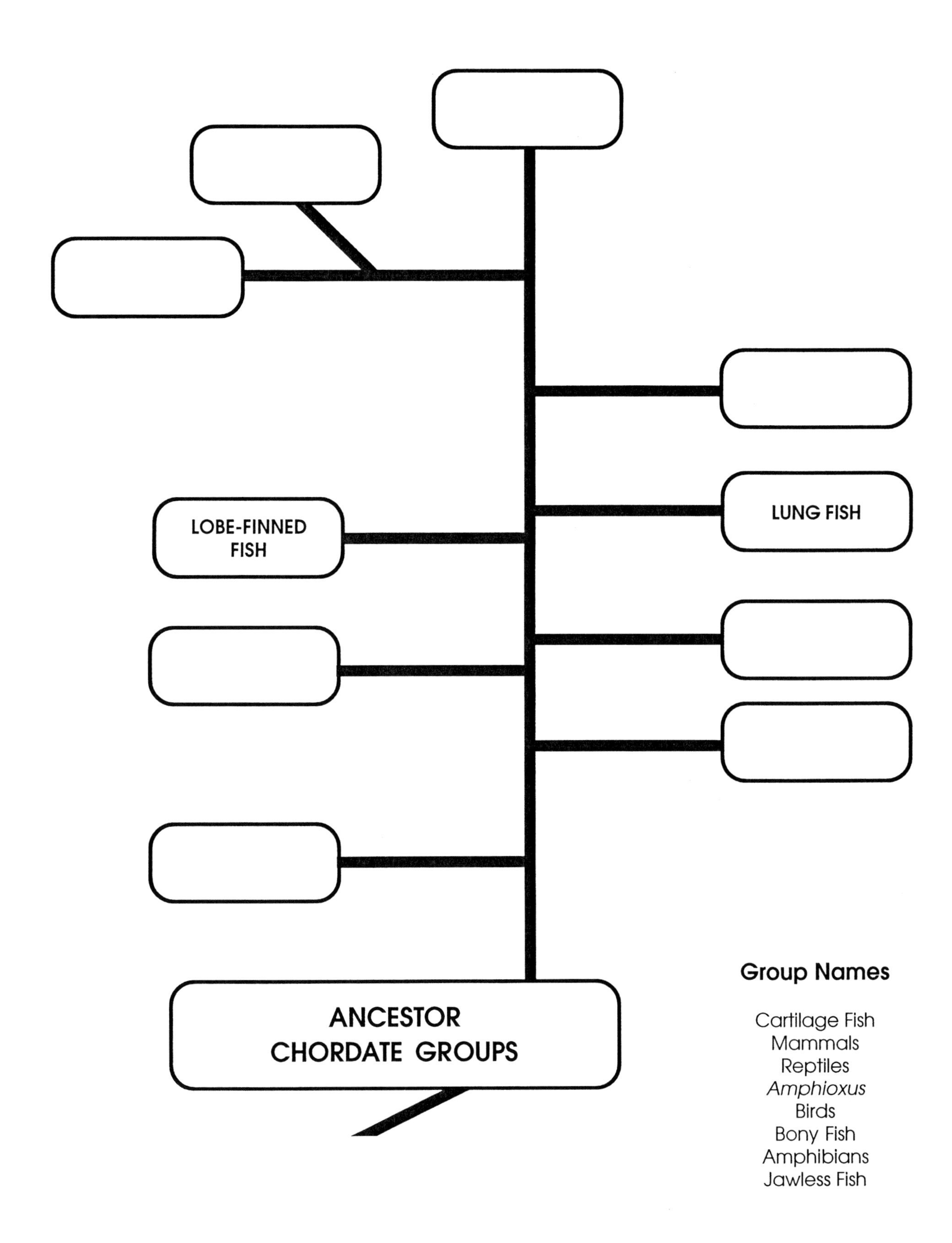

Matching Vertebrates

Make the correct match between the two columns.

Group Names		Defining Features
Chordate (*Amphioxus*) ____	A	First animal with both legs and lungs.
Jawless Fish ____	B	One of two groups that can regulate body temperature.
Cartilage Fish ____	C	Has hair. It is the only group with internal development of the embryo supported by a placenta.
Bony Fish ____	D	A small (5 to 7 cm) fishy-looking creature (but not a fish) that has a notochord, gill slits, and a dorsal nerve cord.
Lung Fish ____	E	First animal with internal fertilization. First to lay an energy rich and protective egg to support the next generation.
Lobe-Finned Fish ____	F	First fish with a lung.
Amphibian ____	G	First fish with limbs.
Reptile ____	H	The most successful fish today.
Bird ____	I	The first fish with a skeleton and a jaw.
Mammal ____	J	A fish creature that can be a foot long or so today. This group has a vertebral column, but no skeleton beyond that. It dominated the seas 400 million years ago and some were very large.

Dimensional Realms

(Tutorial)

"Isn't it amazing? The bumblebee violates the laws of physics when it flies!" Most of us have heard someone say this before, or we have been told about the super strength of ants and the jumping prowess of fleas. These descriptions suggest that some organisms can break the laws of nature, and that it may be human folly to think that we can understand nature.

Actually, stories about animal super-powers reveal very little about the laws of physics or bumblebees, ants, and fleas. The flying structures of bumblebees are exactly what physics would predict for organisms their size. Furthermore, fleas aren't super-bugs—they jump by using legs that can be cocked like springs. This is no more amazing than a human fired out of a circus cannon. All of these feats can be explained. But, why does nature appear to be so different when we compare small creatures with big ones?

The following Exercises will help you to understand why the laws of physics "appear" to change from small-scale events to large-scale events. The concept is called *dimensional realms,* and it explains how different is the world of bumblebees compared to the world of humans.

Exercise #1 "Different Realms—Different Forces" 276

Exercise #2 "The Small Organisms" 278

Exercise #3 "The Large Organisms" 281

EXERCISE #1

"Different Realms—Different Forces"

Always look for the next largest thing—and the next smaller.
—Eliel Saarinen (20TH century architect)

RANGE OF STRUCTURES IN THE UNIVERSE

Scale/Size	Example
10^{25} meters	Farthest known galaxy from Earth. Close to the size of the universe.
10^{21} meters	Milky Way galaxy.
10^{13} meters	Our solar system.
10^{9} meters	Moon's orbit around the earth.
10^{7} meters	Earth.
10^{4} meters	An average city.
10^{2} meters	Largest buildings, boats, and trees.
→ 1 meter	Human scale.
10^{-2} meters	Insects.
10^{-5} meters	Cells of your body.
10^{-8} meters	DNA and large proteins.
10^{-20} meters	Atoms.
10^{-23} meters	Nucleus of the atom.
10^{-26} meters	Quarks: the building blocks of nuclear particles.

Dimensional realms are like different worlds from each other.

A ***dimensional realm*** describes a situation of scale (or size range) in which certain physical forces dominate all objects in that realm, and other forces are much less important.

There are many dimensional realms in the universe, and each realm has its unique set of controlling physical forces. For example, the dimensional realm of small organisms (like bumblebees) is very different from the realm of large organisms (like humans).

EXAMPLES OF DIMENSIONAL REALMS
Subatomic particles
Atoms
Molecules
Cells
Small organisms
Large organisms
Ecosystems
Planet

Physical forces interact differently in each dimensional realm.

The dimensional realms on the previous page could be subdivided into narrower categories if you were studying very detailed changes in physical forces acting in the universe. The concept of dimensional realms encompasses more information than we can consider in this course, so you will focus on two realms—small organisms (smaller than your thumb) and large organisms (larger than your thumb).

Physical Forces

The basic differences between the dimensional realms of large and small organisms is revealed if you compare the relative importance of four physical forces in the everyday lives of these organisms.

The difference between the realms of the large and the small organisms is based on the four physical forces.

- ***Molecular forces*** are those that act between molecules. There are two important types—attraction forces and repelling forces. Attraction forces hold molecules together. Repelling forces push molecules away from each other.
- ***Viscosity*** is the thickness of the medium in which the organism lives. It results from the strength of the attraction forces between the molecules that make up that medium. Air has a very low viscosity. Water has a higher viscosity. Wood has an even higher viscosity.
- ***Inertia*** is the force of an organism as it moves through its environment. The bigger an organism is, the more inertia it has, and the easier it can move through air or water.
- ***Gravity*** is the force of attraction between an organism and the planet. The bigger the organism, the stronger is the force of gravity.

? QUESTION

1. What is a dimensional realm?

2. What important controlling factors change from one dimensional realm to another?

3. List the four physical forces that influence the different realms of small and large organisms.

4. List two dimensional realms that include objects smaller than the small organisms, and list two realms that include objects bigger than large organisms.

EXERCISE #2

"The Small Organisms"

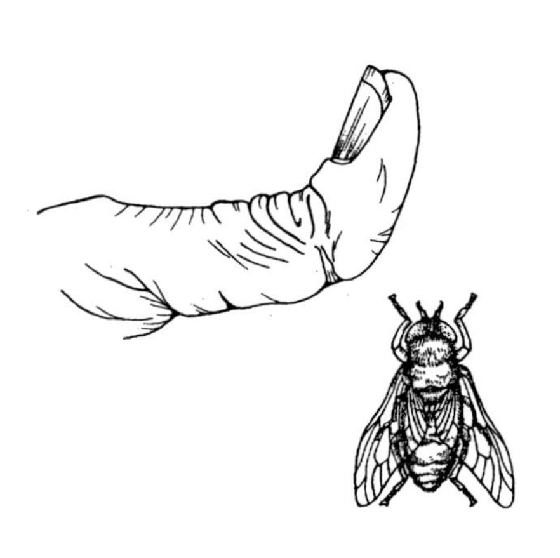

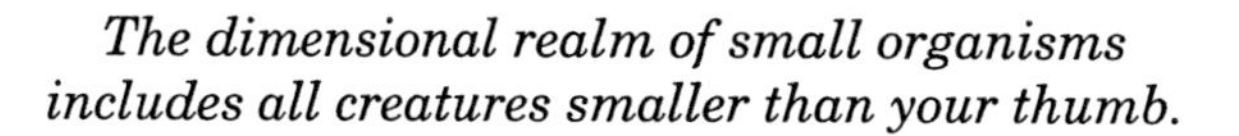

The dimensional realm of small organisms includes all creatures smaller than your thumb.

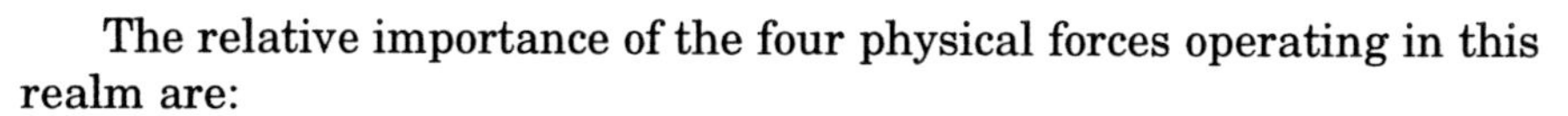

The relative importance of the four physical forces operating in this realm are:

- ▶ Molecular forces dominate the lives of small organisms.
- ▶ Small organisms don't have much inertia.
- ▶ The viscosity of air or water is high compared to the inertia of small organisms.
- ▶ Small organisms aren't affected much by gravity.

In summary, molecular forces and viscous forces play the controlling roles in the dimensional realm of small organisms. Furthermore, inertia and gravity have almost no effect on the lives of small organisms. Using this information about the relative importance of physical forces in the realm of the small organisms, you should be able to explain the unique behaviors of the organisms that are presented on the following pages.

Flying

? QUESTION

1. If a dead mosquito and a dead bird were dropped from the Eifel Tower at the same time, the mosquito would hit the ground long after the bird. This is because the force of ____________________ on the mosquito is weak compared to the thickness (viscosity) of air.

2. It's hard for a mosquito to fly through the air because its force of ____________________ is weak compared to the viscosity of air.

3. A mosquito trying to fly through air is quite comparable to you trying to swim through water. What could you buy at the sporting goods store that would help you swim along?

4. How is the shape of the wings of small organisms (bees, flies, mosquitoes, etc.) related to their ability to fly? Answer this as though small insects have the same problem moving through air as you have moving through water. (Refer to your answer to question #3.)

5. How would you best describe the flying behavior of a small organism? (circle your choice)

Glides through the air or Swims through the air

Because the bumblebee is smaller than your thumb, it needs a paddle-shaped wing in order to fly.

6. Some very small insects have wings like this: Explain the functionality of this wing design.

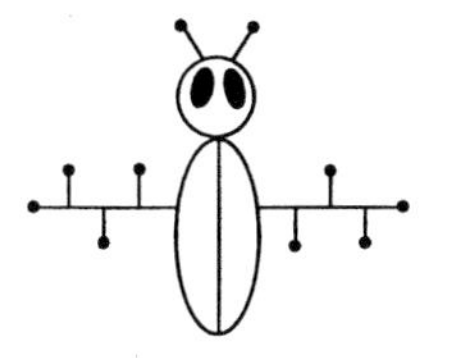

Now you know why bumblebees don't have aerodynamically-shaped wings like birds or airplanes. Bumblebees are smaller than your thumb. They live in the realm of the small organisms, and their wings are designed to help them *paddle* through the air.

Swimming

? QUESTION

1. Flies can't dive after food like a pelican because they don't have enough ____________________ force to break through the surface tension of water.

2. The gliding distance for a protozoan (one-celled organism) is equal to the diameter of an atom. A trout can glide five or more body lengths. Using your understanding of the physical forces controlling the dimensional realm of the small organisms, explain the extremely short gliding distance of a protozoan.

3. Think back to your observations of swimming paramecium in the lab class. Their movement illustrates just how difficult it is for a very small organism to get through the water. What is the swimming pattern in paramecium? (**Hint:** What design would you use if you had to get a piece of steel through wood?)

Walking on Water

? QUESTION

Insects that can walk on water are very small. The surface of water is a film of water molecules holding tightly onto each other, thereby creating surface tension. Using your understanding of the physical forces that control the realm of small insects, explain how insects are able to walk on water.

Walking on the Ceiling

? QUESTION

Some very small organisms have feet with special hair-like or cup-like modifications that stick lightly to surfaces in the environment (like weak adhesive tape). These organisms can walk on walls and the ceiling as long as the molecular attraction between their feet and the ceiling is greater than the strength of ____________________ (a physical force).

On Being Small

The advantages of being small include:

- It's easy to hide.
- You don't need much to eat.
- You reproduce and evolve quickly.
- Gravity poses no problems on your structural design or movement.

But there are some disadvantages. Small organisms can't move as far or as fast as large organisms. Viscous forces and molecular forces dominate their lives. And small organisms can't use inertia or gravity to their advantage like large organisms.

Since the majority of life-forms are small, there must be an advantage.

For example, ants could never live like humans. They can't control changes in their environment like fire. They don't have enough brain cells for memory functions. An ant hammer would not produce enough inertia to drive an ant nail into wood. The pages of an ant book could not be turned because the molecular forces holding the paper pages together are too strong for ants to overcome. Ants couldn't pour liquids because water sticks together as drops, and water drops are too big for them to handle.

The dimensional realm of small organisms is dominated by molecular forces, and the roles of inertia and gravity are practically negligible in the lives of these organisms. As you shall see in the next Exercise, the realm of the large organisms is very different from the world of the small.

EXERCISE #3

"The Large Organisms"

The dimensional realm of large organisms includes all creatures bigger than your thumb.

The bigger they are, the harder they fall.

The relative importance of the four forces operating in this realm are:

- The lives and structure of large organisms are dominated by gravity.
- Large organisms have a lot of inertia when they move.
- The viscosity of air and water are low compared to the inertia of large organisms.
- Molecular forces don't dominate the lives of large organisms.

In summary, inertia and gravity play controlling roles in the dimensional realm of large organisms. In fact, as size increases from mouse to elephant, gravity and inertia become even more important. Molecular forces and the viscosity of air or water do not affect the lives of large organisms to the same extent that these forces affect small organisms.

Size, Weight, and Strength

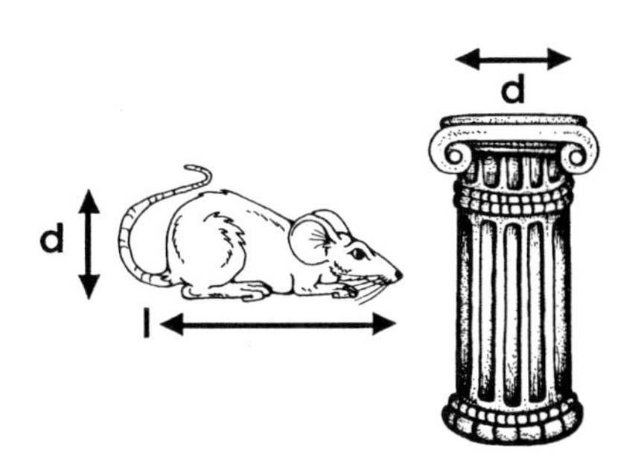

The size of an organism can be described by using length (l) and diameter (d). The weight of an organism is directly proportional to its volume. Volume can be estimated by multiplying the length x the diameter x the diameter (ld^2).

The weight of an organism must be supported by its strength. The strength of an organism's supporting structure (bone and muscle) is directly proportional to the cross-sectional area of the structure. The thicker a leg is, the stronger it is. Cross-sectional area can be estimated by multiplying the diameter x the diameter (d^2).

Procedure

Fill in the table below by making the appropriate calculations. You will discover how weight and strength do not change equally as the organism increases in size.

Example	Size	Weight (ld^2)	Strength (d^2)
mouse	l = 2 d = 1		
rat	l = 4 d = 2		
cat	l = 8 d = 4		

? QUESTION

1. What structural problem develops as the size of the organism increases?

2. To correct for the above problem, how must the structure of larger organisms change? (Draw a picture comparing the leg bone of a mouse and an elephant to illustrate your answer.)

Flying in Birds

The weight of a bird in flight must be supported by lift, which is created by air flow over the surface of the wings. Lift is directly proportional to the area of the wings (area = length x width).

? QUESTION

1. If you enlarged a small sparrow-like bird to a bigger and bigger size, which would increase faster? (circle your choices)

   body size or wing area   weight or lift

2. What eventual consequence is created by continuing to enlarge a bird?

   Name a bird that exhibits that consequence. ____________________

3. There is an evolutionary solution to the problem presented in the previous two questions. The solution is for the large bird to change its basic body form. How is the body form of a big flying bird different from the form of a small bird? Draw the outline of a small bird and a large bird to illustrate your answer.

Small Bird	Large Bird

Evolutionary Changes in Size

When times are good, life gets big.

In general, large organisms are more specialized than small organisms, and large organisms are more efficient at utilizing their food resources. During fairly stable environmental conditions, large organisms have the advantage over small ones. However, during environmentally unstable times, big organisms tend to die off. A big animal needs more to eat than a small animal. This is a serious problem when there is a shortage of food. Also, larger species evolve more slowly because their generation time is longer. They can't adapt to environmental changes as fast as small organisms.

? QUESTION

1. Which physical forces are the most important in the dimensional realm of large organisms (bigger than your thumb)?

2. What are the advantages and disadvantages of being a large organism?

3. What are the advantages and disadvantages of being a small organism (smaller than your thumb)?

4. How would you characterize the environment during the time when the dinosaurs were most successful?

5. Birds and mammals survived the extinction that killed the dinosaurs. Describe the size of birds and mammals at the time that dinosaurs became extinct, and explain how their size may have allowed them to survive.

In Conclusion

You now have a general idea why small organisms and large organisms have lives that are so different. They live in different dimensional realms! It is natural to apply your assumptions about human life to a smaller realm. But, understanding the controlling factors that influence each realm explains why the appearance of organisms changes dramatically from one realm to another. These general principles apply throughout the universe.

The life of a cell is not the same as the life of a whole organism. The life of a bumblebee is not the same as the life of a bird. It's all a matter of understanding dimensional realms!

Embryology

Embryology is the study of early developmental phases in plants and animals. An ***Embryo*** is any stage after the egg becomes fertilized, but before the developing organism looks like the adult form (about 8 weeks for humans). In advanced vertebrates, any stage after the embryo and before birth is called a ***fetus***.

A century ago, there were two contrasting theories about development. One view (*preformation*) held that a miniature version of the adult existed in the egg and grew into the adult form. This theory is incorrect. The second theory (*epigenesis*) stated that new structures and body systems are continually being created during embryonic development. This description is correct. All vertebrates begin as a single cell, but develop amazing degrees of complexity by the time they are born.

Scientific research has revealed two very important embryological concepts. First, the study of embryos shows how simple life-forms could have evolved into complex life-forms. The developing traits provide biologists with the opportunity to observe how genetic mechanisms operate to create an organism. The second concept is that there are developmental processes common to all vertebrates. This means that studying the embryos of simpler organisms can help us to understand development in human embryos.

In today's lab, you will learn the general terminology and descriptions of embryonic stages, and you will compare the embryos of various organisms at similar stages found in human development.

Exercise #1 "Ontogeny Recapitulates Phylogeny" 285
Exercise #2 "Eggs, Sperm, and Fertilization" 288
Exercise #3 "Methods of Feeding the Embryo" 294
Exercise #4 "Stages of Development" 295
Exercise #5 "Disruption of Normal Development" 300

EXERCISE #1

"Ontogeny Recapitulates Phylogeny"

The title of this Exercise has been a recurring TV quiz-show question. The quote is from Ernst Haeckel, a German scientist at the turn of the 20TH century. He thought that embryonic development replayed the entire evolutionary history of a species. In other words, watching the development of a frog would be like seeing a movie of the evolution of vertebrates leading to the frog. Since Haeckel's time, science has discovered that both evolution and development are far more complicated than his original idea suggested.

The human embryo does proceed through various levels of complexity similar to those of earlier vertebrates. There is certainly as much change in the human embryo from conception to birth as there is change in the fossil record from one-celled organisms to complex land vertebrates. The situation faced by the human embryo is similar to the challenges presented to vertebrate groups that evolved from a water environment to dry land conditions. But, we do not first become a fish, then a frog, followed by a reptile, and finally a mammal. The developmental "movie" is much more blurred than that.

The embryonic development of an animal provides clues to its evolutionary history.

Procedure

Read the brief descriptions of human embryonic stages below, and then refer to the "Ontogeny Recapitulates Phylogeny" chart on the next page. Write the name of the appropriate human embryonic stage that compares in complexity to organisms in evolutionary history.

	HUMAN EMBRYONIC STAGES
Name	**Description**
Zygote	A diploid cell formed by the union of egg and sperm.
Morula (3–4 days)	A small solid ball of identical cells.
Blastocyst (1 week)	Cells begin differentiating into tissues.
25 Day Embryo	The heart begins to beat, but there are no organ systems yet (i.e., circulatory systems, etc.).
4 Week Embryo	The *notochord* has formed. This is the very beginning of a skeletal system that will develop in later weeks.
4–5 Week Embryo	Organ systems have developed including a primitive type of kidney called the *mesonephros.* This kidney will soon degenerate and be replaced by a new one.
5–6 Week Embryo	Organ systems are developing very rapidly. A new type of kidney appears called the *metanephros.* This will eventually grow into the adult human kidney.
6–8 Week Embryo	There is a rapid development of a brain lobe called the telencephalon which grows into the *cerebrum* in later weeks.
8 Week Fetus	This stage is now called the *fetus* because it looks like the adult form.

There is certainly as much change in the human embryo from conception to birth as there is change in the fossil record from one-celled organisms to complex land vertebrates.

ONTOGENY Developmental Stages	**RECAPITULATES** Repeat	**PHYLOGENY** Evolutionary History
________________		Pig and Primate Embryo (developed cerebrum of brain)
		↑
________________		Chick Embryo (has dry land kidney called *metanephros*)
		↑
________________		Frog Embryo (has aquatic kidney called *mesonephros*)
		↑
________________		Early Vertebrates (beginning of internal skeleton)
		↑
________________		Flatworms and Roundworms (definite organs present)
		↑
________________		Jellyfish (cells organize into tissues)
		↑
________________		Colonial Protozoa (cluster of identical cells)
		↑
________________		Eukaryotic Cells (unicellular with organelles)
		↑
(no comparable stage in human development)		Prokaryotic Cells (no cell organelles)

? QUESTION

1. Define *embryo*.

2. Define *fetus*.

3. What week of development marks the beginning of the fetus stage in humans?

4. What is meant by "ontogeny recapitulates phylogeny"?

5. Is it true that the human embryo first looks like a fish, a frog, and then a reptile before reaching the fetal stage? Explain your answer.

EXERCISE #2

"Eggs, Sperm, and Fertilization"

Eggs

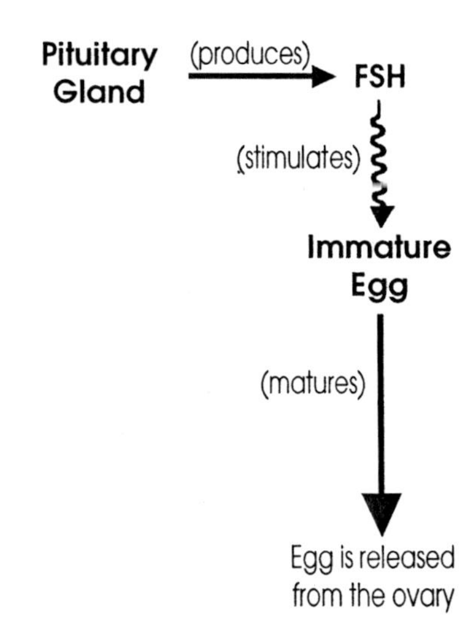

By the fifth week, a human female embryo already has pre-egg cells multiplying in her ovaries. These pre-egg cells begin meiosis and develop into a cell stage called the *oocyte*. Each oocyte is surrounded by a small cluster of cells called the ***follicle***. There are about a million oocytes in the ovary of a female baby at birth. These immature eggs are stopped at Prophase I of meiosis until she reaches puberty. This is a significant vulnerability because the chromosomes of oocytes are exposed to many environmental chemicals during childhood. Chromosomes can be damaged by these chemicals.

The oocytes mature into eggs beginning at puberty. Usually one egg is released (called ***ovulation***) each lunar month. This monthly process, called the ***menstrual cycle***, is initiated by follicle stimulating hormone (***FSH***), which is produced by the pituitary gland (one of the glands of your endocrine system).

As the egg matures, the other cells surrounding it divide and grow. Most of this growing cellular mass remains in the ovary after the egg is released. It is called the ***corpus luteum***. There is a fatty substance stored in the corpus luteum that contains the hormones ***estrogen*** and ***progesterone***. These hormones are released into the bloodstream and stimulate the growth of the inner uterine wall (***endometrium***) for possible implantation of an embryo.

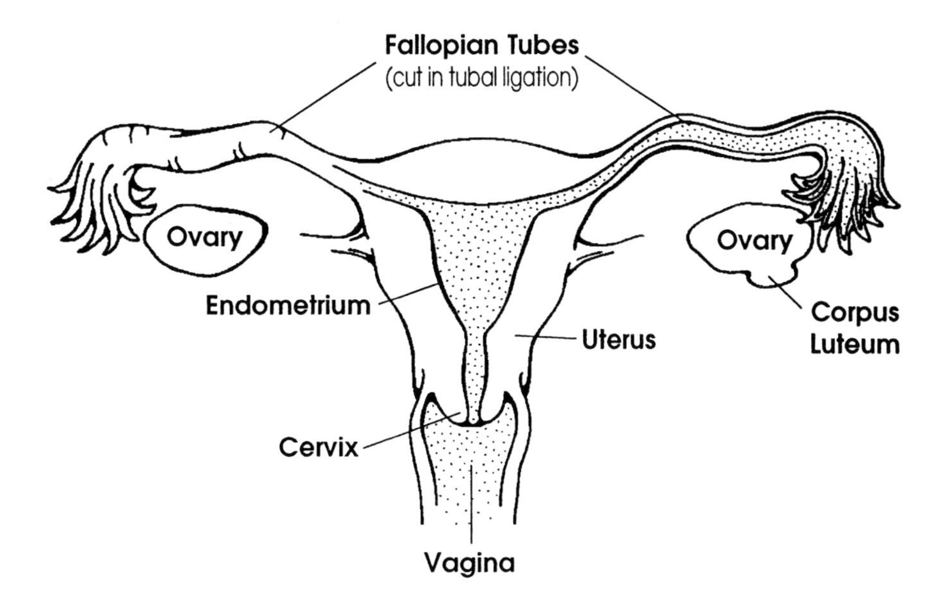

The monthly menstrual cycle is fairly complex but can be summarized as:

Day 1 Menstrual flow begins from previous month's cycle. One egg begins to mature in the ovary for release in the next cycle.

Day 14–15 Ovulation. (Ovary releases an egg.) The corpus luteum produces hormones that stimulate the endometrium of the uterus to grow.

Day 28 If the egg isn't fertilized, then the uterus lining is shed with the menstrual flow.

? QUESTION

1. Health officials warn: "The eggs of your daughters can be damaged by certain chemicals you are exposed to during pregnancy." How can the eggs of your daughter be damaged when she won't start menstruating until 12 or 13 years after her birth?

2. Which endocrine gland produces the hormone that stimulates an egg to mature?

3. Where does the corpus luteum form, and what hormones does it produce?

4. What day of the menstrual cycle does ovulation usually happen?

5. The menstrual flow is a shedding of the ____________________ lining.

Materials

- A compound microscope.
- Three microscope slides: Immature Egg
  Mature Egg
  Corpus Luteum

Procedure

[1] Work with two other lab groups. Put a different slide on each microscope, and use low magnification to get an overview of the sectioned ovaries.

[2] The immature eggs are bigger than most other cells in the ovary. A maturing egg cell is larger and has a ring of cells surrounding it. The spherical mass of cells is called the follicle. The corpus luteum is even larger than the follicle, and is a solid mass of cells without an egg inside. (The egg has been released.)

[3] Look back and forth among the three slides until you can see the difference in the structures.

[4] Draw each structural stage in enough detail so that you can find it again (perhaps on a test) with the help of your picture. *Return the slides when you are finished.*

Immature Egg	Mature Egg	Corpus Luteum

Sperm

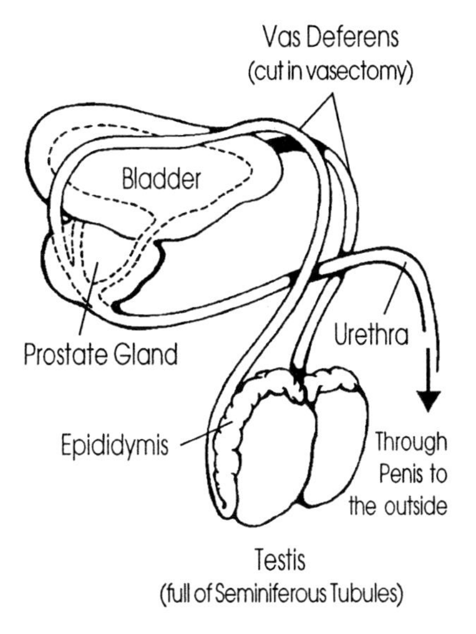

The reproductive system of the human male embryo is noticeably differentiated by the eighth week. However, the male fetus does not begin meiotic division of sex cells as does the female fetus. The female has immature eggs that are stopped at Prophase I of meiosis. The male doesn't begin sperm production until puberty.

It takes about three weeks for sperm to be produced. If a male is exposed to damaging chemicals, only the current "batch" of sperm is affected. After a month, sperm production could be back to normal. Of course, chromosome damage does sometimes happen in the male gametes as in the female. But, consider how little time sperm chromosomes are exposed to potential damage by environmental factors compared to egg chromosomes.

Sperm are produced (200–500 million daily) along the inside of an extensive system of tubes (***seminiferous tubules***) in the testes. They are stored in enlarged tubes called the *epididymis.*

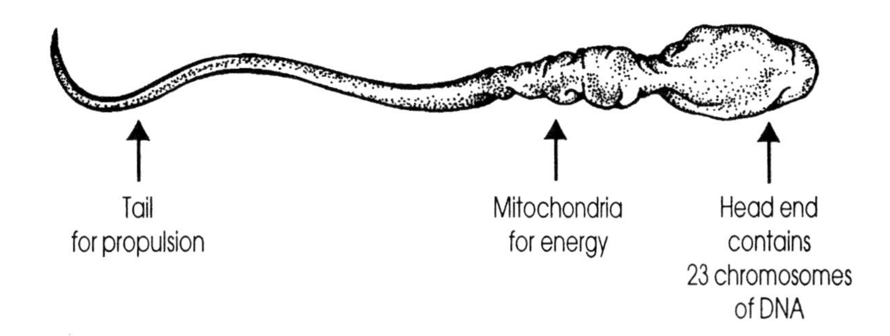

? QUESTION

1. Explain how sperm are less vulnerable to chromosome damage than eggs.

2. Where specifically are sperm produced?

3. Where are they stored?

Materials

- Two microscope slides: Testis
  Human Sperm

Human Sperm

Procedure

[1] Examine the cross-section of a testis, and find the seminferous tubules filled with various stages of sperm production.

[2] Draw a picture of the testis. Label the seminiferous tubules and sperm.

[3] Examine the slide of human sperm. Identify the head, mitochondrial region, and tail. Make a quick sketch and label it.

Testis

Sperm

[4] *Return the slides.*

Fertilization

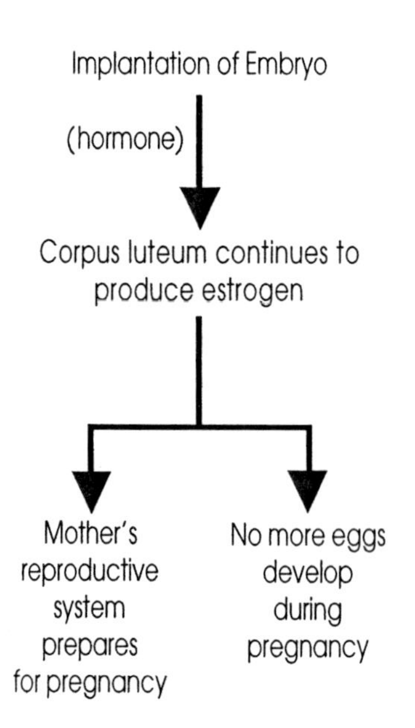

The human egg is released from the ovary with a layer of follicle cells stuck to its outside. Sperm cannot fertilize the egg until this layer of cells has been removed. Each sperm cell produces a small amount of enzyme that dissolves the jelly-like glue holding the small cells to the egg. Much of this enzyme is necessary to separate the follicle layer from the outside of the egg.

Eventually, enough of the protective layer surrounding the egg is dissolved, and one of the sperm penetrates the egg. Once that happens, the egg membrane reacts by forming a new layer of jelly around the egg, thereby preventing any more sperm from entering it. If more than one sperm fertilizes an egg, the resulting embryo dies.

Fertilization usually occurs as the egg moves through the fallopian tube towards the uterus. The embryo continues to develop for about one week before it implants in the wall of the uterus. After implantation, a hormone from the growing embryo signals the corpus luteum (in the ovary) to produce large amounts of estrogen, which then stimulates the mother's reproductive system to prepare for pregnancy. Estrogen also prevents new eggs from developing during pregnancy. (This is why estrogen can be used for birth control.)

? QUESTION

1. It is necessary for a male to have a high sperm count to be fertile. If only one sperm is necessary to fertilize the egg, then what function is served by the other sperm?

2. What event signals the mother's reproductive system to prepare for pregnancy?

3. What prevents more than one sperm from fertilizing the egg?

Materials

- A depression slide and coverslip.

Procedure

If your instructor has been able to get mature sea urchins, then this is the time to watch fertilization actually happen.

1. Your instructor will inject the urchins with a salt solution that stimulates the release of their gametes. The urchins are then placed upside-down in a beaker of sea water. The gametes will flow out of the animal.
2. Sperm looks milky. The eggs are granular and usually have a slight pink or yellow color.
3. Put a drop of the eggs into a depression slide (no coverslip). Carefully place the slide on your microscope and examine the unfertilized eggs using the 10x objective lens.
4. The sperm must be diluted with sea water because too many sperm causes abnormal fertilization. If the sperm collection beaker looks slightly milky, then dilute the sperm further. Add a drop of the sperm to your egg slide. Cover the slide with a coverslip, and immediately observe the events under the microscope. A ***fertilization membrane*** usually forms within 2 minutes.
5. When you've seen the membrane form, put this slide aside, and recheck it every 30 minutes. Don't leave the slide on the microscope with the light on because fertilized eggs will soon overheat. If everything goes well, the first ***cleavage*** (division) should happen in about an hour. Watch for it!
6. Draw pictures of the following:

Unfertilized Egg	Fertilization Membrane	Cleavage

EXERCISE #3

"Methods of Feeding the Embryo"

There are two strategies for embryonic development: fast, with the embryo soon on its own; or slow, with the embryo fed by yolk or placenta.

Quick-Development Self-Feeding

Invertebrates usually give no special care to their embryos. The only source of food during the first stages of development is provided by the cytoplasm in their eggs. After fertilization, there is rapid cell division and development to allow the embryo to feed on its own. As one example, the sea urchin embryo develops to a self-feeding stage in 24 hours or less. There are many small pieces of organic matter and micro-organisms in water for young embryos to eat. (You will see sea urchin embryos during Exercise #4.)

Yolk

Nutrition in the form of ***yolk*** is another way of giving food to the embryo. Fish are examples of organisms that have small amounts of yolk in the egg. Birds and reptiles have much more yolk in their eggs. Larger amounts of yolk provide longer possible times for development. This is especially necessary for the more advanced land vertebrates.

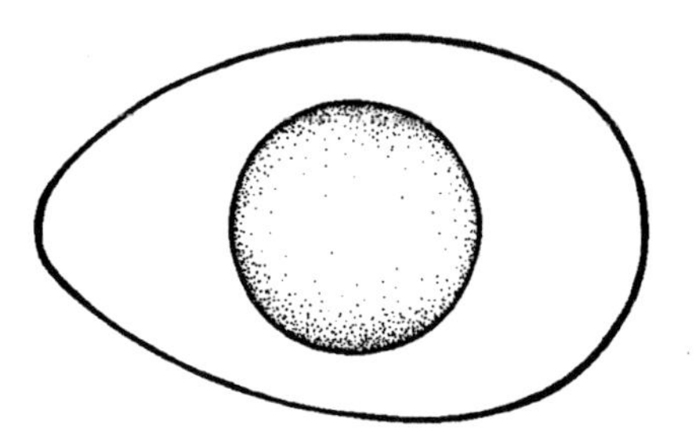

Reptile & Bird
Egg Cell
(very large yolk)

Frog
Egg Cell
(more yolk than urchin)

Sea Urchin & Human
Egg Cell
(very little yolk)

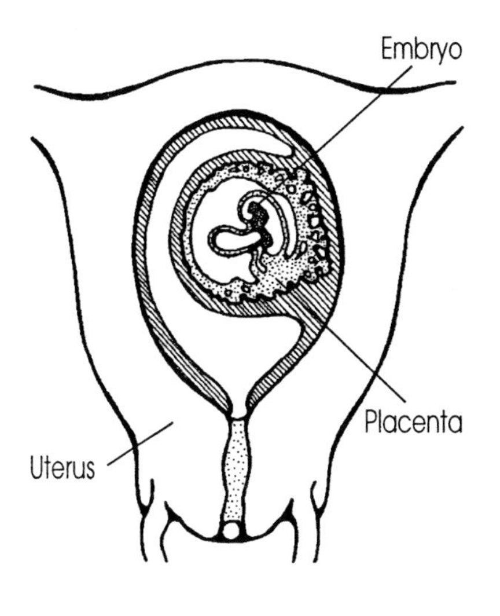

Placenta

Mammals provide nutrition to their embryos through a specialized integrated feeding structure called the ***placenta***. It is formed partly by the embryo and partly by the mother. Nutrients, wastes, and blood gases are exchanged between the embryo's blood system and the mother's blood system.

The human embryo, from the zygote to blastocyst stage, must survive on the cytoplasm of the original egg cell. The egg divides but cells don't grow in size. In about one week the developing human embryo begins to implant into the uterine wall for the next stage of feeding (the placenta).

The placenta begins as a few specialized cells on the outside of the implanting embryo. These special cells digest their way into the wall of the uterus. Soon after implantation, finger-like projections begin the formation of a more efficient structure. By the second month, the placenta is well developed, and continues to grow with the fetus.

? QUESTION

1. List three methods for feeding early embryos by different organisms.

2. Why do complex animals require more specialized embryonic feeding mechanisms?

3. Why do most land animals require a more developed embryonic feeding mechanism than aquatic animals?

EXERCISE #4

"Stages of Development"

During this Exercise, you will examine 10 stages of vertebrate development using various organisms from sea urchin to pig embryos to illustrate each stage. We will skip many of the details, and focus only on a few structures as examples of the events.

These animal examples are very similar to human developmental stages.

Cleavage, Morula, and Blastocyst

The first divisions in development are called ***cleavage*** because the zygote divides (or cleaves) into two cells, then four, then eight, etc. Eventually, a small solid ball of cells forms called the ***morula***. Cells of the morula continue to divide, producing a hollow ball of cells called the ***blastocyst***. The human blastocyst develops in about one week after fertilization. You can get an idea of what these early stages look like by observing early embryos of sea urchin and frog.

Materials

- Two microscope slides: Early Sea Urchin Development
  Frog Blastula (stage 8)

Procedure

[1] Examine the early developmental stages of the sea urchin.

[2] Compare the sizes of the earliest embryos undergoing cleavage. Are the stages from zygote to morula about the same size? ____________

Are these early cells growing or just dividing? ________________

[3] Draw pictures of these stages:

Zygote	Early Cleavage	Morula

[4] Examine the early frog embryo.

[5] The frog blastocyst (also called *blastula*) is a partial hollow ball of cells. You should be able to see several different cell types. These cells are changing into new kinds of cells for the next stage of development.

[6] Draw a picture of the frog blastocyst, and show the different cell types.

[7] *Return the slides.*

Frog Blastocyst

Early Chick Embryos

Materials

- Two microscope slides: 21-Hour Chick Embryo
  28-Hour Chick Embryo

Procedure

1. Work with another lab group. Put the two slides on different microscopes so that you can look back and forth between them.
2. The 21-hour embryo shows early development of the nervous system. The long ***neural fold*** becomes the spinal cord in a later stage. Perhaps you can see that the head end is slightly larger.
3. The 28-hour embryo has a more developed neural fold, and you can see that the brain end has enlarged. Also, there are several paired blocks of tissue along the ***neural tube***. These blocks are called ***somites***. The somites develop into the internal organs, skeleton, and muscles of the adult body.
4. Draw pictures of the two embryonic stages, and label the neural fold, neural tube, head end, and somites.
5. *Return the slides.*

21-Hour Chick Embryo

28-Hour Chick Embryo

Later Chick Embryos

Materials

- Two microscope slides: 56-Hour Chick Embryo
  72-Hour Chick Embryo

Procedure

These chick embryo stages are comparable to three and four-week-old human embryos.

[1] Put the two slides on different microscopes for comparison. These are side views of the embryos, which are shaped like a question mark (either forward or reversed depending on how the slide was made).

[2] You should be able to see changes in the brain development between the two embryonic stages. The brain consists of several lobes. In the 72-hour stage there is a lobe in front of the lobe with the eye in it. That front lobe is the ***cerebrum***.

[3] The somites are easy to see. How many are there? ________________

[4] There is a structure bulging from the middle of the embryo, and positioned below the brain. This is the developing heart. In the older embryo, the bottom chamber of the heart is larger. That larger chamber is the ***ventricle***.

[5] Draw a picture of the 72-hour embryo showing brain and spinal cord, eye, heart, and somites. Label your drawing.

[6] *Return the slides.*

72-hour Chick Embryo

Pig Embryo

Materials

- The slide of the 10-mm pig embryo.

Procedure

The 10-mm pig embryo is comparable to a human embryo stage at six to eight weeks.

1. You should recognize a well developed nervous system, including backbone, large brain, but smaller eyes than in the chick embryo.
2. Can you find the ***limb buds***? In the human embryo these buds will develop into arms and legs.
3. Just below the heart is a large dark structure, the ***liver***.
4. Draw a picture of the pig embryo showing brain, eye, vertebral column, heart, liver, and limb buds. Label your drawing.
5. *Return the slides.*

10-mm Pig Embryo

EXERCISE #5

"Disruption of Normal Development"

Most people have a profound misunderstanding of what can go wrong during development.

Considering all of the events that must go exactly right during embryonic development, it is easy to see that events can go wrong. The general principle for understanding developmental disorders is:

The earlier the disruption,
the more serious are the consequences.

Some researchers estimate that one-half of conceptions do not make it to birth. Women are usually unaware of the earliest embryonic failures, which are also the most common. However, there are several disruptions of later development that are very serious and reveal themselves at birth (4–5% of births), or as late-term miscarriages.

Gene Mutations

Gene mutations can destroy the normal production of essential enzymes controlling basic metabolic processes in the fetus.

Most states require the immediate testing of all babies for PKU (phenylketonuria), which is a deficiency of an enzyme responsible for the metabolism of the amino acid *phenylalanine*. This disorder is similar to other gene mutations—no product is formed, and there is a buildup of the substrate.

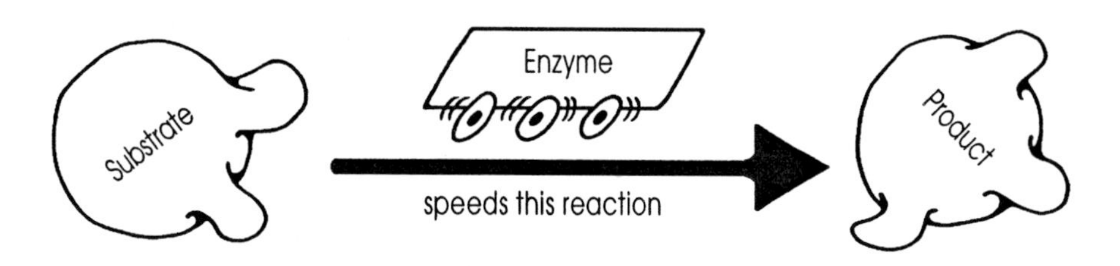

High concentrations of these substrates are usually toxic or disruptive, and often the product is essential for other processes. Tay Sachs, galactosemia, and tyrosinosis are other metabolic disorders resulting from gene mutations.

Chromosomal Aberrations

When mistakes occur during meiosis of sperm or egg production, abnormal numbers of chromosomes in the embryo are possible. These events are almost always lethal. However, some chromosomal aberrations (X, XXY, and three #21 chromosomes) occur in about 1% of births. These babies have serious health problems. The *trisomy* of #21 chromosomes produces Down's syndrome, the most common form of mental retardation. The other chromosomal aberrations also result in similar problems.

Possible Sites of Implantation

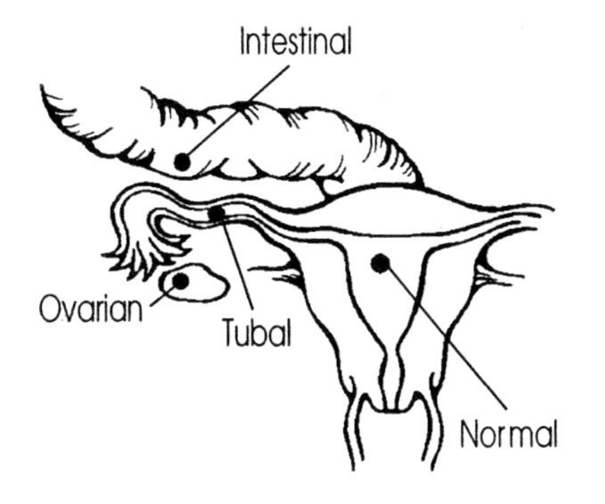

Ectopic Pregnancy

Occasionally, the embryo implants in the wall of a fallopian tube, or on the outside of the uterus or intestine. There are two serious consequences of this abnormal event. First, the child will have to be delivered by Cesarean section. The second serious problem is that the placenta attaches to the wall of the wrong organ. That organ can be damaged or destroyed by placental development. This condition is so dangerous to the mother's life, that termination of pregnancy may be her only chance of survival.

Incomplete Development of Organ

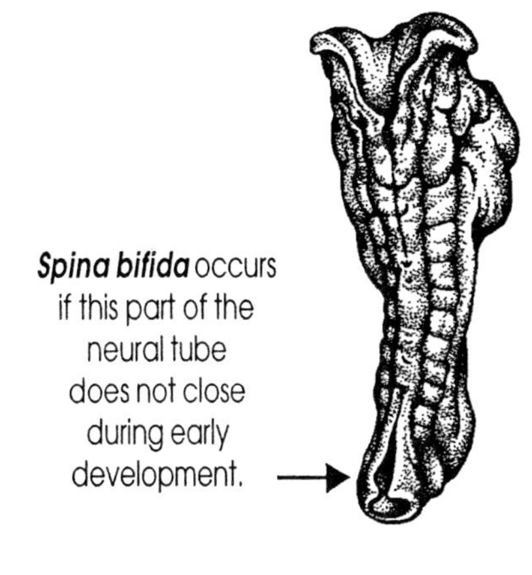

Spina bifida occurs if this part of the neural tube does not close during early development.

Sometimes an organ does not develop completely during early development. The cause can be a foreign chemical like thalidomide, or a disease like measles. Common examples are spina bifida (neural tube doesn't close), or holes between the heart chambers.

The general principle of developmental disorders applies especially in these cases. The earlier the problem happens, the more serious are the consequences.

Bad Habits During Pregnancy

We have very little control over the previously described disruptions of normal development. In fact, many early miscarriages are the result of the mother's hormonal system "warming up" for future reproductive success.

However, there are disruptions in development that we *can* control—bad habits during pregnancy. Science makes a certain direct statement about bad habits during pregnancy:

"Drink alcohol, smoke cigarettes or marijuana, use drugs, or have poor nutrition, and your child will be below normal in mental abilities, and may be physically deformed."

Fetal Alcohol/Drug Syndrome

This fetal damage is predictable and results from the mother's drinking or drug habit. Fetal Alcohol/Drug Syndrome includes characteristic facial deformations, a degree of mental retardation, and defects in internal organs (especially the heart and liver).

Cigarette Smoking

The majority of babies born by mothers who smoke, have a smaller placenta, lower birth weight, and lower overall health. The effects on the fetus are not usually as extreme as with alcohol or drug use by the mother, but the birth results are 100% predictable.

Poor Nutrition

The usual problem with poor nutrition in this country is not a lack of Calories, but an unbalanced diet (not enough protein, vitamins, and minerals). The effects of poor nutrition are lower birth weight, lower overall health, and below-average mental ability.

If the nutritional problems during pregnancy are severe, as seen in overpopulated parts of the world, all of the detrimental effects during embryonic and fetal development are greatly magnified.

? QUESTION

1. What is the general principle for understanding developmental disorders?

2. Perhaps as many as __________ of all embryos don't survive to birth.

3. What is the percent of births with developmental defects?

4. What is the most serious potential risk in an ectopic pregnancy?

5. List the two direct biochemical effects of most gene mutations.

6. What is the usual result of chromosomal aberrations?

7. What are the common effects on embryonic development caused by alcohol, drugs, cigarettes, and poor nutrition?

Diet Analysis

Americans spend more money on high-Calorie fast foods and low-Calorie diet foods than anyone else in the world. In addition, more books have been written on topics related to body weight than any other health issue. The next time you are in the supermarket, notice how much shelf space is devoted to ailments of the digestive system.

This week's lab will reveal some insights gained from using a scientific approach to the study of your diet.

Exercise #1 "Energy Balance of the Body" 303
Exercise #2 "Nutrients in Food" 305
Exercise #3 "Calories from Nutrients" 307
Exercise #4 "Dieting Strategies" 309
Exercise #5 "Energy Expenditure During Daily Activities" 311
Exercise #6 "Special Considerations" 313

EXERCISE #1

"Energy Balance of the Body"

Constant Body Weight
Energy Input = Energy Output

The energy balance of the body is a simple principle that explains weight changes. **Remember:** ***If energy input*** (food eaten) ***equals energy output*** (food metabolized for activity), ***then the body weight remains constant.***

If more energy (food) goes into the body than is used by the metabolic processes, then your body weight will *increase* (fat storage). Likewise, if less energy goes into the body than is used, body weight will *decrease.*

Energy Input

The total chemical energy in a food can be determined by measuring the amount of heat given off when that food is burned. However, several complications arise when human foods are analyzed. Some of the chemical energy in food cannot be digested. A primary component of plants is *cellulose* (cell walls). Our stomachs do not produce the enzyme to break this starch into sugars for energy. Termites and cows, however, do have this ability, which is why they can get more energy from eating plants.

Another complication in calculating the energy input for humans is that some of the digestible nutrients in our food aren't completely absorbed by the small intestine. Absorption varies from person to person. It may depend on the concentration of digestive enzymes or on the speed that foods are moved through the digestive tract. This is similar to other physiological differences in metabolism. For example, some people have inherited an ability to easily store fat. This may have been necessary for our ancestors. They needed to store fat during times of abundant food, and lived off that fat during periods of food shortage.

Although variations in physiology complicate the calculations of energy input from foods, the energy input values listed in the following table are accurate enough to begin the estimation of energy balance in your body.

Important Energy Estimates *
1 g of carbohydrate = **4** Calories of energy input
1 g of protein = **4** Calories of energy input
1 g of fat = **9** Calories of energy input
4200 excess input Calories = 1 pound gain in fat (9.3 x 454 g)
4200 deficit of Calories = 1 pound loss in fat

* These estimates are "rounded off" for easier calculation.

Energy Output

Energy output is the term used to represent all of the energy required to maintain the metabolic processes and activities of an organism. The most accurate and direct way of determining the amount of energy used by the body during an activity is to measure the ***amount of heat given off***. Physics tells us that *heat is released whenever energy is transformed from one form into another* (such as nutrient energy into physical work). A technical problem with using this method of analysis is that the person must be inside an insulated container surrounded by a known quantity of water. The temperature of the water increases as heat is released by the person. Although this procedure is very accurate, it is also expensive and difficult.

Another method of measuring the body's energy output is the ***oxygen consumption technique***. If the oxygen requirement of a resting person is known, then that value can be compared to the increased amount of oxygen used during a particular physical activity. This approach is easier and less expensive than the heat method. Most general studies of energy expenditure are based on oxygen consumption.

? QUESTION

1. One Calorie of fat provides more input energy than one Calorie of protein. (T or F)

2. One gram of fat provides more input energy than one gram of carbohydrate or protein. (T or F)

3. An average person requires about 1600 Cal to maintain constant weight. If this person eats no food (water only), then how much fat can they lose in a week? (Assume that only fat is being metabolized during the fast. Actually, your essential proteins and sugars are also used as an energy source during a fast, and their depletion can be a serious health threat.)

4. What important law of physics tells us that heat production is an accurate estimate of energy expended by the body?

5. You will be using several foods as examples during this lab. Let's see how you evaluate them before doing the Exercises. Put a check mark if you think the food is high Calories, high protein, or high fat.

Food	High Calorie	High Protein	High Fat
Tuna Sandwich			
Milk (8 ounces)			
After School Snack: 1 cup of peanuts 6 crackers 1 oz of cheese			
Cheeseburger			

6. Why does the burning of some foods provide an inaccurate Calorie value of that food for humans?

7. If the normal resting energy output is 80 Calories per hour and the energy output during moderate exercise is 200 Calories per hour, then what is the energy output for the exercise?

EXERCISE #2

"Nutrients in Food"

Food Composition Tables are included in the appendix of most nutrition books. These reference tables give approximate amounts of carbohydrates, protein, and fat in the foods listed. On the last page of this lab there is an abbreviated table you can use for all the Exercises.

Basic Calculations

A tuna sandwich is the example we will use for calculating the amounts and percentages of nutrients in food. A sandwich is a mixture of ingredients, and the following information was obtained from the reference tables.

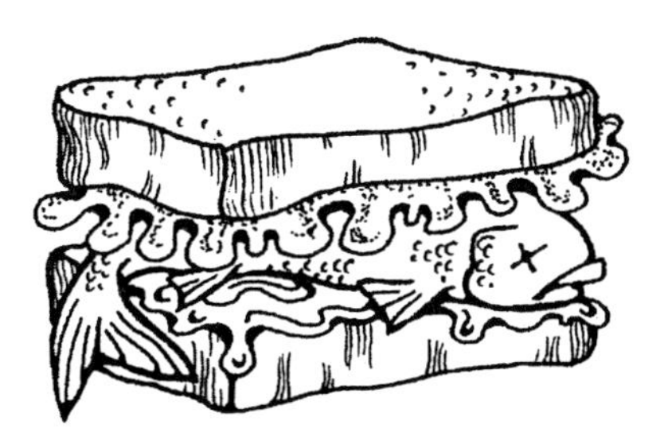

Tuna Sandwich	Carbohydrate	Protein	Fat
2 Slices of Bread	24 g	4 g	1.4 g
1 Tbsp. of Mayonnaise	trace	trace	11 g
Lettuce	trace	trace	trace
2 ounces of Tuna	0 g	15 g	1 g
Totals	24 g	19 g	13.4 g

Step 1

The calculations for nutrient analysis *based on weight* take some time but are simple to do. The first step is to determine the ingredients in a food. This was done for you in the tuna sandwich example.

Step 2

Use the nutrient information in the Food Composition Table at the end of this lab to determine the total weight (in grams) *for each category of nutrient* (carbohydrate, protein, and fat). This was done for the tuna sandwich.

Step 3

Add the weights of all three nutrients (**C** + **P** + **F**). For the tuna sandwich the total weight of carbohydrate plus protein plus fat is:

$$24 + 19 + 13.4 = 56.4 \text{ g.}$$

Step 4

Calculate the *percent* of each nutrient in the food using the following formula:

$$\frac{\text{(Weight of a Nutrient Category)}}{\text{(Total Weight of All Nutrients)}} \times 100 = \underline{\quad ? \quad} \%$$

$$\frac{24}{56.4} = 42\% \ \mathbf{C} \qquad \frac{19}{56.4} = 34\% \ \mathbf{P} \qquad \frac{13.4}{56.4} = 24\% \ \mathbf{F}$$

Procedure

Let's see if you can calculate the weight percentages of each nutrient in the entrées below.

Entrées	C	P	F
1. 8 oz glass of Whole Milk			
Total Weight of Nutrient			
Percentages by Weight			
2. After School Snack:			
1 cup of peanuts			
6 crackers			
1 ounce of cheese			
Total Weight of Nutrient			
Percentages by Weight			
3. Cheeseburger			
Total Weight of Nutrient			
Percentages by Weight			

EXERCISE #3

"Calories from Nutrients"

In Exercise #2 you calculated the ***weight percentages*** of nutrients in a tuna sandwich. The ***percentages based on Calories*** of each nutrient is a more accurate description of your diet.

Tuna Sandwich
24 g C 19 g P 13.4 g F 293 Calories

Basic Calculations

Step 1 The Food Composition Tables are used to determine the weight (in grams) of each nutrient in the tuna sandwich.

Step 2 The weight of each nutrient category must be multiplied by the ***Caloric Conversion Factor*** for that nutrient. Calculate the Caloric value of each nutrient in the tuna sandwich.

Tuna Sandwich				Caloric Conversion Factor	Caloric Value	
24	g	C	x	4 Cal per gram →	________	
19	g	P	x	4 Cal per gram →	________	
13.4	g	F	x	9 Cal per gram →	________	
				Total =	________	Cal

Step 3 The final step is to calculate the percentage of Calories contributed by each nutrient. *Do the calculations for the tuna sandwich.*

$$\frac{\text{Caloric Value of a Nutrient}}{\text{Total Calories in the Food}} \times 100 = \text{\% Based on Calories}$$

$$C = \frac{\_\_\_\_\_\_}{293} \times 100 = \_\_\_\_\_\_ \%$$

$$P = \frac{\_\_\_\_\_\_}{293} \times 100 = \_\_\_\_\_\_ \%$$

$$F = \frac{\_\_\_\_\_\_}{293} \times 100 = \_\_\_\_\_\_ \%$$

Procedure

Calculate the percentage of Calories for each nutrient in the entrées you analyzed in Exercise #2.

Entrées	Caloric Value	% of Total Calories
Milk C = g x 4 → P = g x 4 → F = g x 9 →		
After School Snack C = g x 4 → P = g x 4 → F = g x 9 →		
Cheeseburger C = g x 4 → P = g x 4 → F = g x 9 →		

? QUESTION

1. Refer to your answer to question #5 in Exercise #1. How would you describe these foods now?

   Tuna Sandwich = ____________________

   Milk = ____________________

   After School Snack = ____________________

   Cheeseburger = ____________________

2. Calculate the Caloric percent of fat in a 3 ounce piece of sirloin steak. (Use the following data: P = 20 g; C = trace; and F = 27 g.)

   ____________________ % fat

3. A green leafy salad is considered as being low in fat and low in calories. What happens to the fat content when you add a tablespoon of salad oil or creamy dressing?

EXERCISE #4

"Dieting Strategies"

There are many different ways of cutting Calories out of the diet. We will analyze three common dieting strategies.

Eat All You Want of This or That

There are *two* variations of this approach to dieting. One method says that you must eat a certain amount of a particular food *before* eating your meals. For example, the ***Eat-Six-Grapefruits-Per-Day*** diet instructs you to eat two grapefruits before each of the three meals of the day.

The second variation of this diet approach says that you can eat all you want from a narrow list of foods. For example, the ***Eat-All-of-the-Hard-Boiled-Eggs-and-Oranges-You-Want*** diet.

Procedure

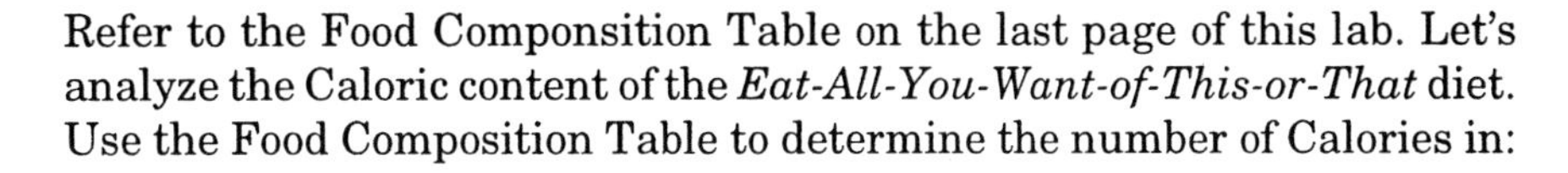

Refer to the Food Componsition Table on the last page of this lab. Let's analyze the Caloric content of the *Eat-All-You-Want-of-This-or-That* diet. Use the Food Composition Table to determine the number of Calories in:

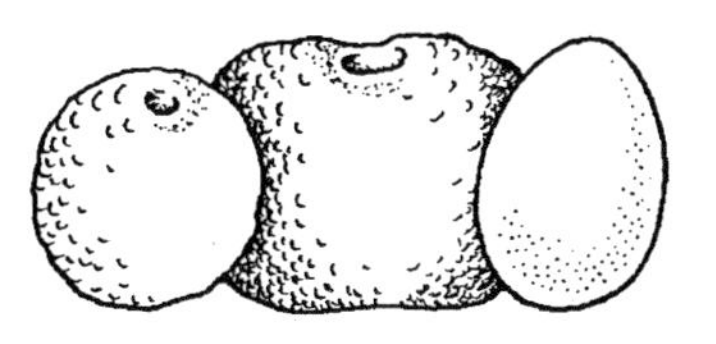

1 orange = ______________ Cal

1 grapefruit = ______________ Cal

1 egg = ______________ Cal

? QUESTION

1. Assume that a person is in energy balance if they keep their food intake to 1600 Calories per day. How many total Calories are in six grapefruits?

   ______________ Cal

2. How much food (compared to normal) are you likely to eat after having two grapefruits before each of your three meals? (Your answer should reveal the "trick" of how this diet works.)

3. Let's assume that most people on the second variation of this diet wouldn't eat more than 3 eggs and 2 oranges for each of the three meals of the day.

   9 eggs = ______________ Cal

   6 oranges = ______________ Cal

   The *Eggs-and-Oranges* diet is how many Calories less than the 1600-Cal energy-balance diet?

4. It takes about a 4200-Calorie deficit to lose a pound of body fat. How many days do you have to stay on the *Eggs-and-Oranges* diet until you lose a pound of fat?

   ______________ days

Special Prepared Meals

This approach to dieting works reasonably well while you are on the diet, and its success is the result of *two* factors. If you are paying for special prepared foods, then you are motivated to stay on the diet longer and not violate it. Also, someone other than you has planned a reasonably good-tasting meal that is not high in Calories.

? QUESTION

1. Why do most people hesitate in using this type of diet?

2. What is the weakness of this approach to dieting after you have stopped the diet?

Cut Something Out

There are *two* variations of this approach to dieting. One version is subtle and is illustrated by the ***Butter-and-Jam-on-Toast*** example. The second version is an extreme approach exemplified by the ***Pritikin-type*** of diet during which you eliminate most fat-containing foods.

Butter and Jam on Toast

We start with a person who has a well established habit of eating consistent meals. However, recently he has noticed that he is 4 pounds heavier than two years ago. Each morning he eats the same breakfast.

1 bowl of oatmeal
1 piece of toast
1 pat of butter
1 tablespoon of jam

? QUESTION

1. Use the Food Composition Table to determine the Caloric value of the butter and jam.

   1 pat of butter = ________________ Cal

   1 tablespoon of jam = ____________ Cal

2. If only the jam were eliminated from his toast each morning, how many days would he be on the diet until he lost one pound of fat?

   ____________ days

3. If both the butter and jam were eliminated from his toast each morning, how many days until he would lose one pound?

   ____________ days

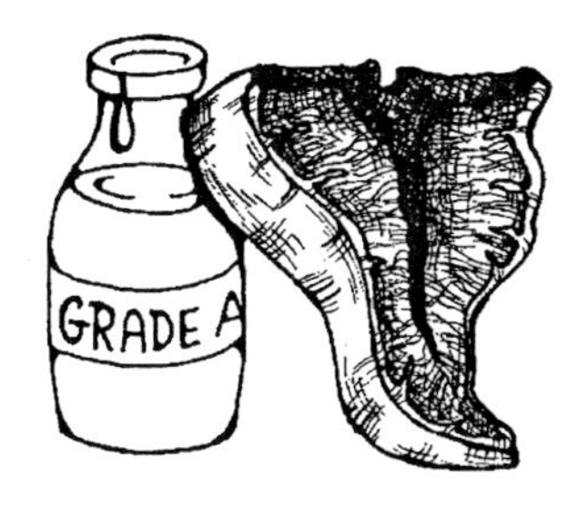

No Milk or Steak

The second version of cutting something out of the diet is illustrated by the next case. Again, we start with someone who has a consistent diet (this keeps the example simple). This person normally eats a 6-ounce piece of sirloin steak and one 8-ounce glass of milk for each lunch. He has been instructed to replace his normal lunch with a low-fat 400-Calorie pasta and vegetable meal.

? QUESTION

1. Refer to Exercise #3 for the Caloric value of:

   8 ounces of milk = ______________ Cal

   6 ounces of sirloin steak = ______________ Cal

2. How much Caloric reduction resulted from the switch from the *steak and milk* meal to the low-fat pasta and vegetable meal?

   ______________ Cal per day

3. How many days are required to lose one pound of fat on this low-fat lunch diet?

   ______________ days

4. What are the advantages of using the *Cut-Something-Out* diet compared to using the other two types of diets?

EXERCISE #5

"Energy Expenditure During Daily Activities"

The accurate measure of a person's ***energy expenditure*** during each of their daily activities is a technical challenge for exercise physiologists. Detailed energy analyses of competitive athletes during performance training have become very imporant for coaches and trainers. Other studies have established estimates for the average person during typical daily activities. Using the estimates, you can predict what will happen to the body energy balance when changes in your job or lifestyle occur. In addition, you can estimate how much more activity is required for you to lose a certain amount of weight.

Procedure

[1] List your general activities during a typical 24-hour period. Record these activities in the following Energy Expenditure Table.

[2] Use the information in the **Caloric Conversions for Activities** table to complete your calculations for energy expenditure. Match each of your daily activities with one listed in the conversion table that most closely approximates the effort expended during your activity. *Be sure to multiply the Caloric Conversion by your weight in kilograms.*

ENERGY EXPENDITURE TABLE

Daily Activities	Caloric Conversion x Your Weight	# of Hours Spent Doing Activity per 24-Hour Day	Calories Expended Doing Activity per Day
Total Calories Expended in 24 Hours =			

CALORIC CONVERSIONS FOR ACTIVITIES*

Activity	Energy Cal/Hr/Kg
Sleeping	0.9
Lying Still	1.0
Sitting	1.4
Standing, Reading, Writing	1.5
Driving a Car	1.9
Light Exercise	2.4
Walking (3 miles per hour)	3.0
Carpentry, Metalwork, etc	3.4
Bicycling (moderate speed)	3.5
Moderate Aerobic Exercise	4.1
Fast Aerobic Exercise	6.4
Slow Running (5 miles per hour)	8.1
Swimming (2 miles per hour)	8.9
Speed Walking (7 miles per hour)	9.6
Walking Up Stairs	15.7
Rowing in a Race	17.0

* These are approximations based on exercise books that you are likely to find in a library.

? QUESTION

1. Pick one of your usual daily activities. Calculate how many hours of that activity are required to burn 500 Calories per day.

   Description of Activity: ______________________

   Caloric Conversion for Activity:

   ______ Cal/hr/kg x Your Weight ______ = ______ Cal/Hr

   Hours of the Activity to Burn 500 Calories Per Day = ______

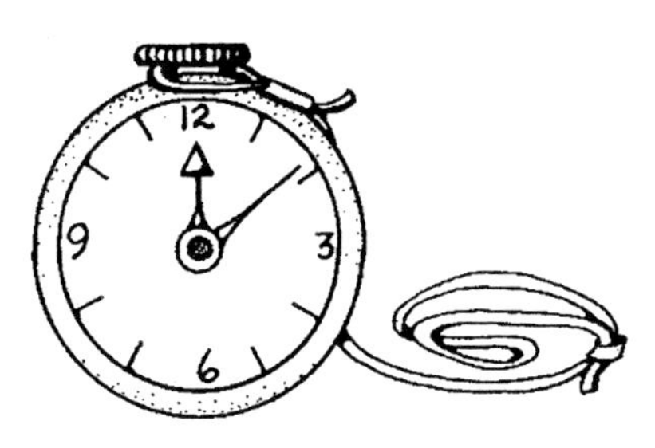

2. A 12-oz beer has about 170 Calories and a 12-oz cola has about 130 Calories. How many minutes of the activity in question #1 are required to burn off a can of cola or beer?

   12 oz cola = ____________ minutes of activity

   12 oz beer = ____________ minutes of activity

EXERCISE #6

"Special Considerations"

Scientific investigations have revealed much about diet analysis and energy balance in the human body. Few of these insights are discussed in popular magazine articles. Yet, some of the discoveries offer possible explanations for the differences in weight gain among people. The extra energy output by ***foot bouncing*** and by ***brown fat*** is investigated in this Exercise. You will also discover some unexpected ***effects of exercise on appetite***. These "special considerations" are only a few of the ideas about diet and exercise that are available to you in scientific books.

Foot Bouncing

Either you or someone you know is a foot bouncer. He or she just can't sit still. The foot bounces or the legs move back and forth. Two interesting questions about these people have been tested: (1) Do they burn more Calories than people who sit still? and (2) Do they move more after a big meal? The answer to both of these questions is yes, and we can estimate how much extra energy is expended by their "foot bouncing."

Procedure

[1] Estimate the number of hours that you spend sitting during a typical day. Include sitting in class, watching TV, studying, talking with others, etc.

____________ hours per day

[2] Look at the Caloric Conversions for Activities table in Exercise #5, and determine the Caloric *difference* between "sitting" and "driving a car." We will assume that the extra movements while driving a car are equivalent to the "foot bouncing" movements.

____________ Cal/hr/kg extra for "foot bouncing"

[3] Calculate the daily Caloric expenditure for "foot bouncing" by multiplying the number of hours spent sitting X the caloric conversion X your weight.

_____________ Cal

[4] Assume that the average "non-foot-bouncer" person burns 1600 Cal per day. What percent more Calories is being burned by the foot bouncer?

_____________ %

Brown Fat

Some of the fat cells in your body are specialized to help maintain body temperature. These special ***brown fat*** cells burn fat and produce extra heat. This allows you to keep warm on a cold day. Some people give off much more heat than others after eating a big meal. Their brown fat may be burning the extra fuel from a big meal. Therefore, some of those extra Calories won't be stored as fat. The heat produced by these people suggests that their metabolic rate is at least 10% higher than normal. In contrast, people with less brown fat may be storing the extra food energy as fat instead of metabolizing it.

? QUESTION

1. Find the energy expenditure for "sitting" in the Caloric Conversions for Activities table. Use this estimate for the normal basic metabolic rate. If "brown fat" increases the energy output to 10% above normal, what is 10% of the sitting energy?

   _____________ Cal/hr/kg

2. If this extra heat-production effect lasted for 10 hours in a 60-kg person, then how many extra Calories would the "high-brown-fat" person burn in a day?

   _____________ extra Calories

Appetite and Physical Activity

Scientific studies have revealed many factors influencing appetite. For example, if blood sugar or body temperature drops below a threshold level, you get hungry. The amount of chewing during a meal and the stretching of the stomach also reduce appetite. And, of course, we know that early childhood training and daily habits play a role in our desire to eat.

Who would believe that light exercise would decrease appetite?

Scientific investigation discovered an unexpected effect of exercise on appetite. It was known that when people worked harder they ate more food. But no one thought to study sedentary people who began a light-activity schedule. The surprising discovery was that the food intake (appetite) actually *decreased*. The implications of these findings are important! Consider the conclusion in reverse: ***A person who stops light activity and then becomes sedentary will increase their food intake.*** This explains why it is so easy to gain weight in our modern, low-activity lifestyle.

This summary graph provides many useful insights concerning the effects of physical activity on appetite. Refer to it when answering the questions.

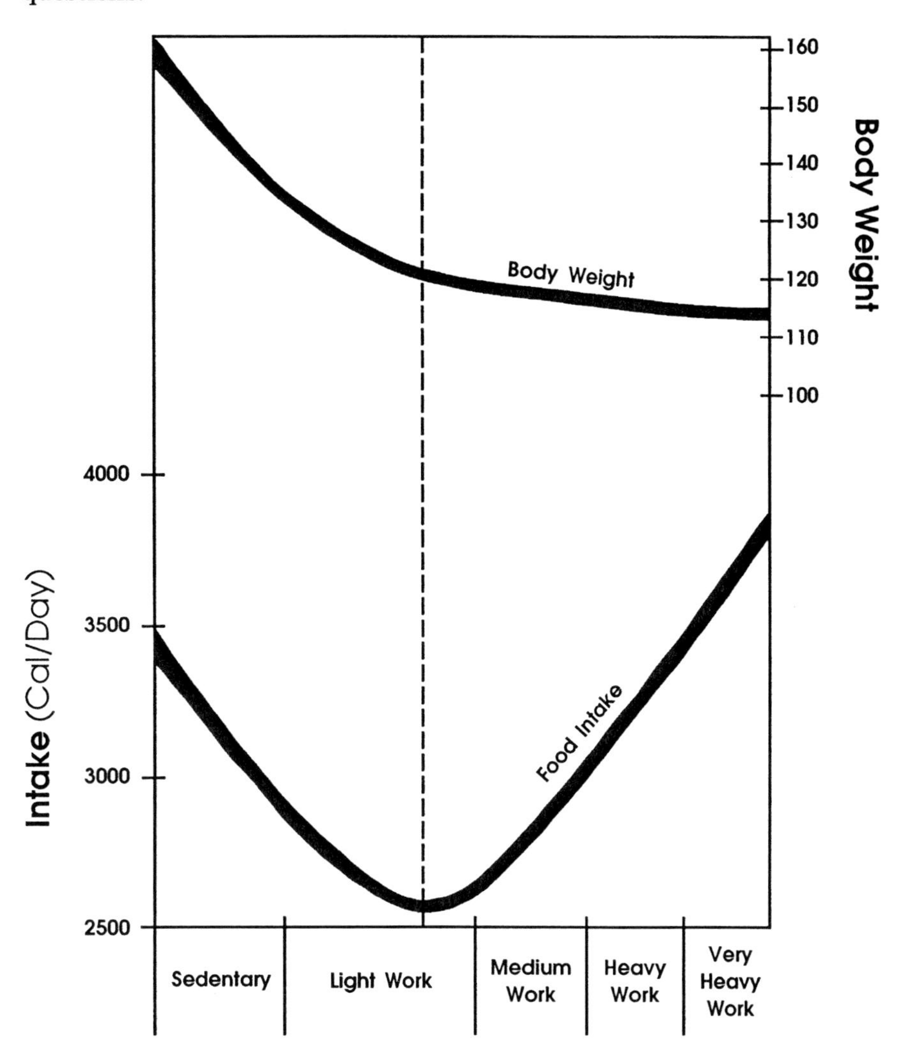

BODY WEIGHT AND CALORIC INTAKE AS FUNCTIONS OF PHYSICAL ACTIVITY

? QUESTION

1. How much appetite reduction (in Caloric intake per day) occurred when the study subjects shifted from sedentary to light work?

_______________ Cal reduction per day

2. **Remember:** A person who shifts from sedentary to light work is also burning more Calories because of the *increased activity*. What is the difference between the Caloric expenditures of sitting and the light work? (Refer to Exercise #5.)

Difference between sitting and light work = _______ Cal/hr/kg

Your weight = _______________ kg

Total extra Calories burned per day = _______________ Cal
(Assume you do the light work for 8 hours)

3. How many days will it take you to lose 10 pounds? Consider both the reduced food intake and the increased physical activity.

_______ days

4. The graph shows that the people tested lost about 35 pounds when they shifted from sedentary to light work. The time for this weight reduction was between 4 and 6 months. Do your calculations, based on what you learned in this Exercise, generally agree with the study findings? (Multiply the number of days it would take you to lose 10 pounds (question #3) times 3.5 to determine how long it would take you to lose 35 pounds.)

______________ days to lose 35 pounds

Conclusion

We hope that you have gotten value from the scientific approach to diet analysis as presented in this lab. Our discussions are only a beginning of many considerations about diet that have been discovered. There are important questions remaining to be answered. Visit your library for more information. And by the way, have you ever considered a career in nutritional physiology?

FOOD COMPOSITION TABLE

FOOD	C (grams)	P (grams)	F (grams)
Bread (1 slice)	12	2	0.7
Butter (1 pat)	trace	trace	4
Catsup (1 tablespoon)	4	trace	trace
Cheese (1 ounce)	1	7	9
Cracker (1 saltine)	3	0.4	0.5
Egg (1 medium)	trace	6	6
Grapefruit (1 medium)	24	2	trace
$\frac{1}{4}$ pound Hamburger Meat	0	28	23
Jam (1 tablespoon)	14	trace	trace
Lettuce (1 leaf)	trace	trace	trace
Mayonnaise (1 tablespoon)	trace	trace	11
Orange (1 medium)	16	1	trace
Peanuts (1 cup)	27	37	65
Salad Greens (1 cup—no dressing)	7	2	trace
Salad Oil (1 tablespoon)	0	0	14
Sirloin Steak (3 ounces)	trace	20	27
Tuna (2 ounces)	0	15	1
Whole Milk (8 ounce cup)	12	9	9

The Heart

The heart is an incredible pump. It can beat 2.5 billion times in a lifetime, pumping 35 million gallons of blood. Furthermore, the heart is capable of varying its output between 50 ml and 250 ml per stroke. It can contract at rates from 60 to 160 beats per minute. You can't buy a better pump!

This week's lab will include some aspects of heart structure and function. Also, you will learn to measure blood pressure, and discover some medical implications related to heart function.

Exercise #1 "The Heart as a Pump" 317
Exercise #2 "Heart Sounds" 321
Exercise #3 "Heart Rate and the Pulse" 323
Exercise #4 "How to Measure Blood Pressure" 326
Exercise #5 "Medical Implications" 328

EXERCISE #1

"The Heart as a Pump"

The heart must control the direction of blood flow, maintain pressures, separate pulmonary and systemic circuits, yet minimize wear and tear while accomplishing these tasks.

When you are resting, your heart pumps about 5 liters of blood per minute. This is about the same rate as a slow flow of water from the bathroom faucet when you brush your teeth. During strenuous exercise your heart can pump 30 liters of blood per minute. This is about equal to the water flow when you fast-fill the bathtub. The heart is capable of this wide range of performance because of its structure.

Two Pumps in One

The heart is actually two pumps—a ***right pump*** and a ***left pump***. The right pump delivers blood to the lungs; this route is called the ***pulmonary circuit***. The left pump pushes blood to the rest of the body; this route is called the ***systemic circuit***.

Notice that this diagram is drawn as if the heart is facing you. This means that the right side of the heart is on the *left* side of the drawing. All anatomy diagrams are drawn in this view. Remember this whenever you look at a medical picture.

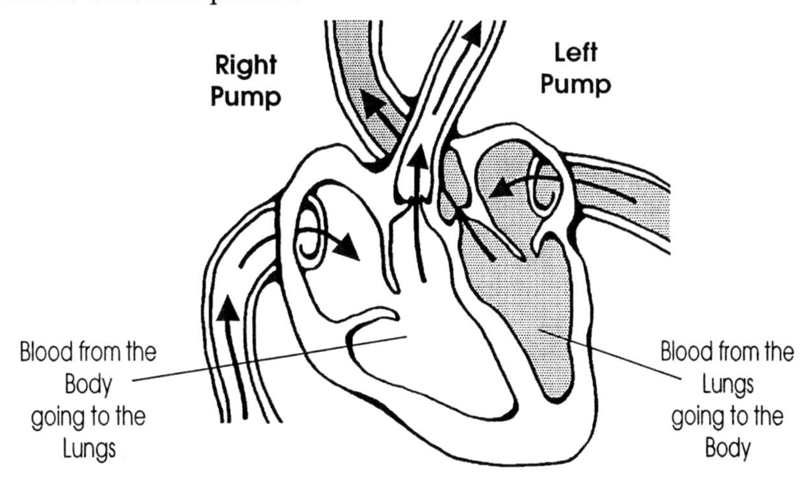

Two Chambers per Pump

The atria are the storage chambers; the ventricles are the pumping chambers.

There are *two* chambers in each of the two heart pumps. The top one is a temporary storage chamber called the ***atrium***, and the bottom one is a pumping chamber called the ***ventricle***. Blood from the body tissues flows into the atrium of the right heart pump. This blood is then pushed through a valve and enters the right ventricle. The ventricle does the hard work of pumping blood out of the heart. The right ventricle pumps blood to the lungs where it is *oxygenated*. While the ventricle is pumping blood out of the heart, the atrium fills with blood entering the heart. This efficient design allows the atrium to quickly refill the emptied ventricle, resulting in a fast-pumping heart.

Oxygenated blood from the lungs enters the atrium of the left heart pump. This blood is then moved into the left ventricle, which pumps the blood to all of the body tissues.

? QUESTION

1. Which chamber has to do the most work? (circle your answer)

Atrium or Ventricle

2. Which chamber would have a thicker muscle wall? (circle your answer)

Atrium or Ventricle

3. Which pump has to do the most work? (circle your answer)

Right Ventricle or Left Ventricle

4. Which pump would have a thicker muscle wall? (circle your answer)

Right Ventricle or Left Ventricle

5. The right heart pump moves blood to the ______________________.

6. The left heart pump moves blood to the ______________________.

Heart Valves

A heart valve is designed to plug an opening when blood moves in the wrong direction.

Four heart valves are strategically located to prevent backflow as blood moves through the heart. These valves are like one-way doors—they only open in one direction. There is a ***chamber valve*** between each atrium and ventricle. These two chamber valves ensure that blood will not flow back into the atria when the ventricles contract.

Blood is pushed out of the ventricles and into the two big arteries leaving the heart. There is a valve in each of these arteries. The ***artery valves*** prevent backflow into the ventricles once blood has been pumped into the arteries. The four heart valves ensure that blood moves in only one direction through the heart circuit. Each of the heart valves has its own special name, but we'll leave those details to an anatomy class.

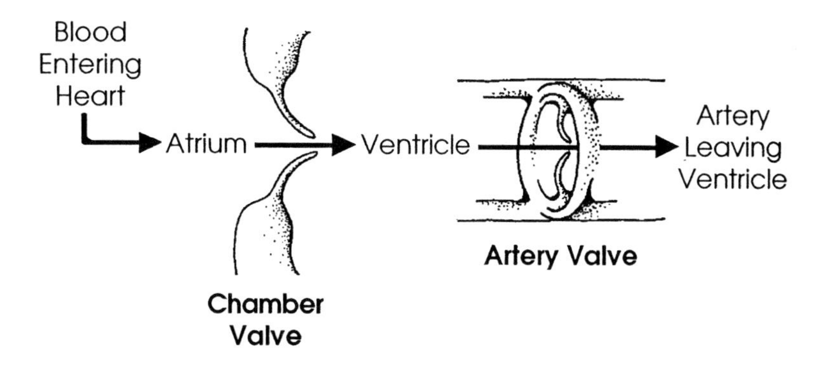

A heart valve is designed to plug an opening when blood moves in the wrong direction. Think of a valve as being something like a parachute that is attached to the heart or artery wall. If blood moves in the wrong direction, the "parachute" (valve) fills with blood and expands to plug the opening. When the blood moves in the correct direction, the valve collapses like an upside-down parachute. This allows the blood to easily pass through the valve.

EXAMPLE

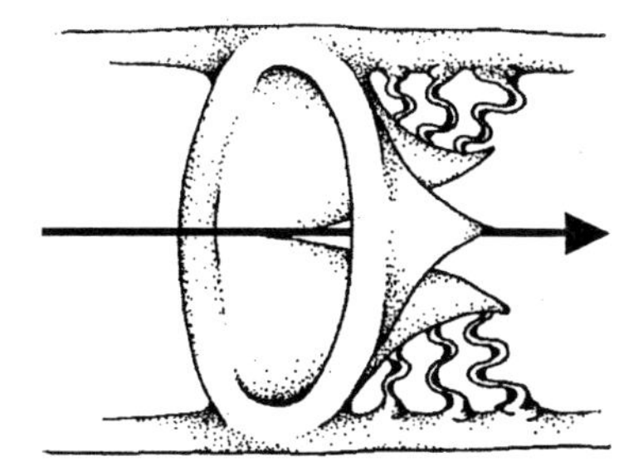

Blood moving in the correct direction pushes the valve aside, and blood enters the ventricle.

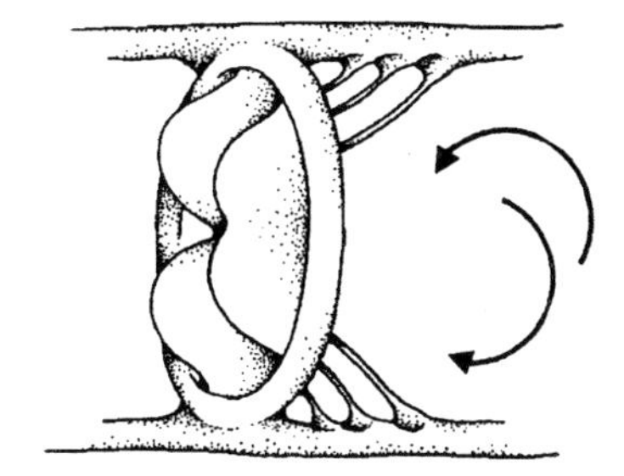

Blood moving in the wrong direction fills the valve which plugs the opening so that blood cannot re-enter the atrium.

The heart valves are *flexible* so that they can fill with blood as shown above. There are *special cords* attaching the valve to the heart wall. These cords operate similar to the ropes of a parachute.

? QUESTION

1. What would happen to blood flow if one of the valve "cords" broke? Be specific.

2. What would happen to blood flow if one of the valves was scarred by disease, narrowing the opening?

3. If you had a moderate heart valve problem, what would the heart have to do to compensate?

4. On which side of the heart would a moderate heart valve problem have more consequence to your health? Explain your answer.

Materials

- A sectioned sheep heart from the display table.

Procedure

1. See if you can identify the four chambers of the heart. **Remember:** One of the ventricles should have a thicker muscle wall.
2. Find a heart valve and feel the valve to determine its flexibility. Can you find the valve "cords"?
3. Show your instructor when you can identify all of these structures.
4. Draw a simple sketch of the dissected heart. This will remind you of what you saw in case you are tested on it later.

Dissected Heart

EXERCISE #2

"Heart Sounds"

The heart sound is often described as "lub-dub." You might think that the two parts of this sound come from the separate contractions of the upper and lower chambers of the heart. That is not correct. Actually, these sounds are more closely associated with the closing of the heart valves.

The first sound, ***lub***, happens when the blood vibrates after the ***chamber valves close*** between the atria and ventricles. The second sound, ***dub***, occurs when the blood vibrates just after the heart ***artery valves close***. The sounds are created by vibration waves. With the aid of a stethoscope, a physician can hear these heart sounds and determine if there has been any damage to the valves.

Materials

- A stethoscope.

Procedure

[1] Clean the earpieces of the stethoscope with a cotton ball soaked in alcohol. *Always repeat this procedure whenever another person uses the stethoscope.*

[2] Fit the stethoscope earpieces in your ears so that they are comfortable and point slightly forward in the ear passage. (Your ear passage points forward before it turns inwards to the ear drum.)

[3] Move the bell of the stethoscope around the *left* side of your chest starting at the lower center notch of your rib cage.

As the stethoscope is moved around the heart area, you will hear the "lub" sound better at some places and the "dub" sound better at other places. An experienced physician or nurse can position the stethoscope to hear each heart valve and determine whether there is an abnormal sound. Abnormal sounds indicate possible valve damage or other circulation problems.

This part gets a little tricky, so read carefully.

Most people expect that the two big heart arteries would exit from the bottom of the ventricles, but they don't. These arteries come out of the top of the ventricles, and arch upwards above the heart. Refer to the diagram below, and notice the location of the two heart arteries. The artery valve allowing blood flow out of the ventricle is next to the chamber valve controlling flow from the atrium into the ventricle.

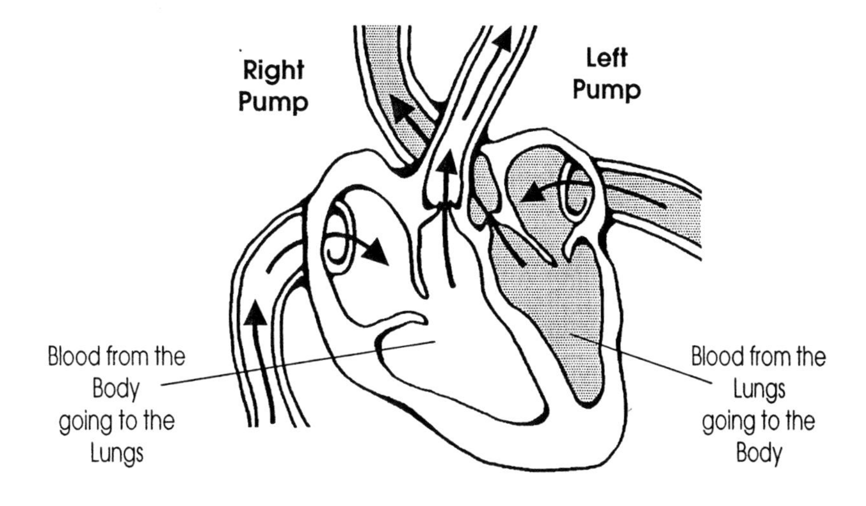

A "lub swish" sound could mean that the mitral valve (left chamber valve) is damaged.

The **"dub"** sound of the heartbeat resonates *upward*, which is the direction that the heart arteries leave the heart. In which position of the stethoscope (A or B) do you hear mainly a "dub" sound? ____________ This would indicate that you are located near the top of the heart where the heart artery valves make their sounds.

Continue to move the stethoscope to *both sides* of this position until the heart sounds quiet. You are mapping the general location of the top of the heart. Mark the extent of this "dub" zone on the rib cage diagram. The sound zone is bigger than the heart because the sound spreads outward.

The **"lub"** sound of the heartbeat resonates *downward* from the chamber valves, so you hear it best at the bottom of the heart. In which position of the stethoscope (A or B) do you hear mainly a "lub" sound? ________________ This would indicate that you are located nearer the bottom of the heart where the sound of chamber valves is loudest.

Continue to move the stethoscope to the *left* of this position until the loud "lub" sound quiets. You are mapping the general location of the bottom of the heart. Mark the extent of this "lub" zone on the rib cage diagram.

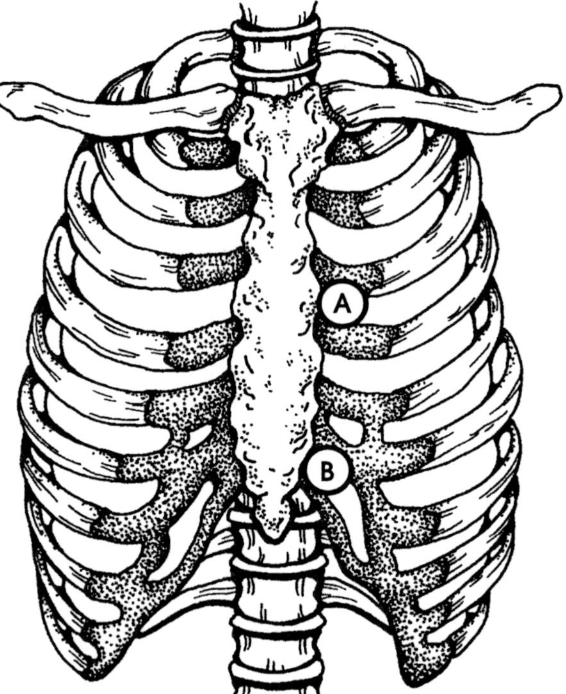

In Conclusion

Draw an arrow from the center of the "dub" zone to the center of the "lub" zone on the rib cage diagram. That arrow is the general orientation of the heart. The orientation is . . . (circle your choice)

vertical (straight up and down).

to the left of vertical
(bottom of the heart points to the left).

to the right of vertical
(bottom of the heart points to the right).

This sound technique of inferring the orientation of the heart is less accurate than the EKG method used in hospitals. The EKG can reveal whether the heart has enlarged on its left or right side. Enlargement is an indication of a heart or circulation abnormality.

EXERCISE #3

"Heart Rate and the Pulse"

The heart of a resting person contracts somewhere between 60 and 80 times per minute. These contractions can be counted with the aid of a stethoscope, which you will do in a few minutes. Heart rate can also be determined by counting the number of pulse waves that pass by a spot on an artery. Counting these waves by touch is how you measure the ***pulse***.

The Pulse Wave

The ***aorta*** is the large artery that supplies the blood to all other arteries that feed the body tissues. Each heart contraction forces a volume of blood (50–250 ml) into the aorta. First, the aorta is "ballooned" out by blood, then the elastic artery wall snaps back an instant later.

The recoil causes the adjacent area of the aorta to balloon out and snap back. The alternate expansion and recoil of the aorta wall "pulses" outward from the heart to the other arteries of the body. You feel these waves passing by whenever you press a finger on an artery.

The ***carotid pulse*** is felt when you press your fingers against the side of your throat. The ***radial pulse*** is felt when you press your fingers on the thumb side of your upward-turned wrist.

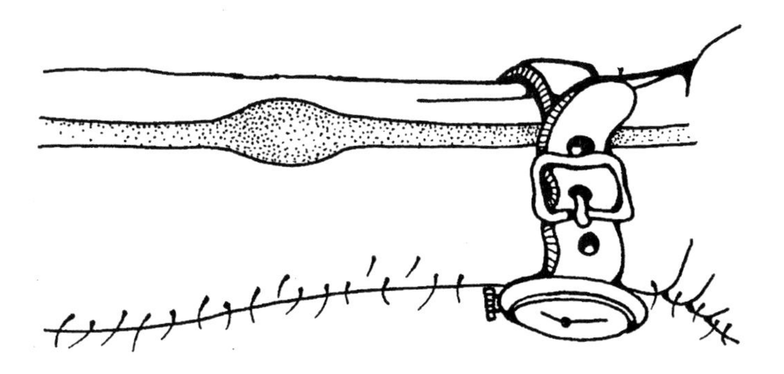

Procedure

[1] Use the stethoscope to count your heartbeats for 30 seconds.

Stethoscope Heart Rate = ________ beats per minute.

[2] Count both your radial pulse waves and carotid pulse waves for 30 seconds.

Radial Pulse = ________ waves per minute.

Carotid Pulse = ________ waves per minute.

? QUESTION

1. Which had the stronger pulse waves? (circle your choice)

   Carotid or Radial

2. Which is closer to the heart? (circle your choice)

   Carotid or Radial

3. Which would produce a stronger pulse wave? (circle your choice)

   A smaller heart or A bigger heart

4. Arteriosclerosis hardens the artery wall with scar tissue. If arteries have been partially injured by arteriosclerosis and the arterial wall is less flexible than normal, then would the pulse be stronger or weaker than normal?

5. If a person has arteriosclerosis, what happens to the blood pressure near the end of the arteries? (circle your choice) **Hint:** Is some of the energy of the heart contraction "used up" by the pulse wave?

   a. it is the same as normal

   b. it is higher than normal

   c. it is lower than normal

   What would be the consequences?

6. Half of the people have a smaller-than-average sized heart and half have a larger-than-average sized heart. In which group would you expect the *heart rate* to be higher? Explain your answer.

7. In general, females have a higher heart rate than males. What explanation can you give for this difference?

Heart Rate and Longevity

Ever so slowly, the heart wears out.

Research in comparative physiology suggests that the average mammal heart beats about 1.5 billion times before it wears out. Although there are exceptions to this heart longevity rule, it seems to be generally true whether you're a mouse or an elephant. A mouse's heart beats about 10 times faster than an elephant's heart, and a mouse lives about $\frac{1}{10}$ as long.

Based on the mammalian average for total heart beats, the modern human species is predicted to live about 35 years. Humans score above most other species for longevity. This is probably because we are smarter and can avoid more hardships than the average mammal. However, we also have a limit—somewhere around 2.5 billion beats—if we are lucky enough to survive disaster and illness. How you spend these heartbeats is partly determined by the activities in your lifestyle.

Let's assume that the longevity rule is generally true for humans. A woman who is already doing enough daily activity to keep her heart healthy, asks the question, *"If I train in a very strenuous sport for 4 hours a day beyond my normal activity, then how much am I shortening my life by doing this sport?"* Assume that her normal heart rate of 70 is elevated to 120 during the heavy training.

? QUESTION

1. How many *extra* heartbeats does she use per day of sport training? ____________

2. Her normal heart rate of 70 per minute means that she would use 100,800 heart beats on a normal day without sport. If an extra 100,800 beats shortens her life by one day, then how many days of sport training does it take to shorten her life by one day? ____________

3. How many years of sport would shorten her life by one year? ____________

4. Let's assume that this person is considering 8 hours of strenuous sport per day. How many years of this sport activity would it take to shorten her life by one year? ____________

Before we ascribe too much importance to a higher heart rate, remember that females generally live 10% longer than males even though females have a 10% higher resting heart rate. Obviously, other important factors affect longevity.

5. Smoking elevates the heart rate about 10% above normal; so does drinking 2 to 4 cups of coffee per day. If you were a smoker or a coffee drinker for 40 years, how many years of longevity might be lost due to the increased heart rate alone (not taking into account the obvious health risks of tobacco and caffeine)? ____________

6. Negative stress can elevate the heart rate 10–20% above normal. How many lost years of longevity might result from a 20-year stress-filled job that elevated heart rate 20% above normal? ____________

EXERCISE #4

"How to Measure Blood Pressure"

Knowing how to measure your blood pressure is one of the best health-maintenance tools you can have. We offer this Exercise to promote your good health.

Blood Pressure

Blood pressure in body arteries is created by the contraction of the left ventricle. As you would expect, the pressure is highest when the chamber contracts. Pressure during heart contraction is called the ***systolic pressure***. When the ventricle relaxes the blood pressure drops. However, instead of dropping to zero, the blood pressure is partially maintained by the recoil of artery walls that are stretched by blood pumped out of the heart. The lower artery pressure during the relaxation of the ventricle is called the ***diastolic pressure***.

Medical books state that the typical resting blood pressure is 120 over 80. The 120 refers to the systolic pressure and the 80 refers to the diastolic pressure.

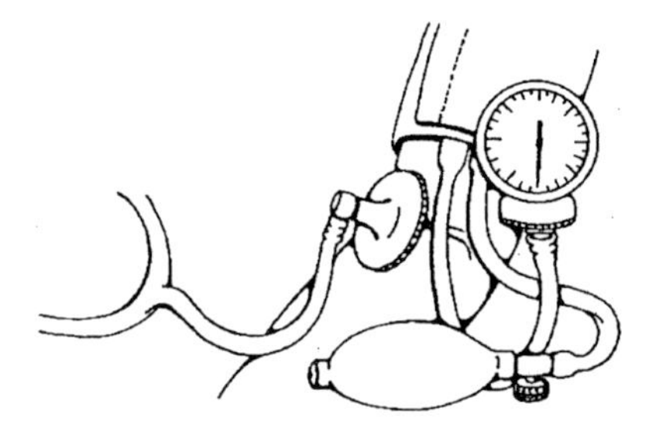

Measuring Blood Pressure

The method we will use to measure blood pressure is fairly simple. It is easier to show you how to measure blood pressure than to explain all of the details in writing. Therefore, most of the instructions will come from your lab teacher.

Read through the general steps of the procedure and the hints that follow before you begin practicing.

Materials

- A blood pressure cuff.

Step 1

Fasten a pressure cuff around your upper arm. Place the stethoscope diaphragm over the brachial artery (*inside bend of elbow*). Pump the cuff full of air until all blood is stopped in the brachial artery. The thumping heart sound that you hear through the stethoscope will fade and disappear as the cuff pressure is pumped above the systolic pressure.

Step 2

Release the pressure on the cuff by slowly opening the valve. Listen to the brachial artery with the stethoscope.

Step 3

When you first hear a "thumping" sound, read the gauge on the pressure cuff. This is the *systolic pressure*. The blood is just starting to squirt past the cuff during the contraction of the heart.

Step 4

Continue to release the pressure on the cuff until the thumping sound disappears. Read the pressure gauge. This is the *diastolic pressure*. The blood flows past the cuff during both the contraction and relaxation of the ventricle. The sound disappears when the flow changes from a pulsating squirt to a constant flow.

Hints

- Don't pump the pressure cuff over 150 until you've practiced the technique several times.
- Don't keep pressure on your arm for more than 30 seconds.
- Let your arm rest for at least 2 minutes after each reading before taking another measurement. This is especially important while you are learning the technique.
- Take your time. Learn this procedure well. It is important that you can measure your own blood pressure. Checking your blood pressure every few months provides you with a thorough understanding of your normal physiology. When an abnormal change occurs, you can seek medical advice.

Procedure

1. Record your blood pressure while sitting: ________________ B.P.

2. With the blood pressure cuff still on your arm but not pumped up, run in place for 30 seconds. Measure your blood pressure after exercise: ________________ B.P.

Cold Water Test

This test is used to determine the effect of a sensory stimulus (cold) on blood pressure. The normal reflex response to a cold stimulus is a slight increase in blood pressure (both systolic and diastolic). In a normal individual, the systolic pressure will rise no more than 10 mm Hg, but the increase in a ***hyper-reactive*** individual may be 30 to 40 mm Hg. We will discuss the implications of these responses later.

Procedure

1. The subject should be seated comfortably.
2. Immerse the person's free hand in ice water (approximately 5 °C) to a depth well over the wrist.
3. After waiting 30 seconds, measure the blood pressure.

   Your normal resting blood pressure is ________________.

   Your blood pressure after cold water is ________________.

   Are you a normal or a hyper-reactive individual based on the cold-water test? ________________________

Implications

There is some evidence that people showing a hyper-reaction to a cold stimulus may have a greater chance of developing high blood pressure later in life. Perhaps there is some minor defect in the physiology of these people. Or they may have inherited a more reactive nervous system—favored in hunter-gatherer times—but easily overstimulated by our modern chaotic life. We just don't know.

Before giving too much importance to the results of the cold water test, remember that lack of exercise, high salt intake, unhealthy diet, and stressful situations are known causes of high blood pressure, and are mostly under your control.

EXERCISE #5

"Medical Implications"

There are many medical implications related to what you have learned about the heart, and several of these health issues are discussed in this Exercise. **Remember:** You should always ask your physician to explain health-related problems in a way that you can easily understand. And go to the library. Inform yourself.

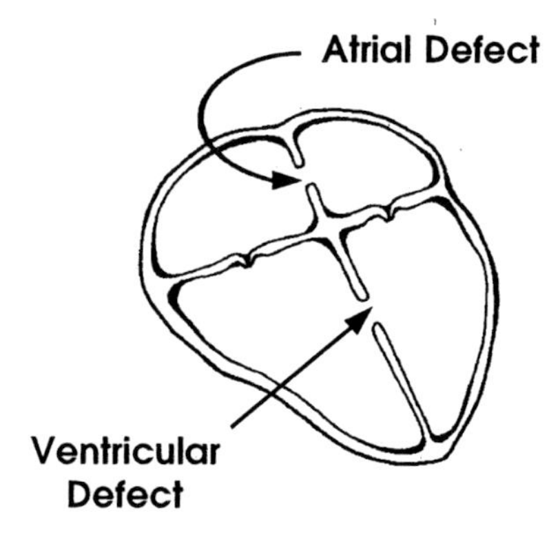

Abnormal Holes in the Heart

The heart in a fetus has a hole between the right and left atria. This opening allows fetal blood to partially bypass the lung circuit since the lungs aren't needed to get oxygen during life in the womb. Normally, the atrial hole closes shortly after birth. A birth defect results if this hole does not grow closed. Another birth defect occurs when there is an abnormal opening between the right and left ventricles. Both of these heart abnormalities can have serious health implications if left uncorrected.

? QUESTION

1. What important molecule is carried by blood entering the left heart pump and is not in the blood entering the right heart pump?

2. If the fetal hole between the right and left atria does not close, what happens to the blood in these two chambers?

3. Which ventricle operates under the most pressure (does the most work)?

4. What would happen to the pressure in the two ventricles if there was a hole between them?

5. What would the heart do to compensate for the pressure problem created by a ventricle hole? **Hint:** People with this abnormal hole must have it repaired while they are young or they won't live long.

Coronary Arteries

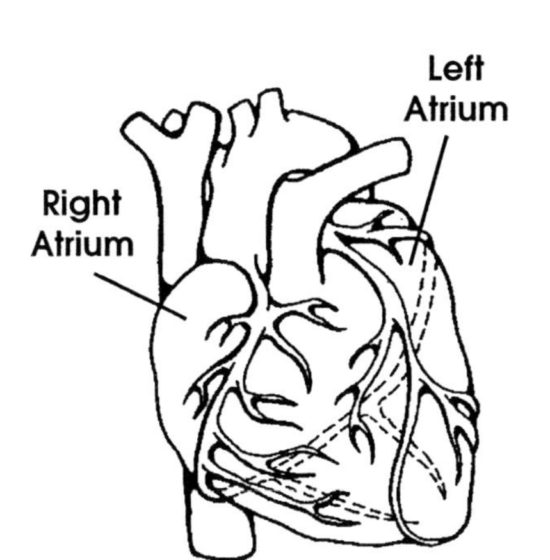

The heart muscle works very hard and it must be supplied with oxygen and nutrients just like any other part of the body. The vessels that supply blood to the heart muscle are called the ***coronary arteries***.

There are two important circulation patterns that you can see in this diagram. The right atrium is fed only by the right coronary artery, and the left atrium is supplied only by the left coronary artery. However, each coronary artery supplies blood to parts of *both* ventricles. The ventricles do more work than the atria, and must be supplied with more blood.

Another important aspect of vessel structure in the coronary arteries is the connection between the arteries. Connections between arteries are called ***collateral circulation***. These connections are alternate routes of blood flow to tissue if one path is blocked. Some parts of the heart have no collateral circulation. Other parts have only very small-diameter collateral vessels because they are unused. Also, there are different amounts of collateral vessels among people. Can you find a collateral vessel in the heart diagram above? Color that vessel.

Normal Circulation

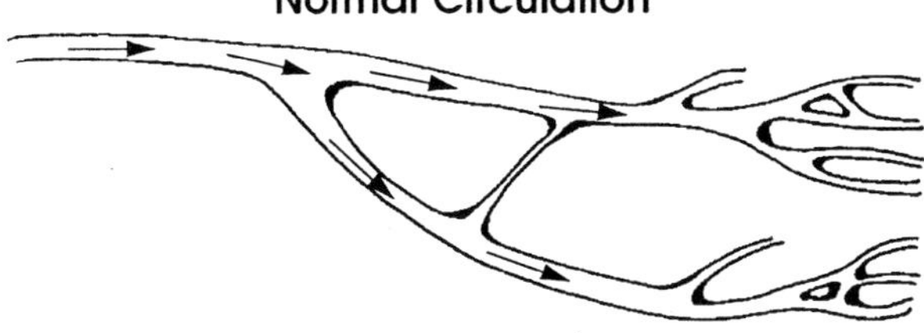

Collateral circulation provides detour around blockage.

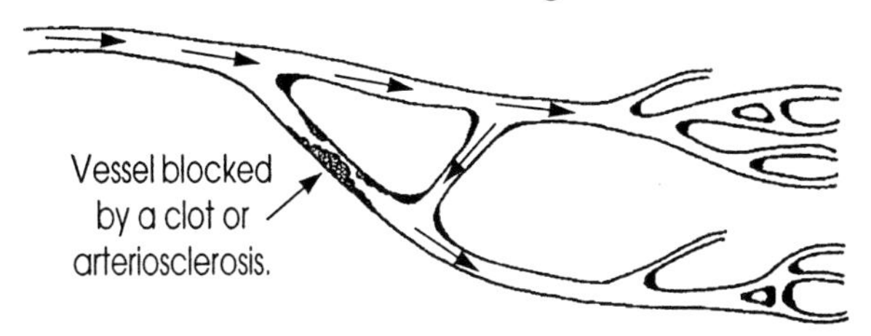

? QUESTION

1. A patient is told that she has a narrowing of the right coronary artery. Which chambers of her heart are going to be the most affected by this disorder?

Which chamber on the right side has to do the most work and could be the most serious health concern?

2. A group of patients were told that they had plugged arteries in their hearts. In addition, they had all suffered a similar size of heart attack. All of these patients survived. Some of them had parts of their injured hearts return almost to normal after several months. The other patients had no such luck. Explain these differences in terms of coronary circulation.

3. A heart attack on which side of the heart would probably cause the most serious immediate risk to the person?

EKG

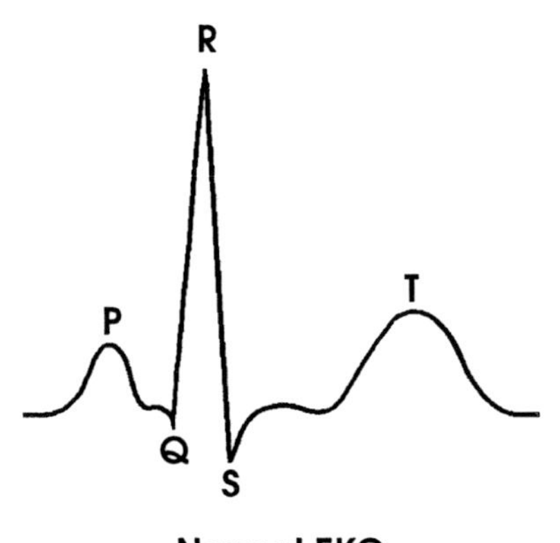

Normal EKG

The ***electrocardiogram*** (called ***EKG*** or ECG) is a recording of the small electrical currents produced by the contracting and relaxing heart muscle. These electrical patterns indicate whether there is a normal or abnormal functioning of the heart. A normal EKG is shown to the left.

The ***P wave*** is a recording of the electrical activity in the atria, and it is especially important in diagnosing problems in the heart's natural *pacemaker*.

The ***QRS wave*** is a recording of the activating current in the ventricles. This current travels along a special conducting pathway 10 times faster than it would be transmitted through normal heart muscle. The result is that all the muscle cells in both ventricles are stimulated at the same time, causing these chambers to contract quickly and strongly.

The ***T wave*** occurs just after the ventricles contract, and is a recording of the normal recovery phase of the ventricles. This is a period when the muscle cells perform various biochemical reactions that prepare the ventricles for the next contraction.

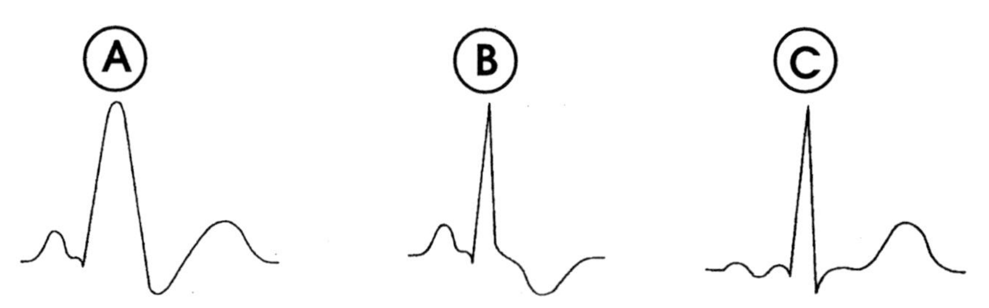

? QUESTION

1. A person who drinks a lot of coffee complains that his heart beats irregularly. Which one of the above EKG waves reflects this problem? _______________ Explain your answer.

2. A person has a greatly enlarged heart from the overwork created by long term high blood pressure. Which abnormal EKG wave reflects this problem? _______________ Explain your answer.

3. A person with poor coronary circulation to the heart muscle has some heart injury, but may not have suffered death of the heart muscle. Which abnormal EKG wave reflects this problem? _______________ Explain your answer.

4. Why is a heart attack sometimes called a "coronary"?

Senses and Perception

The human must be aware of and adjust to environmental demands. Our ***senses*** provide information about the physical environment. Our ***perception***, developed by the experiences of trial and error, evaluates sensory information, allowing us to make the appropriate adjustment to the environment. Working together, these two processes have contributed to the success of the human species.

Human senses result from specialized receptors located in various parts of the body. These *receptors* are activated by only one kind of *stimulus* (sound, touch, light, chemicals, etc.). The information from one receptor is kept separate from that sent by another sense organ. In the brain, particular areas are specialized for processing and interpreting the information pertaining to each sense.

Today's lab is designed to demonstrate some of the sensory and perceptual mechanisms of your nervous system.

Exercise #1 "Touch" 331
Exercise #2 "Temperature Sensation" 333
Exercise #3 "Hearing" 333
Exercise #4 "Smell" 335
Exercise #5 "Taste" 336
Exercise #6 "Vision" 337
Exercise #7 "Reflexes" 339

EXERCISE #1

"Touch"

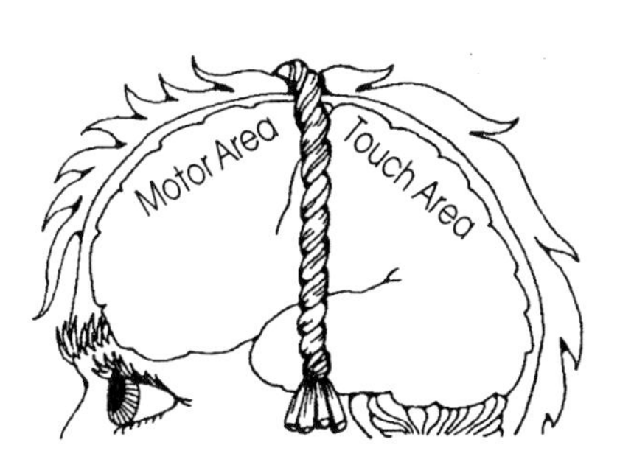

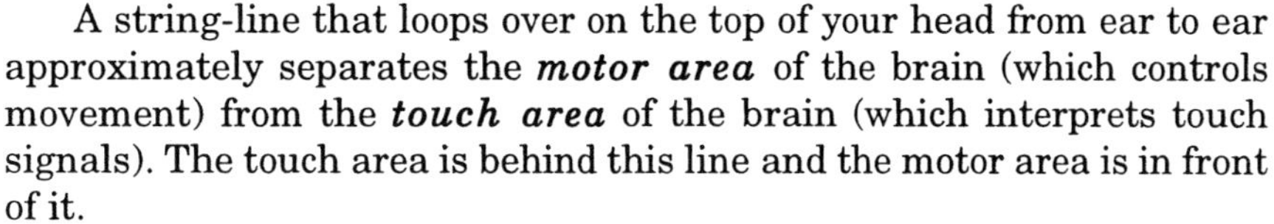

A string-line that loops over on the top of your head from ear to ear approximately separates the ***motor area*** of the brain (which controls movement) from the ***touch area*** of the brain (which interprets touch signals). The touch area is behind this line and the motor area is in front of it.

You will determine the density of touch receptors in several areas of your body. Using that information, you can test the idea that *areas of greater sensitivity contain more touch receptors.*

Materials

- A horse hair.
- A metric ruler.
- A compass.

Procedure

[1] Be sure to perform all the tests on each person in your lab group.

[2] Mark off a 1 cm x 1 cm square on your fingertip, back, and another area of your body that you would like to test. You might choose an area that itches often or feels strange when touched.

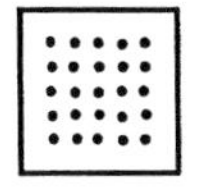

[3] Close your eyes while your lab partner lightly touches you 25 times with the horse hair. The touching should be done in a grid-like pattern that covers all of the square you have marked.

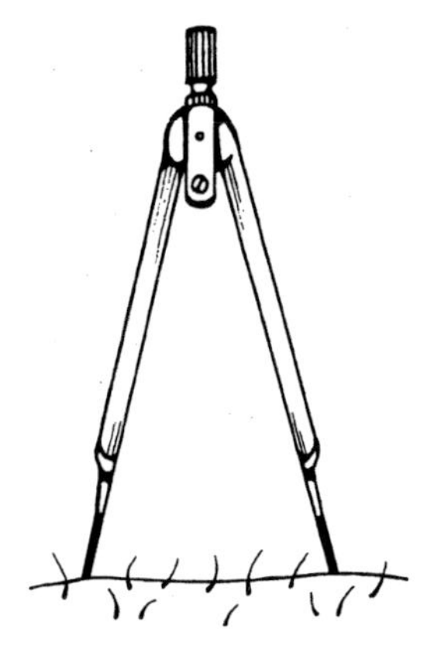

4. Each time you feel the touch of the horse hair, say so. Record the number of positive responses in the Touch Experiment Table on the next page.
5. Next, use the points of a compass to *lightly stimulate* the subject's skin in the area of the marked boxes. The compass points must be blunt and not poke through the skin. (File the points if necessary.) Start with the points close together, then increase their distance apart until the subject definitely feels *two distinct points.* Be sure that the two points are applied simultaneously each time, and retest to see if there is error due to imagination.
6. Measure the distance between the two compass points when the subject clearly perceives two points. This is called ***two-point discrimination***. Record the results for each area of the body that you mapped for touch receptor density. Include the results from everyone in your lab group.

TOUCH EXPERIMENT TABLE

Part of the Body	# of Positive Responses During 25 Touches in 1 cm^2	Two-Point Discrimination (in cm)
Fingertip		
Back		
(Other) ____________		

? QUESTION

1. Which test area had the greatest density of touch receptors?

2. Which test area had the best two-point discrimination? Explain.

3. How is *two-point discrimination* related to *density* of touch receptors?

4. When you have an itch somewhere on your back, why does it take so much scratching before you finally find it?

EXERCISE #2

"Temperature Sensation"

During this Exercise you will determine whether your body detects the *actual* temperature of the environment or only the *change* in environmental temperature.

Materials

- A large beaker of cold water (10 °C).
- A large beaker of hot water (50 °C).
- A large beaker of 30 °C water.

Procedure

[1] If the water beakers are already set up at the demonstration table, then check and adjust the temperatures using water from the hot plate or ice cubes in order to maintain the three temperature conditions listed above.

[2] Place the index finger of one hand into the cold water, and the index finger of the other hand into the hot water for 15 seconds.

[3] After 15 seconds, quickly place both fingers into the 30 °C water. Record the sensations.

Cold-water Finger = ____________________________.

Hot-water Finger = ____________________________.

? QUESTION

What seems to be the most important factor related to your perception of skin temperature? (circle your choice)

actual temperature or *change* in temperature

EXERCISE #3

"Hearing"

The ear is divided into three parts: *outer*, *middle*, and *inner* ear. When sound waves enter the ear, the ***eardrum*** (between the outer and middle ear) is shaken and special small bones vibrate. These ***middle ear bones*** transmit the sound vibrations into the inner ear where the ***auditory nerves*** leading to the brain are activated. The area of the brain that is specialized for interpreting sounds is next to the ears.

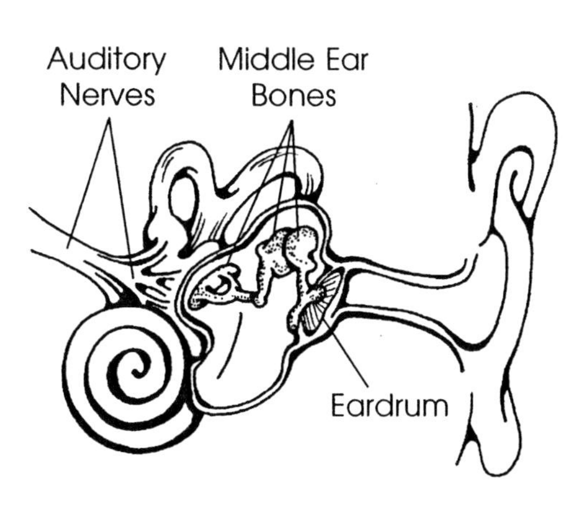

Materials

- Some cotton for ear plugs.
- A set of tuning forks.
- A meter stick.

Procedure

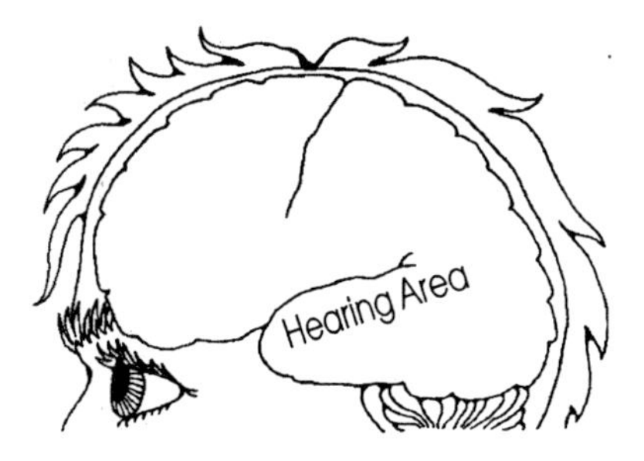

[1] Do this test in a quiet room. Have the subject close one ear with cotton and close his eyes. Strike the tuning fork against the table and hold it in line with the open ear. Move the tuning fork away from the ear until the subject just loses the ability to hear it. Measure the distance. Repeat the test again to validate your first measurement. Record the hearing distance for the other ear. *Be sure to strike the tuning fork with equal force each time you do the test.*

[2] Repeat the test with each of the six tuning forks of different tones to determine if you have hearing loss in any of the six ranges. If one of your ears has a hearing loss at a particular tone range, then do the next test.

[3] This next test should not be performed in a quiet room. Place the handle of a vibrating tuning fork on the midline of the subject's forehead.

A person with normal hearing will localize the sound as if it were coming from the midline. If one ear has defective middle-ear function (ear bones), then the sound will be heard much better in the defective ear than when the tuning fork is not in contact with the forehead. If there is an affliction of the auditory nerve in the ear, then touching the tuning fork to the forehead won't improve hearing in the defective ear.

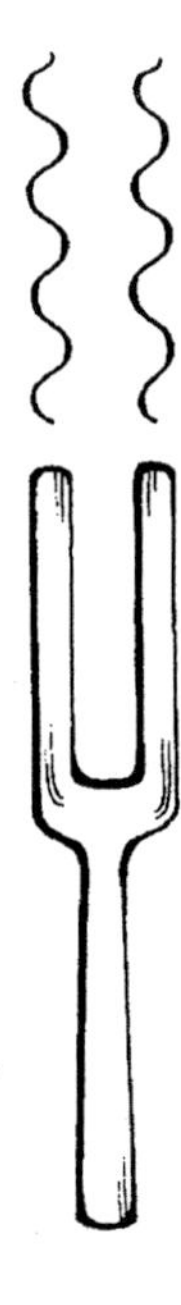

RESULTS OF HEARING TESTS

Sound Frequency (cycles per second)	Farthest Distance sound heard from Left Ear	Farthest Distance sound heard from Right Ear
128 cps fork		
256 cps fork		
512 cps fork		
1024 cps fork		
2048 cps fork		
4096 cps fork		

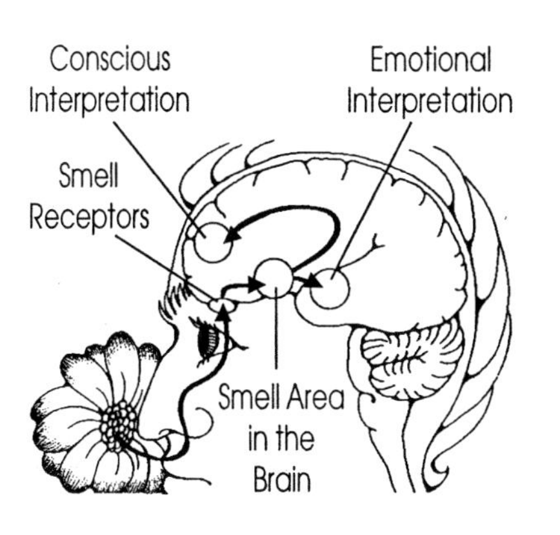

EXERCISE #4

"Smell"

Recent studies show that smell is much more important in human behavior than was previously thought. Some researchers suggest that the evolutionary specialization of the mammal forebrain began with the sense of smell. The exact role of smell in our lives is not understood. This sense seems to be more closely linked to emotional memories than to the conscious activities of our brains. As you experiment with the various odors in the exercise below, describe the type of *emotional reaction* you have to each.

Materials

- A smell kit.

Procedure

1. Close your eyes. Have your lab partner pass an open odor vial about 3" under your nose for a couple of seconds. Repeat the test if necessary.
2. *First*, determine if you can smell the odor. *Second*, determine if you can correctly identify the smell. *Third*, describe any special memories associated with the smell.
3. Record the results of your test in the Odor Recognition Table.

ODOR RECOGNITION TABLE

Vial Number	Detects Smell	Identifies Smell	Memories Associated with the Smell
1	____	____	______________________
2	____	____	______________________
3	____	____	______________________
4	____	____	______________________
5	____	____	______________________
6	____	____	______________________
7	____	____	______________________
8	____	____	______________________
9	____	____	______________________
10	____	____	______________________
Totals	____	____	

? QUESTION

1. How many of the smells were associated with emotional memories?

2. List three examples of how specific smells might be used to sell you a product.

    a.

    b.

    c.

EXERCISE #5

"Taste"

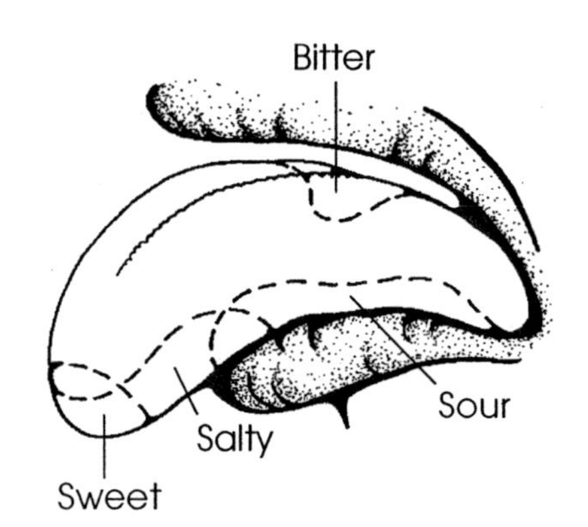

The tongue has at least four different taste receptors (***salty***, ***sweet***, ***bitter***, and ***sour***). However, the taste of many chemicals is also influenced by your interpretation of their *smell*. In this Exercise, you will examine different aspects of your ability to taste.

Genetically Determined Taste

Your ability to taste certain chemicals is not influenced by previous eating habits, but is determined by whether or not you have inherited the *gene* controlling the taste response to that particular substance. This is an important lesson to remember when certain foods don't taste bitter to you, but other people complain about them (especially your children). They may have the gene to taste it, and you don't!

Materials

- The special taste papers for PTC and thiourea.

Procedure

[1] Take one of the taste papers, and touch it to your tongue. You will immediately know if you are a taster!

[2] Do the same test with the other taste paper.

[3] *Put the used taste papers in the trash.*

? QUESTION

1. Are you a taster for thiourea?

2. Are you a taster for PTC?

3. If you are a non-taster and you want to be a high-class chef, what might you do to compensate for this genetic limitation?

Sugar Taste Threshold

This experiment should give you insight about why some people prefer more sugar in their foods.

Procedure

[1] Go to the demonstration table and determine your sugar taste threshold.

[2] Dip a strip of tasting paper into each solution, and record whether or not you can detect a sweet taste. *Discard the used taste papers.*

[3] After testing all of the solutions, go to the front desk for the key to the sugar concentration of each solution.

[4] Record your results in the summary chart on the front chalkboard.

? QUESTION

1. What was the lowest threshold for tasting sugar?

2. What was the highest threshold?

3. Do the people with a high taste threshold also like to add more sugar to their food? (Perhaps you could determine this by asking your classmates how much sugar they add to their coffee.) Your lab may have a setup for making a "food coloring" tongue print to count the number of taste buds in different people.

EXERCISE #6

"Vision"

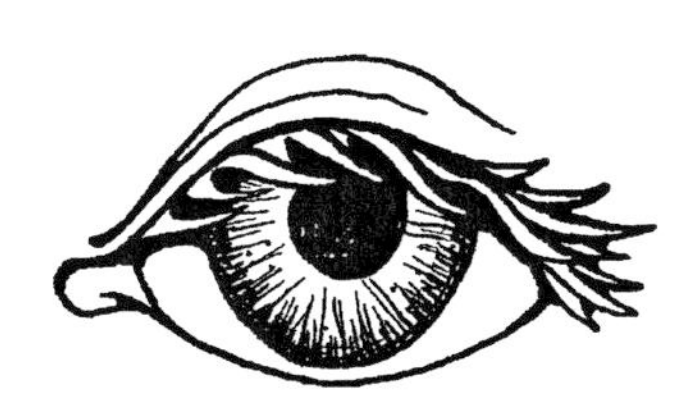

Human beings are primarily visual animals. Vision is the dominant sense you use to relate to the environment. There is a lot of scientific literature on visual perception, and we encourage you to investigate this information when you have time to do so. Most human behavior has been shown to be strongly influenced by visual perception. What you don't know can be used against you!

Preferred Eye

This Exercise is designed to reveal which one of your eyes is used for certain visual functions. Your preferred eye is the one your brain chooses to use when both eyes can see the same object.

Procedure

[1] Pick an object that's about 30 feet away. Make a circle with your thumb and first finger of both hands.

[2] Straighten and raise your arms from your waist to a position where the circle surrounds the object. Keep your head and feet positioned straight ahead.

[3] Without further movement, close one eye. Then open the closed eye, and close your other eye.

[4] Which eye has the *same view* as the view with *both eyes* open? This is your preferred eye. **Hint:** When you close your preferred eye, the distant object will move out of the circle formed by your hands.

? QUESTION

1. Which eye is your preferred eye?

2. If you are left-eyed, what problem will you have in shooting a rifle?

3. Why should you use your preferred eye when looking through a monocular microscope?

Eye with Best Vision

Use the classroom eye chart to determine which of your eyes has the best vision (without glasses).

? QUESTION

1. Which of your eyes has the best vision?

2. Talk with other lab students, and discover whether the eye with the best vision is always the same one as the preferred eye.

Results: ______________________________

Eye with Best Depth Perception

There are fairly simple ways of determining which of your eyes has the best depth perception. If this equipment is available, then determine the depth perception for each of your eyes. If this equipment is unavailable, then refer to the information chart to answer the questions below.

INFORMATION		
Vision Tests	**13 Left-Handed People**	**11 Right-Handed People**
Eye with Best Depth Perception	9 left eye 3 right eye 1 same in both eyes	2 left eye 8 right eye 1 same in both eyes
Preferred Eye	7 left eye 6 right eye	5 left eye 6 right eye
Eye with Best Vision	1 left eye 1 right eye 11 same in both eyes	2 left eye 1 right eye 8 same in both eyes

? QUESTION

The eye with best *depth perception* is most closely associated with . . . (circle your choice)

Preferred Eye or Preferred Hand or Eye with Best Vision

EXERCISE #7

"Reflexes"

Some people have a very good excuse for being "jumpy."

The brain is capable of sending both ***facilitatory*** (speed up) and ***inhibitory*** (slow down) signals to the reflex centers in the spinal cord. People differ with respect to the balance of these opposing effects. You see this difference by watching how reactive a person is (very calm vs. quick reacting).

A reading reflex test can be used to determine which reflex type you are. *Concentrating on reading should reduce the effect your brain normally has on your reflex centers.* If reading reduces your reflex response, then normally your brain must be stimulating reflexes (you are a quick-reacting person). If reading increases your reflex response, then your brain normally inhibits reflexes (calm reacting). Your brain's reflex emphasis can change.

No person is 100% one type or the other all of the time.

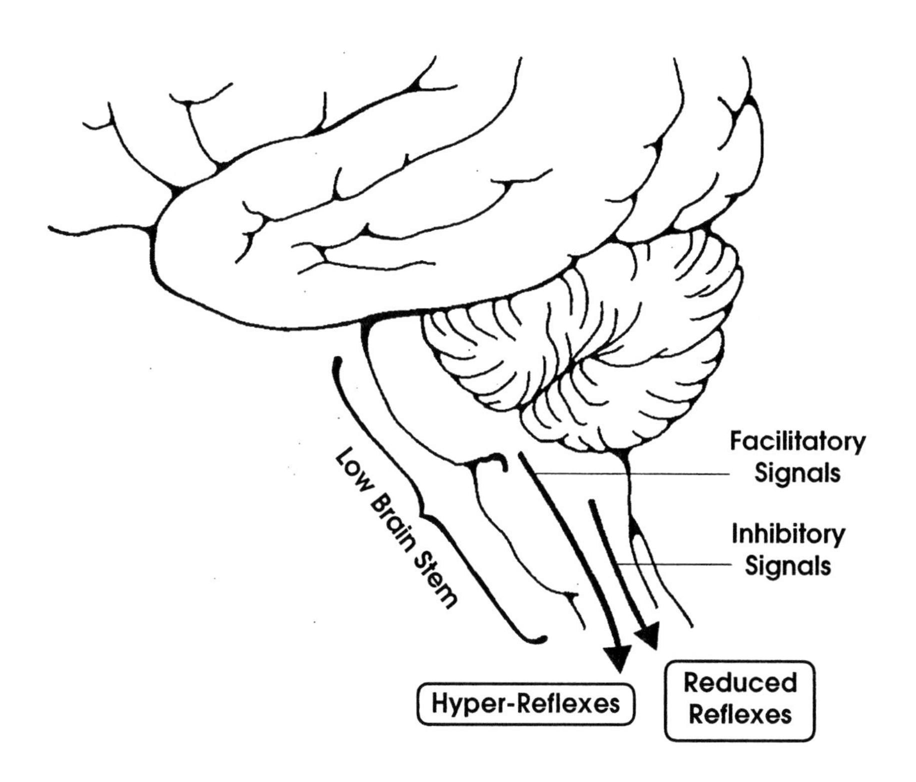

Materials

- A patellar hammer.
- A meter stick.

Procedure

[1] Before beginning the test, ask your lab partner to evaluate whether you are the calm or quick-reacting type.

Lab Partner's Opinion: ________________

Your Opinion: ________________________

[2] Sit on a table so that your legs hang freely over the edge. Have your lab partner hit the patellar ligament (just below the knee) with the reflex hammer. *Don't hit too hard.* This may take some practice. Measure the amount of leg movement several times in order to get an average estimate of the reflex intensity.

Normal Reflex: ________________________

[3] Next, read from a textbook while your lab partner measures the amount of reflex leg movement. Is the reflex more intense or less intense during the reading conditions?

? QUESTION

1. Under normal circumstances, does your brain activate or inhibit spinal reflexes?

2. So, which reflex type are you?

3. Does this agree with how you evaluated yourself before the test?

4. Compare your conclusions with those of other students in the class. What did you discover?

Patterns in Nature

Nothing is what you can't tell you're in because it looks the same no matter how you look at it.
—K.C. Cole, Discover Magazine

Curious people have noticed that nearly everything in nature has a form, and the different forms seem to be variations on several basic patterns: *meander, spiral, explosion,* and *branching.* What causes these patterns to occur? Do they occur throughout both the living and non-living parts of nature?

You will begin investigating these questions by considering how space fills with different patterns of form.

Exercise #1 "Space" ... 341
Exercise #2 "Comparison of Patterns" ... 344
Exercise #3 "Rules of Branching" ... 347
Exercise #4 "Cracking and Packing" ... 348
Exercise #5 "Film—*The Shape of Things*" ... 350

EXERCISE #1

"Space"

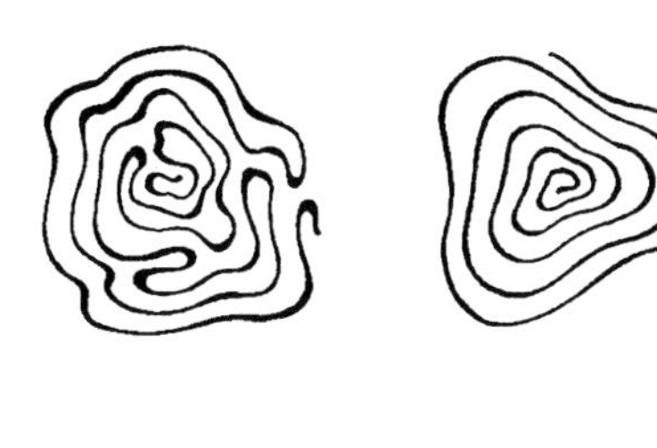

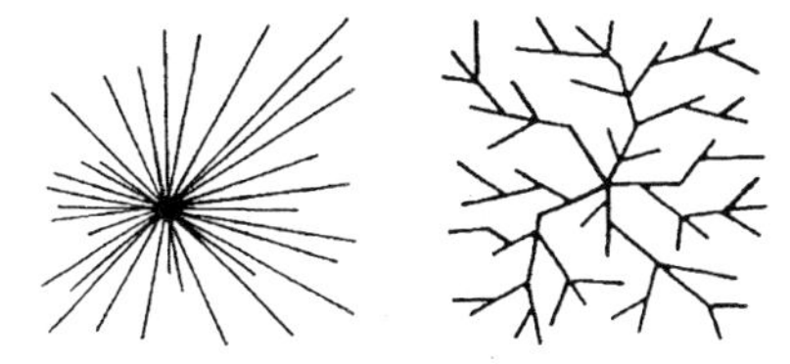

It is a common belief in Western Civilization that space is "nothing," and that you fill it with "something." This idea describes space as a passive emptiness—a set of coordinates where something is located.

Modern physicists think that space itself may be an array of bubbles, or string-like particles, or loops of matter that are millions of times smaller than an atom. This "ocean" of space may be the basic fabric of the universe.

If you imagine space as an "ocean," then several analogies become apparent. In the ocean, there are no waves without the ocean itself. And all life in the ocean arises from the materials present in the ocean. Everything "in" the ocean is actually an expression of some "part" of the ocean.

If space is like the ocean analogy, then the patterns of growth and form in nature are expressions of the basic properties of space. That certainly explains why the same patterns appear everywhere and in all kinds of matter. Whatever the correct understanding of space may be, you can investigate patterns in nature, and develop general principles that describe them. Let's try to understand the following two basic questions about patterns.

What are the patterns of growth and form in organisms?

and

What are the values of each growth pattern in organisms?

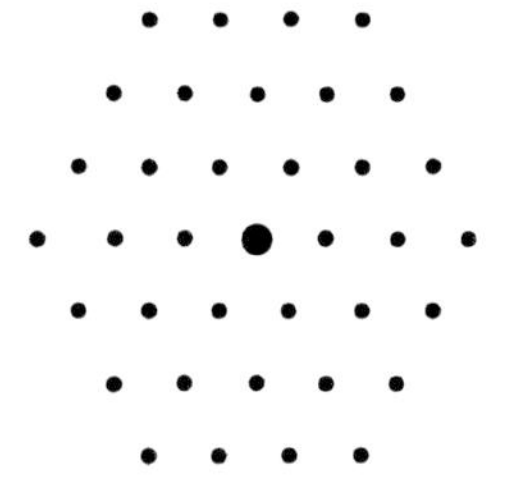

Methods for Describing Patterns

In today's lab, space is represented as dots on the page. The center dot is where a growth pattern begins. As an organism grows, it will "fill" all of the space represented by the dots.

There are four steps you must use in order to describe a growth pattern.

Step 1

Start the growth pattern of the organism at the center dot in space.

Step 2

Connect all of the dots using one of the patterns (meander, spiral, explosion, or branching). As you draw a pattern, use only lines that are 1 cm in length (as shown in the example) to connect any two dots.

Beginning a Meander Growth Pattern

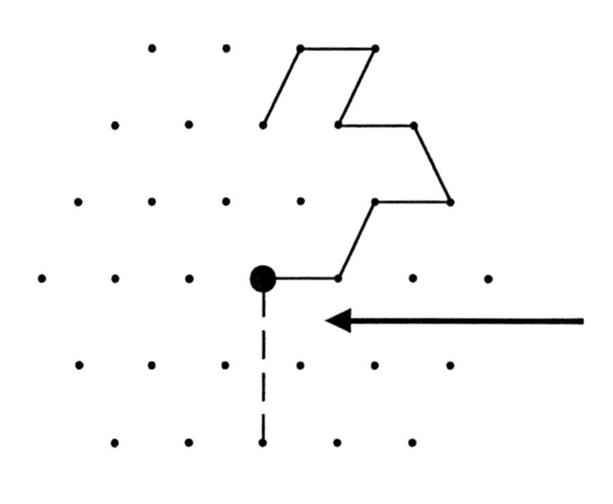

Remember: You are connecting dots on the page to represent the filling of space by a particular growth pattern. Therefore, you cannot draw a line through the same dot twice, or let any two lines cross. (That would be like saying that an organism can fill the same space twice.) Also, make sure that your pattern goes through each of the dots (i.e., fills the space completely).

Step 3

Calculate the ***total length*** of the lines of the pattern. The distance between two dots is one centimeter. In the meander example above, the length is 8 cm so far.

Step 4

Calculate the directness of the pattern. ***Directness*** is defined as the average distance from the starting point (center dot) to each of the other dots by following the line (or lines) of the pattern.

In the meander example, the average distance between the center and each of the eight dots can be calculated by:

$$\frac{1+2+3+4+5+6+7+8}{8} = \frac{36}{8} = 4.5 \text{ cm (directness)}$$

Hint: The numbers on the top of the equation are the individual distances between the center and each of the dots. When measuring directness, you must follow the lines of the pattern. Because there are eight dots along the meander line, you determine the average distance by dividing the sum of the top numbers in the equation by the number of dots (8, so far).

Total Length = The growth efficiency of an organism.

The best designs are low in total length and directness.

The ***total length*** of a pattern is the length of all lines in the pattern necessary to connect all of the points in space. A lower total length indicates a more efficient design of space filling than a high total length.

Directness = The interconnectedness of the parts of an organism.

The ***directness*** of a growth pattern is the average distance from the center dot (starting point) to the other dots in the pattern. A lower directness value means that there is a *short* distance from one part of the organism to the rest of the organism. A higher directness value means that there is a *long* distance from one part of the organism to the rest of the organism.

In Exercise #2, you will compare both the total length and directness of the four universal patterns in nature.

? QUESTION

1. List four universal patterns in nature.

2. Summarize the theory that space is not "nothing," but may be the basic structure of "everything" in the universe.

3. What does the *total length* of the growth pattern tell you about an organism?

4. What does the *directness* of the growth pattern tell you about an organism?

EXERCISE #2

"Comparison of Patterns"

Draw the following patterns, and calculate the total length and directness of each. Remember, the pattern can go through each single dot only once.

Meander

Total Length = ________

Directness = ________

Spiral

Total Length = ________

Directness = ________

The lines of the explosion pattern must be drawn with a straight edge outwards from the center. Draw as many lines as it takes to go through all of the points. You will have to draw a few lines longer than one unit in order to connect the center to some of the dots.

Explosion

Total Length = ________

Directness = ________

Branching

Total Length = ________

Directness = ________

Comparisons of Total Length and Directness

Compare the total length and directness of the four patterns. If a pattern has a low directness value, then give it a "good" rating in the comparison table. Otherwise, assign the pattern a "bad" rating. Give the highest total length a "bad" rating.

Pattern	Total Length (efficiency of growth)	Directness (interconnectedness)
Meander		
Spiral		
Explosion		
Branching		

? QUESTION

1. Which two patterns are best in terms of directness?

2. Which two growth patterns do plants use? Give an example of each.

   Why do plants need interconnectedness (directness)?

3. Frogs lay eggs, as do many other organisms. Which two egg-laying patterns would take the least amount of energy by the frog? (By the way, these are the two egg-laying patterns in nature.)

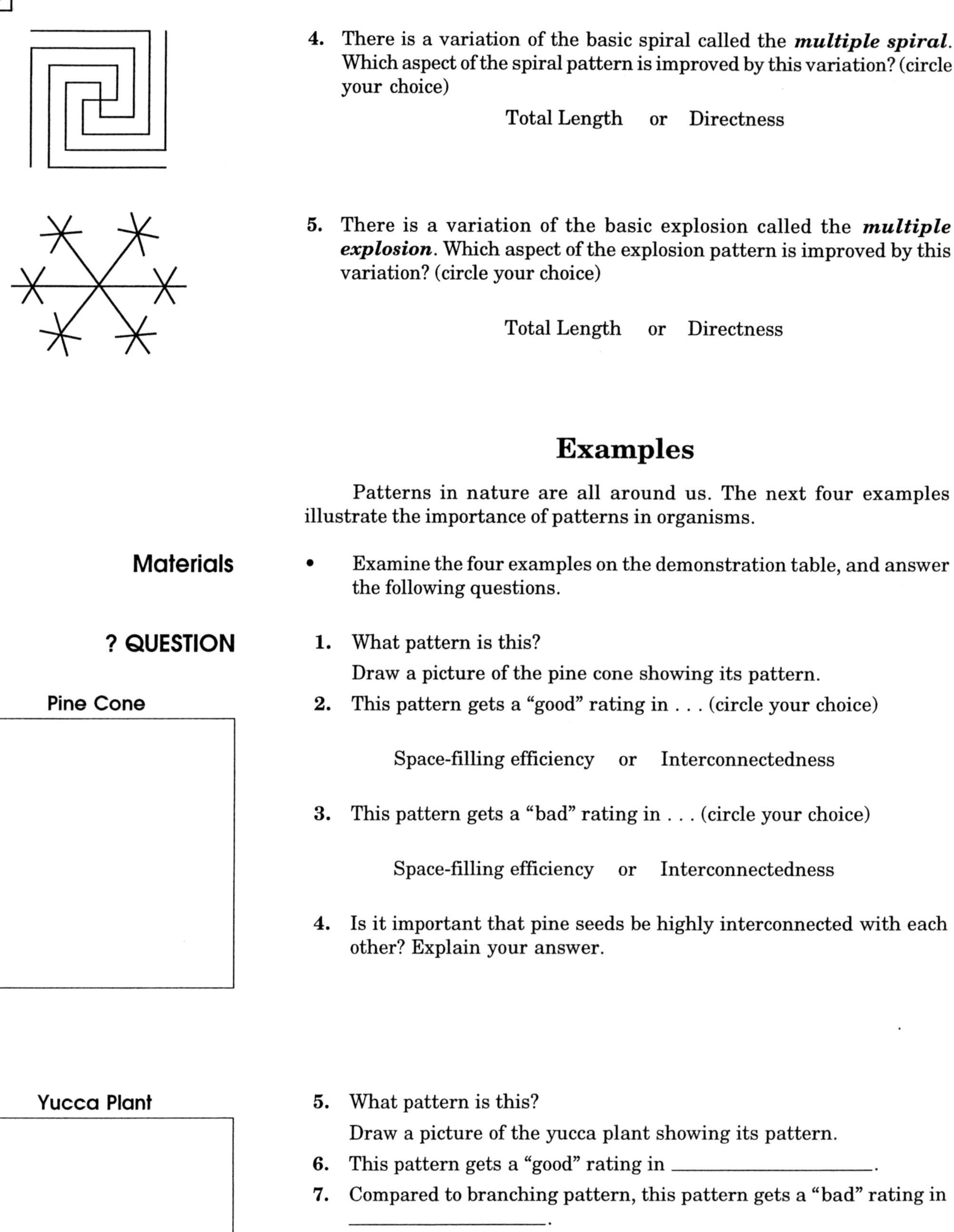

4. There is a variation of the basic spiral called the ***multiple spiral***. Which aspect of the spiral pattern is improved by this variation? (circle your choice)

Total Length or Directness

5. There is a variation of the basic explosion called the ***multiple explosion***. Which aspect of the explosion pattern is improved by this variation? (circle your choice)

Total Length or Directness

Examples

Patterns in nature are all around us. The next four examples illustrate the importance of patterns in organisms.

Materials

- Examine the four examples on the demonstration table, and answer the following questions.

? QUESTION

Pine Cone

1. What pattern is this?
   Draw a picture of the pine cone showing its pattern.
2. This pattern gets a "good" rating in . . . (circle your choice)

   Space-filling efficiency or Interconnectedness

3. This pattern gets a "bad" rating in . . . (circle your choice)

   Space-filling efficiency or Interconnectedness

4. Is it important that pine seeds be highly interconnected with each other? Explain your answer.

Yucca Plant

5. What pattern is this?
   Draw a picture of the yucca plant showing its pattern.
6. This pattern gets a "good" rating in ____________________.
7. Compared to branching pattern, this pattern gets a "bad" rating in ____________________.
8. As a general rule in nature, would you expect plants with this leaf pattern to be big or small? Explain your answer.

Feather

9. What pattern is this?

   Draw a picture of the feather showing its pattern.

10. This pattern gets a "good" rating in both total length and directness. A low total length means that the feather is an efficient design. What functional feature of the feather is created from a "good" directness rating?

Human Blood Vessels

11. If you look at a picture of the human circulatory system, what general pattern do you see for all of the vessels collectively?

    Draw a picture of human blood vessels showing the general collective pattern.

12. What are the two values of this pattern?

13. If you follow just one vessel at a time, its pattern is a ____________.

14. What is the value of this single-vessel pattern?

    The pattern of leaf veins in plants is remarkably similar to the pattern of blood vessels in animals. Why should you expect this?

EXERCISE #3

"Rules of Branching"

In most circumstances, branching is the best space-filling pattern. Researchers have studied branching patterns in trees, blood vessels, lung tissue, lightning, and many other natural forms. They have discovered a number of almost universal rules describing branching. Three of these rules are presented next. After reading them, it will be your job to find examples of each rule to show to your instructor.

Note: In plants, the controlling variable responsible for the branching rules is the high resistance to flow of water and nutrients created in smaller diameter branches. Because of this high resistance, small branches must come off the trunk at 90° which is the shortest distance to the leaves.

Rule 1

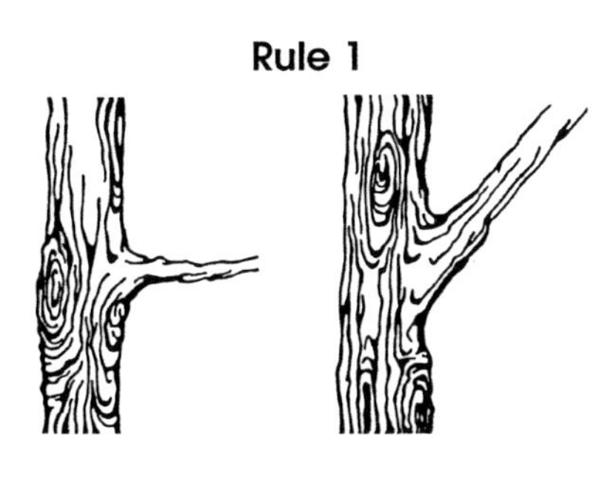

Angle of Branching vs. Diameter of Branch

The angle between the trunk and the branch is dependent on the diameter of the branch.

If the branch coming off a main trunk is *small* compared to the trunk, then it will come off at an angle close to 90°. If the branch is *large,* then it will come off noticeably less than 90°. This is true of all branching in nature.

Rule 2

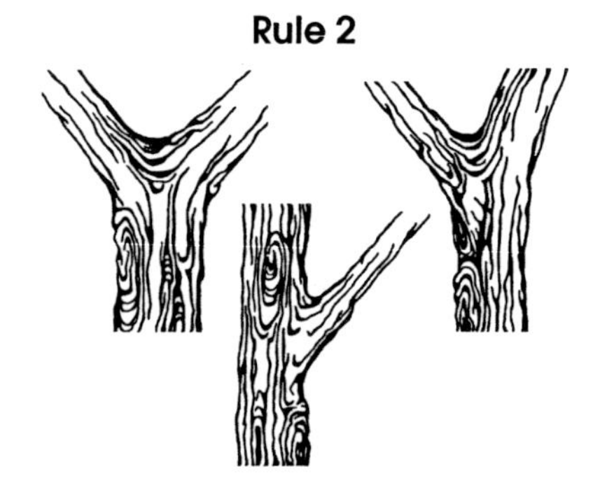

Deviation of Trunk vs. Diameter of Branch

The closer the branch size is to the size of the main trunk, the more the trunk above the branch deviates from the vertical axis.

If the branch is equal in size to the main trunk, then the main trunk will deviate from its original axis at an angle equal to that of the branch. Branches smaller than the trunk deflect the main trunk only slightly, or not at all if they are very small.

Rule 3

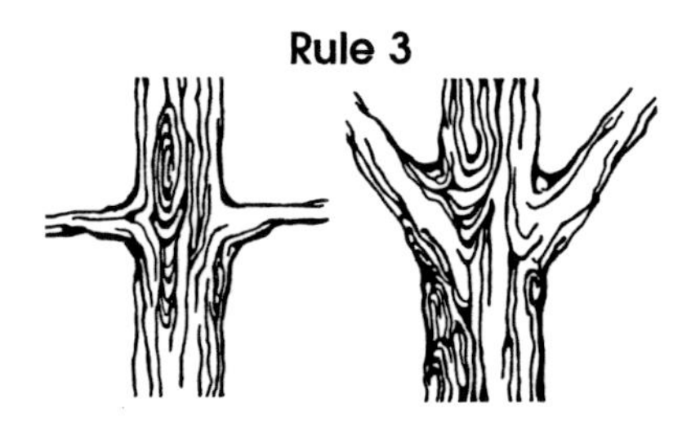

Branching on Opposite Sides of the Trunk

If two branches come off on opposite sides of the same place on the trunk, then the trunk above the branches does not deviate from the vertical axis. However, the angle of those branches still follows Rule 1.

Procedure

It's your turn to find examples of these three rules, and show them to your instructor. Give a brief description of each example, and make a sketch in the margin to help you remember each branching rule.Places you might look are: bushes and trees on campus, veins of leaves (use a dissecting microscope), and pictures of the human circulatory system.

	Example	Check-Off
Rule 1		
Rule 2		
Rule 3		

EXERCISE #4

"Cracking and Packing"

When spherical objects pack together in nature, or larger masses break into pieces, we see two typical patterns: the three-way 120° joint and the right-angle 90° joint.

120° Joints

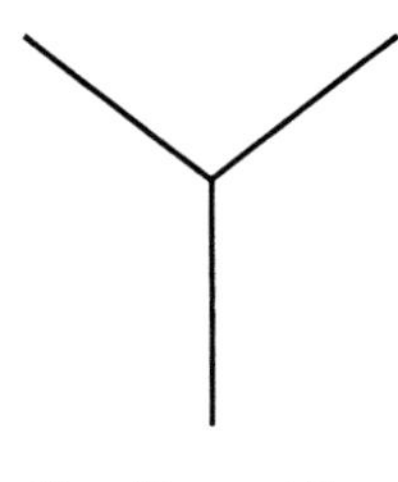

The Three-Way 120° Joint

Packing and cracking patterns occur in a wide range of natural objects. Examples of the 120° joint can be found in soap bubbles, molecular crystals, animal scales, dried mud, crowded biscuits cooked in a pan, turtle shells, continental plates, and many others. Engineers call the 120° joint a "minimum network" because the forces creating it are expressed along the minimum system of surfaces, cracks, or partitions. You might say that this joint system is the most efficient manifestation of cracking and packing forces in nature.

Cracking Rule 1

If a rigid substance is subjected to a cracking force that is expressed in all directions, all at once throughout the substance, then three-way 120° joints will form.

Packing Rule 1

If elastic objects are packed together by forces that push equally in all directions, then three-way 120° joints will form.

90° Joints

The Right-Angle 90° Joint

The 90° joint (or four-way joint) can happen in just about any substance if the forces are applied in a certain way. The 90° joint is *not* a minimal network. The forces creating this type of joint are applied in one direction more than from another.

Once the first cracking or packing lines form in response to the dominant force, all of the stress is relieved in that direction. The stress forces that remain are always *perpendicular* to the first dominant force. This remaining stress force will eventually produce other joints (90°) perpendicular to the first joint line. Parallel joint lines form in one direction, and then a series of joint lines form perpendicular to the original set.

Examples of 90° joints can be found in all substances in nature. You can find them in geologic formations, tree bark, concrete, or packed cells as long as the stress forces are applied more in one direction than another.

Cracking Rule 2

Nature joins a new crack to an old crack only at right angles—90°.

Packing Rule 2

If the forces packing objects together are much stronger in one direction than another, then 90° joints form.

Procedure

It's your turn to find at least two examples of each joint type. Your instructor may have provided some places to start looking (see the demonstration table). Make simple sketches of each example in the space below.

Check-Off by Instructor		
120° Joint	1	
	2	
Right-Angle 90° Joint	1	
	2	

EXERCISE #5

"Film—*The Shape of Things*"

Perhaps you will have the opportunity to watch this film during the last hour of the lab. It is an excellent visual review of everything you've learned today. While watching the film, write down examples of each pattern you observe.

Patterns are everywhere you look.

Meander	
Spiral	
Branching	
Explosion	
3-Way 120° Joint	
Right-Angle 90° Joint	

The patterns and forms that we observe in the world provide deep insight into the most basic forces operating in the Universe. There is much more to patterns than modern people pay attention to. Further information on these topics can be found in the "Selected Readings."

Selected Readings

The following references offer expanded discussions about some of the topics mentioned in *Laboratory Investigations*. One book we highly recommend that every student read is *Finite and Infinite Games*, a vision of life as play and possibility, by James P. Carse. This book is short and can be read in a couple of hours. It presents a very simple yet profound idea about the differences in the ways that people experience their lives, depending on how they choose to play "The Game." We hope that you view your life as an adventurous game, and enjoy exploring some of our Selected Readings.

Adams, Douglas and Mark Carwardine. 1991. *Last Chance to See*. Harmony Books, New York.

Burke, James. 1985. *The Day the Universe Changed*. Little, Brown and Company.

Campbell, Joseph. 1972. *Myths to Live By*. A Bantam Book by Viking Press Inc.

Carse, James P. 1986. *Finite and Infitite Games*. The Free Press, A Division of Macmillan, Inc.

Cavalli-Sforza, Luigi Luca. 1991. "Genes, Peoples and Languages." *Scientific American* (November): 104-110.

Cook, Theodore Andrea. 1979. *The Curves of Life*. Republication by Dover Publications, Inc.

Coppens, Yves. 1994. "East Side Story: The Origin of Humankind." *Scientific American* (May): 88-95.

Dawkins, Richard. 1996. *Climbing Mount Improbable*. W.W. Norton & Company.

Feynman, Richard P. 1988. *What Do You Care What Other People Think?* W.W. Norton & Company.

Gleick, James. 1987. *Chaos: Making a New Science*. Penguin Books.

Gould, Steven Jay. 1989. *Wonderful Life: The Burgess Shale and the Nature of History*. W.W. Norton & Company, Inc.

Hawking, Stephen W. 1988. *A Brief History of Time: From the Big Bang to Black Holes*. Bantam Books.

Hildebrandt, Stefan and Anthony Tromba. 1984. *Mathematics and Optimal Form*. Scientific American Books, Inc.

Kahneman, Daniel and Paul Slovic and Amos Tversky (edited by). 1982. *Judgement Under Uncertainty: Heuristics and Biases*. Cambridge University Press.

McKean, Kevin. 1985. "Decisions." *Discover Magazine* (June): 22-31.

McMahon, Thomas A. and John Bonner. 1983. *On Size and Life*. Scientific American Books, Inc.

Moore, John A. 1993. *Science as a Way of Knowing: The Foundations of Modern Biology*. Harvard University Press.

Morrison, Philip and Phylis Morrison. 1982. *Powers of Ten*. (About the relative size of things in the Universe.) Scientific American Books, Inc.

Rock, Irvin. 1984. *Perception*. Scientific American Books, Inc.

Saarinen, Eliel. 1985. *The Search for Form in Art and Architecture*. Republication by Dover Publications, Inc.

Stevens, Peter S. 1974. *Patterns in Nature*. Little, Brown and Company.

"The Search for Early Man." 1985. *National Geographic* (November): Volume 165, No. 5.